Exhibitionist Japan

From the second half of the nineteenth century, Japan has been a particularly enthusiastic user of exhibitions. Large-scale international exhibitions, including Osaka 2025, form only the tip of an iceberg comprising over 1,300 industrial, regional, and local exhibitions held in Japan over the past 150 years. In this unique history, Angus Lockyer explores how and why these events have been used as catalysts of development and arenas for fostering modern industry, empire, and nation. He traces their complicated genesis, realization, and reception, demonstrating that although they rarely achieve their stated aims, this does not undermine their utility – Japanese expos have provided a model subsequently adopted around the world. The history of this enthusiasm provides a more nuanced understanding of development in modern Japan, and emphasizes the shared experiences of global modernity.

Angus Lockyer is an independent historian of modern Japan.

Exhibitionist Japan

The Spectacle of Modern Development

ANGUS LOCKYER
Independent Scholar

CAMBRIDGE
UNIVERSITY PRESS

CAMBRIDGE
UNIVERSITY PRESS

Shaftesbury Road, Cambridge CB2 8EA, United Kingdom

One Liberty Plaza, 20th Floor, New York, NY 10006, USA

477 Williamstown Road, Port Melbourne, VIC 3207, Australia

314–321, 3rd Floor, Plot 3, Splendor Forum, Jasola District Centre,
New Delhi – 110025, India

103 Penang Road, #05-06/07, Visioncrest Commercial, Singapore 238467

Cambridge University Press is part of Cambridge University Press & Assessment,
a department of the University of Cambridge.

We share the University's mission to contribute to society through the pursuit of
education, learning and research at the highest international levels of excellence.

www.cambridge.org
Information on this title: www.cambridge.org/9781009544252

DOI: 10.1017/9781009544245

When citing this work, please include a reference to the DOI 10.1017/9781009544245

First published 2025

A catalogue record for this publication is available from the British Library

Library of Congress Cataloging-in-Publication Data
NAMES: Lockyer, Angus, author.
TITLE: Exhibitionist Japan : the spectacle of modern development / Angus Lockyer.
DESCRIPTION: Cambridge ; New York, NY : Cambridge University Press, 2025. |
Includes bibliographical references and index.
IDENTIFIERS: LCCN 2024042993 (print) | LCCN 2024042994 (ebook) | ISBN 9781009544252
(hardback) | ISBN 9781009544269 (paperback) | ISBN 9781009544245 (epub)
SUBJECTS: LCSH: Exhibitions–Japan–History. | Technological innovations–Japan–History. |
Economic development–Japan–History. | Japan–Economic conditions.
CLASSIFICATION: LCC T395.5.J3 L63 2025 (print) | LCC T395.5.J3 (ebook) |
DDC 607/.340952–DC23/eng/20241031
LC record available at https://lccn.loc.gov/2024042993
LC ebook record available at https://lccn.loc.gov/2024042994

ISBN 978-1-009-54425-2 Hardback
ISBN 978-1-009-54426-9 Paperback

Contents

Figures

Acknowledgements

This book has taken too long and received too much support to thank all those who have made it possible. It would have been impossible without the following.

The universities, research institutions, and other organizations, which provided salaries, grants, resources, and institutional homes: CU Boulder, JSPS, Minpaku, SOAS, SSRC, Stanford, Tōdai, UBC, UW, Wake Forest, and the Whiting Foundation.

The colleagues who keep departments, centres, libraries, and archives running, notably Lavi Begum, Ishikawa Atsuko, Fujiko Kobayashi, Dave Martin, Carol Miles, Danielle Salaz, Cyp Stephenson, Janice Walker, Monica Wheeler, Robert Whiteing, and Liza Williams.

The teachers and mentors who modelled, encouraged, and cajoled: Joel Beinin, Paul Berry, Peter Duus, Paula Findlen, Susan Hanley, Jim Ketelaar, Kinoshita Naoyuki, Jeff Mass, Richard Roberts, John Treat, Sarah Watts, Kozo Yamamura, Yoshida Kenji, and Yoshimi Shunya.

The collectors and scholars who discovered the materials and told the stories about Japan and exhibitions that have made this one possible: Hashizume Shinya, Itō Mamiko, Kuni Takeyuki, Sano Mayuko, Tada Osamu, Terashita Tsuyoshi, Yoshimi Shunya, and Yoshida Mitsukuni.

The innumerable colleagues who listened to various parts of this and, especially, the ones who read many more words than those that have made it onto these pages: John Parker, Sarah Teasley, Marcia Yonemoto, Phoebe Young, and the two anonymous readers for the press.

The staff at Cambridge University Press and Straive who made preparing the manuscript and producing the book a piece of cake: Jo Breeze,

Rosa Martin, Sethulakshmy S. Nayar, Balamuthukumaran Pasupathy, and Lucy Rhymer.

The family and friends, who have lightened the load. And Jill, who has made all things possible and life a joy.

Note on Names and References

Hakurankai is the Japanese translation of what was originally called an exhibition in British English, *exposition* in French, and fair in American English. More recently, expo has become the preferred term around the world, for example, Expo 70. I use all of these terms, both to ring the changes and to do justice to the time and place in question. I also use place and date as shorthand, for example, Osaka 1970.

All references to individual exhibitions, except where otherwise noted, are drawn from the extensive database maintained by the Nomura group, www.nomurakougei.co.jp/expo.

References to articles in newspapers comprise title (e.g. *Asahi*), date (y/m/d), as well as edition for regional and/or evening editions (e.g. Chūbū pm).

Japanese names follow the customary order, with family name first, except for people living and/or working in the Anglophone world.

Introduction

I arrived in Japan in 1988 as a wave of expos began to crest, but I did not know it at the time. To be fair, there were none in my immediate neighbourhood. I spent a week in Tokyo. The city had been the main stage when exhibitions started up in the late nineteenth century, but more or less out of the game since the war. The metropolitan government was planning a World City Expo for the following decade, but I could not read the newspapers where it was being discussed. I took the bullet train down to Yamaguchi, where I would spend the next two years. The prefecture had seen its first exhibitions just before and during the Asia-Pacific War, celebrating the 'development of Asia' (*kōa*) and the construction of a railway tunnel; and a couple more in the decade or so after to encourage economic recovery; but the current wave was passing it by. Hiroshima, just up the coast, had embraced exhibitions slightly earlier, using them to cement its relationship with the military before the war and its standing in the peace industry afterward. The summer after I arrived, the city turned to them again, to argue that it was the perfect hub for the island resorts of the Inland Sea. During the day, seven fantasy islands tried to make the claim good, but the site was a concrete wasteland at night, when I went there for a reggae concert. I cannot claim that I knew I was at an expo.

My ignorance persisted. The next spring, I found myself in a coach full of farmers at least twice my age. We were heading from their small temple on a mid-sized island in the Inland Sea to the imposing headquarters of the sect, deep in the mountains on the far side of Japan. We reached the outskirts of Osaka. Had I but known it, we were surrounded by the traces of the exhibitions that had catalysed the city's modern development:

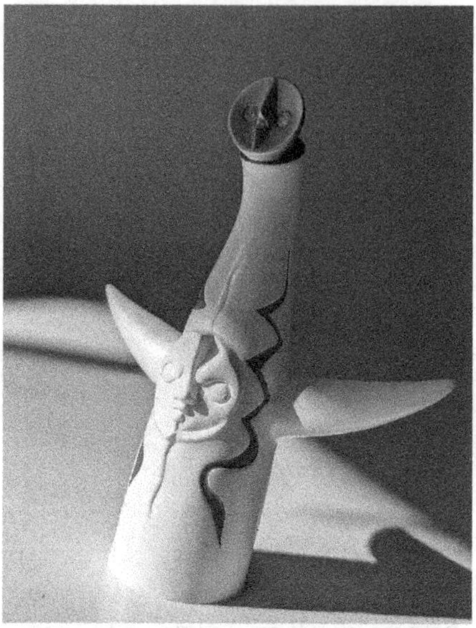

FIGURE 0.1 I first saw the Tower of the Sun in 1990. This small replica has sat on my desk while I've been writing this book.
Source: Jill Berman.

Tennōji, in the south of the city, a temple district transformed into a playground via exhibitions early in the twentieth century; the hot-spring-turned-resort of Takarazuka and the baseball stadia of Nishinomiya, both launched with exhibitions, inciting the interwar suburban sprawl. What I saw, though, on our left, looming over the sound barriers, was the chalky white body, golden face, and outstretched arms of the Tower of the Sun (Figure 0.1). Twenty years earlier, poking through a vast roof, it had been the icon of Expo 70, Japan's first, long-dreamt-of international exhibition. Some of the farmers had probably been. All of them recognized the tower. None of us likely knew that the first ten miles of the expressway, over which we had just rumbled, had been built at the same time, part of the infrastructure sanctioned by the spectacle.

A few years later, I began to put the pieces together. I read Timothy Mitchell in Seattle, who suggested that exhibitions might provide a key with which to unpick the politics of representation that underpinned the modern world.[1] In a corridor in the University of Washington library, I found a bird's eye view of a 1909 expo, which had been held on the then

[1] Timothy Mitchell, 'The World as Exhibition', *Comparative Studies in Society and History* 31.2 (1989), 217–36.

newish campus. It seemed to exemplify the logic that Mitchell had laid out, splaying out the recently discovered riches of Alaska and the Yukon and the promise of the Pacific for American consumption. A little later, at Stanford, I wrote a short paper about how Japan had assimilated this logic at its first industrial exhibition in Tokyo in 1877. The following year, I researched a longer one on the Japan–British Exhibition of 1910. I wandered around what remained of the site in west London. I imagined the exhibits in the cavernous halls over the train tracks, which had presented a Japan ready to leave its pre-industrial past behind. I stumbled into the remnants of a Japanese garden next to a housing estate, which suggested to its English audience that it never quite would. I blew that paper up into a dissertation, to make the same point.

A few years later, I was trying to turn the pockmarked dissertation into a polished book in the Asian Library at the University of British Columbia in Vancouver. It was a good place to do it. The roof had started its life on top of the Sanyo Pavilion at Expo 70. Back then, the pavilion had been designed to embody the 'heart of Japan', complemented by a 'pond of peace'. The exhibit had attracted attention, though, mainly for its vision of the imminent future, embodied by a 'healthy living capsule', which included a human washing machine, demonstrated by a young woman, partially clothed.[2] Now, the roof provided cover as I tried to prove that exhibitions could not but reproduce a late nineteenth-century order of things, whereby the dictates of industry and empire made their remorseless way around the world. I chose not to wonder whether the library roof or the exhibitions might have other stories to tell.

A couple of years later, though, I finally visited my first expo on the outskirts of Nagoya. I had handed in my dissertation at the turn of the century, when its planning seemed to be on the rocks and my elders and betters were convinced that the age of exhibitions was past. I had been confident that it would be easy enough to turn it into a book, which would have the final word. But they and I were wrong, and it will not. I left London on Ash Wednesday, arriving in Tokyo on Maundy Thursday, just before the expo's opening ceremonies, which included the tolling of a bell, a military band, the imperial couple, and a bipedal robot. I broke bread with some priests on Good Friday, then got on the bullet train, arriving in a strangely subdued city late that night. Early the next morning, I got on the city's oldest subway line, then changed onto

[2] *Japan World Exposition, Osaka, 1970: Official Report* (Commemorative Association for the Japan World Exposition, 1972), vol. 1, 452.

FIGURE 0.2 It rained on my first day at an expo. Global House from the Kiccoro gondola at Expo 2005.
Source: Author.

Japan's first maglev train, which trundled up into the eastern hills, suspended over the expressway linking the city and the old pottery town of Seto.

I got off and walked through the wood-lined entrance of the expo and turned right, suspended in mid-air again, this time on a wide, wooden walkway, the Global Loop. I ignored the bombastic corporate pavilions and their robots on my right. I got on the Ferris wheel and got my bearings. I spent some time in a 'growing village'. I wandered down the walkway to one of six 'global commons', where I met the expo's architect, Harada Shizuo, and where the international participants were still setting up shop. I made my way through a 'bio-lung' and the peanut-shaped, bamboo-covered Japanese government pavilion. I ambled through a Japanese garden and into the 'forest experience' zone. I got on a gondola back to the main entrance and another to the second expo site (Figure 0.2), where Japanese and local governments trumpeted the indigenous roots of their environmental credentials and commitment. I got back on the gondola and the maglev and the subway and headed back downtown. I had been one of nearly 50,000 visitors on the more than

400-acre site that day. The organizers had been hoping for three times that number.[3]

It was not quite a holy Saturday, but the strange pause, between the ceremonies and the crowds who would descend later in the year, proved a revelation, nonetheless. It had been clear to me, viewing expos from a distance, that they were simple things, which could be used to tell a familiar story, about the invidious logic of the modern world. Up close, however, in the archive and on the ground, they turned out to be more complex things. True enough, the top note is often one of triumph, celebrating the achievements of state and capital and justifying the discrimination on which they are based. But there is anxiety, too, about their sustainability. There is compromise with, concession to, space for other stories about the world. Most obvious, there has been change over time, which has allowed expos to endure even when the commentariat was insisting that their time was past. Too many years after it got its start, this book has therefore become an attempt to understand why, after a century and a half, expos remain something of a staple in Japan – and therefore, I would argue, a going concern worldwide.

The answer is simple enough. Expos have endured because they are useful. They provide space and time, marked out from the daily grind, which can accommodate the various interests of organizers, exhibitors, and visitors – and so provide a return on their diverse investments. This diversity also means that expos can never serve a single master, mean only one thing, or accomplish one iniquitous end. An expo almost always conceals a welter of disagreement, creates space for more than one possibility, and produces uncertain and various effects, all of which change over time.

When I went back and looked harder, this was evident from the beginning. Japanese exhibits were present at international exhibitions from the very start, soon followed by Japanese participants, who were quick to understand their potential value as a tool of development. Exhibitions both at home and abroad therefore soon became a central concern of state, and the government has remained assiduous in contributing to expos overseas ever since. At home, however, already by the end of the nineteenth century, expos had been taken up by local governments, private companies, and more, who were less concerned about the particular lessons an exhibition might teach and more with the economic effects

[3] *Yomiuri*, 2005/3/27, Chūbu, 1.

it might bring about. This subnational enthusiasm crested twice, in very different circumstances: first in the 1930s, as companies, regions, and colonies sought to counter economic crisis and exploit imperial opportunity; and in the 1980s, as cities and corporations, buoyed by the bubble, reimagined the future. By then, the government too had learned its lesson. International exhibitions, beginning with Expo 70, became a preferred medium with which to channel and justify the investment necessary to further its plans for development. Fifty years later, regional expos are finally dying out, but the enthusiasm for and utility of international exhibitions somehow persist.

This story probably makes some sense if you have been to an expo and experienced the profusion, disorientation, and exhaustion that come quickly on the ground. It likely sounds odd, however, if you assume that expos as a species were defined by the variety that flowered in northwestern Europe, with a viable lifespan of some fifty years (London 1851 to Paris 1900; Chicago 1893 to New York 1939–40). And it may seem plain wrong, if you have read much in the now-extensive literature on international exhibitions.

At the risk of caricature, it is easy when writing about expos to use them to talk about something else: either zooming in on a particular, concrete aspect of the show, or panning out to make them stand for a general, abstract feature of their time and place. The first temptation sees scholars rummaging through an expo or two in order to excavate the history of a specific field. Sometimes this is a discrete sphere of professional practice, for example, art, architecture, or technology.[4] Sometimes, it is a broader area of commercial or administrative concern, for example, entertainment, urban planning, or corporate PR.[5] All of this is fair

[4] For example, for art, Elizabeth Holt, *The Art of All Nations, 1850–73: The Emerging Role of Exhibitions and Critics* (Princeton University Press, 1982) and Kitazawa Noriaki, *Me no Shinden: 'Bijutsu' Juyōshi Nōto* (Bijutsu Shuppansha, 2005). For architecture, Erik Mattie, *World's Fairs* (Princeton Architectural Press, 1998), and Inoue Shōichi, *Senjika Nihon No Kenchikuka: Āto, Kitchu, Japanesku* (Asahi Shinbunsha, 1995). For technology, Yoshida Mitsukuni, *Bankoku Hakurankai: Gijutsu Bunmeishi-teki ni* (NHK Books, 1985), and David E. Nye, *Electrifying America: Social Meanings of a New Technology, 1880–1940* (MIT Press, 1990).

[5] For example, for entertainment, Alan Bryman, *Disney and His Worlds* (Routledge, 1995). For corporate PR, Nanba Kōji, 'Hachijū Nendai no Chihōhaku "Ransai" to Sono Ato', in Fukuma Yoshiaki, Nanba Kōji, and Tanimoto Naho (eds.), *Hakuran no Seiki: Shōhi Nashonariti Mejia* (Azusa Shuppansha, 2009), 5–36. For urban planning, Robert Freestone and Marco Amati, *Exhibitions and the Development of Modern Planning Culture* (Routledge, 2014).

enough. Expos were central in the development of the products and practices, the institutions and spaces, of modern industrial society, together with the visual and material culture to which it gave rise. They were simultaneously the progenitors of museums, trade shows, and theme parks, as well as launch pads for Metabolism, monorail, nylon, and more.[6] The part does not stand for the whole, however. The master plan usually failed to anticipate the use that exhibitors and visitors made of the space. The shiny promise of corporate tech was often gazumped by the seductive distractions of the sideshow. An expo is a complex amalgam, which emerges from the interaction between these elements, not a simple projection from the mind of an architect, animator, or advertiser.

The simplification of this first metonymy also serves as a lure into the second. Here, an expo or two seems to be able to produce and/or reproduce a particular order of people, places, and things. The simplicity of an expo theme, the discipline of a classification, the clarity of a site plan, the seamless presentation of a national exhibit – all make it easy to assume that expos do sterling ideological work, whether crafting a nation, or creating eager servants of state and capital. The spectacle of an expo, by this account, makes its visitors powerless to resist either the diffuse charms of the market, the concentrated appeal of the nation, or the integrated imperative of a neoliberal world. In some accounts, the expo can even become a machine.[7] True enough, from a distance, big events in modern times can look like steamrollers, impossible to resist. Zooming

[6] For example, for museums, Tony Bennett, *The Birth of the Museum: History, Theory, Politics* (Routledge, 1995) and Alice Tseng, *The Imperial Museums of Meiji Japan: Architecture and the Art of the Nation* (University of Washington Press, 2008). For trade fairs, Warwick Frost and Jennifer Laing, *Exhibitions, Trade Fairs and Industrial Events* (Routledge, 2017). For theme parks, Judith Adams, *The American Amusement Park Industry: A History of Technology and Thrills* (Thwayne Publishers, 1991), 19–40.

[7] For the larger claim, Robert W. Rydell, *All the World's a Fair: Visions of Empire at American International Expositions, 1876–1916* (University of Chicago Press, 1985), and *World of Fairs: The Century-of-Progress Expositions* (University of Chicago Press, 1993); Paul Greenhalgh, *Ephemeral Vistas: The* Expositions Universelles, *Great Exhibitions, and World's Fairs, 1851–1939* (Manchester University Press, 1988); and Yoshimi Shun'ya, *Hakurankai No Seijigaku: Manazashi No Kindai* (Chūō kōronsha, 1992). For nationalism, David Raizman and Ethan Robey (eds.), *Expanding Nationalisms at World's Fairs* (Routledge, 2017). For diffuse, concentrated, and integrated spectacle, Guy Debord, *The Society of the Spectacle* (Zone Books, 1994) and *Comments on the Society of the Spectacle* (Verso Books, 1988). For the expo as machine, Vivian Blaxell, 'Preparing Okinawa for Reversion to Japan: The Okinawa International Ocean Exposition of 1975, the US Military and the Construction State', *The Asia-Pacific Journal: Japan Focus* 8.29.2 (2010), https://apjjf.org/vivian-blaxell/3386/article, and Daniela Danna, 'Expo 2015 in Milan: The Power of the Machine', *Journal of Political Ecology* 24.1 (2017), 901–20.

back in, however, reveals something more makeshift, and usually leaking oil. The twentieth century did see spectacles that sought to and may have succeeded in subordinating audiences to more singular ends. Nuremberg and Disney come to mind. But expos were different; and they may have something to tell us about the pretension that spectacle can consistently close the ideological loop. Their official publications tried to present a seamless front, but the official record tells a fractured story – about the negotiation and compromise that dogged the planning of an expo; the reluctance of exhibitors to follow the official line; and the tendency of visitors to wander where they wanted and see what they liked. The simple syllogisms of an expo theme, classification, regulations, and/or master plan disappeared on the ground, inevitably, given the welter of attempts to catch the passing eye and the impossibility of seeing it all.[8] This also does much to explain the evolution of expos over time, from the universal pretension, ordered space, and didactic desire of the first late nineteenth-century generation, to the liberal accommodation and sprawling grounds of their twentieth- and twenty-first-century successors.

Japanese exhibitionism, in other words, provides a useful corrective to our standard accounts of spectacle. It can also refine our understanding of development. Expos tell a story about the latter, trying to make it cohere. Expos put on a show about past, present, and future, pretending that both description and prescription are adequate to the task at hand. Expos have also been used to canalize capital, trying to make it go where the authorities wanted, laying the foundation for the future to come. Capital reproduces itself, however, by finding, flowing into, and widening the cracks. It is no surprise, therefore, that expos have continuously revealed development to be a fractious, uncertain thing, uneven in both its rhythm in time and its spread across space; that their boosters have successively redescribed the grail as industrial promotion, amorphous development, and finally, simply, growth; or that they have turned repeatedly to expos, to recalibrate the prescription, rebalance the culture, and redraw the map.[9] Or at least Japan did, throwing its history too in a slightly different light.

Iain Sinclair, *Ghost Milk: Calling Time on the Grand Project* (Hamish Hamilton, 2012), for the inevitable failure of such grand projects.

8 Compare Laikwan Pang, *The Distorting Mirror: Visual Modernity in China* (University of Hawai'i Press, 2007), 2, on the unstable vision produced by the apparatuses of visual modernity.

9 Sydney Crawcour, '*Kōgyō Iken*: Maeda Masana and His View of Meiji Economic Development', *Journal of Japanese Studies* 23.1 (1997), 69–104, for an example of

The modern Japanese state, as seen at the expo, is usually less imposing than it sometimes could be, and we often believe. Its functionaries worked hard, to compete on the international stage and to make the country fit for purpose. They struggled to do either at expos, however, both overseas and within Japan, even in those times and places when the state's writ was almost divine. There were other ways and other, more disciplinable institutions to insist that its will be done, at least at home. Within Japan, expos reveal, instead, a landscape of cities, regions, and colonies competing for attention and investment. On site, the running was made by the exhibitor with something to show that had not been seen before, and by the professionals, on whom they depended and who knew what it took to attract the crowds. The space was almost always there for visitors to find stories at odds with the official version.

Telling this story also means recognizing the limits of our methods and sources. Expos are not the only modern things to have spawned an archive and a literature that both overwhelms and frustrates. The most comprehensive database of Japanese activity currently lists over 1,300 international and regional expos, in Japan and abroad, between the 1860s and the present. Expo 70 produced an official bibliography with 30,000 items, and 3,000 articles in one newspaper alone.[10] Smaller expos admittedly left a shorter paper trail. All this material can only take you so far, however. Official publications, again, tidy things up, making it seem as if organizers were in charge and expos were indeed well-oiled machines. Media accounts seize on the moments when they were not, but also amplify their claims, about the abstractions of economy, society, and the shape of things to come. Critics and commentators often played the same game, overlooking the complexity on the ground to enable their assertions about the state of the world and what should be done. In other words, it is easy enough to trace how an expo came into being. It is short work to rehearse the story it tried to tell. It is also not that difficult to excavate the local, material, and economic traces it left behind. But it requires a leap of faith to conclude how it might have changed society as a whole, or people's minds. Few expos bothered to ask their visitors what they thought. Journalists sometimes cited a vox pop or two, to animate

promotion. Scott O'Bryan, *The Growth Idea: Purpose and Prosperity in Postwar Japan* (University of Hawaii Press, 2009), for growth and its predecessors.

[10] Nomura, *Hakurankai Shiryō Collection*, for the database. *Nihon Bankoku Hakurankai Kōshiki Kiroku Shiryōshū* (Nihon Bankoku Hakurankai Kinen Kyōkai, 1971), 787–831, for the bibliography. *Yomiuri Shimbun*, for the articles.

the Punch and Judy of a public divided in two, but deadlines discouraged delving into the humdrum or the long term. It is tempting to quilt the surviving threads and scattered patches into whole cloth, oddly but predictably echoing the claims of the organizers, for either endorsement or critique. It is more honest to admit that our accounts, like the expos themselves, are things of fragments, and to restrict them to the sherds that have been left behind.

Finally, this story requires telling in a certain way. A few weeks in the expo archive is enough to recognize the pattern and to reveal that it has been more or less stable over time. Most expos came into being through a roughly similar process, involving iteration, compromise, and concession. The family resemblance was clear when they did, with an extensive site and multiple pavilions, which combined often ineffectual prescription and more compelling distraction. And they succeeded, on their own terms, and therefore endured, by hitting their numbers, if not always attracting vast crowds. It is the resilience of the form, its adaptability to the Japanese environment, to changing circumstance and content, that explains its persistence over time. Doing the story justice therefore means acknowledging both the continuity and the change. On the one hand, Japan saw 1,300 expos that told the same story about a rich past, a challenging present, and a glorious future. The protagonists stayed the same, too: immemorial culture and technological ingenuity had created the heritage on which the country could draw and would guarantee the future it deserved. On the other, the story could be told in different ways, depending on the needs of the time and the place.

In what follows, therefore, I tell the story of Japan at the exhibition more or less as it has panned out over the last century and a half, with the use and utility of expos changing, perceptibly, roughly, every twenty years. Japanese officials first appeared at exhibitions in Europe in the 1860s, shocked by the way in which they revealed the archipelago to be behind the West, but awed by the possibilities of industry on display (Chapter 1). The lesson was soon translated into an early Meiji imperative to promote industry through exhibitions at home and to promote exports at exhibitions abroad (Chapter 2), which established a practice of projecting soft power at exhibitions overseas that continues to this day. Already by the 1890s, though, the state's energies at home were focused elsewhere, and major cities were beginning to use industrial exhibitions to advance and advertise municipal development (Chapter 3). Come the 1910s, expos were beginning to spread to the colonies, which were trying to do the same thing, and to the private sector, where railway companies, department stores, and newspaper companies used them to cultivate customers, thereby providing a burgeoning demand for the services of an emerging professional exhibition industry (Chapter 4).

By the 1930s, local authorities throughout the empire were turning to exhibitions, as they sought to navigate economic uncertainty and exploit new imperial opportunities. Before the war, they had to thread the needle between war and peace, as the state was doing abroad. After, the task became easier, but recovery was more urgent, and expos quickly started up again in response (Chapter 5). By the 1960s, the distortions of development were becoming clear, and the state finally turned back to exhibitions at home as a way of mitigating its effects. Expo 70 in Osaka was a watershed, providing a new model of the way in which exhibitions might service development, and deserves its own in-depth analysis (Chapter 6). It provided a template for international exhibitions thereafter, which were joined in the 1980s by a second regional expo boom, as the country spent the surplus of three decades of economic growth (Chapter 7). The enthusiasm for expos would persist, even after the bubble burst in the early 1990s. Their critics were multiplying, however, and expos beginning to change, admitting that development was not a one-way street (Chapter 8). By the 2010s, finally, the stream of regional expos was slowing to a trickle. Even so, the commitment to international exhibitions remains and the Japanese model continues to spread, as the Conclusion will make clear.

Modern development has relied on spectacle, to persuade us that it is a good thing, to unleash the demand that might bring it about, and to justify the infrastructure by which it might be sustained. The extent to which expos have served development in Japan is clear simply from the statistics, which document their frequency, their popularity, and their return on investment. The extent to which they have changed along the way is quickly apparent from their chroniclers. The extent to which this adaptability is inherent – and the consequences of this for the way we tell the tales of country, spectacle, and the modern times of which they have both been an important part – requires digging behind the story and into the archive. It was already clear, though, in the 1860s. The story starts there.

1

Early Modern Exhibitions

The second year of Bunkyū began on 30 January 1862. At both ends of Eurasia, people had long been putting things on display. For well over a century, big cities, populated with sophisticated consumers and tapped into extensive trade networks, had been providing a demand for and supply of novelty and spectacle. London had upped the stakes a decade earlier, with the first explicitly international exhibition, as a way for Britain to affirm its primacy in the developing world of industry and empire. But the family resemblance to earlier shows and other places was clear. It was not the only place where there was much to see.[1]

In Edo, today's Tokyo, the New Year brought a camel to the west end of Ryōgoku Bridge, which for over a century had been the pre-eminent venue for such attractions.[2] A one-humped predecessor had visited the archipelago in 1821 courtesy of the Dutch in Nagasaki, the culmination of a gradually increasing stream of exotic animal appearances over the previous two hundred years, and had endured a nationwide tour, prompting numerous camel spin-offs, in the form of fans, board games, and the like. The opening of Yokohama for international trade in 1859 brought exotic supply closer to metropolitan demand, with a leopard in 1860 and a tiger in 1861.[3] Both were still in the country as the new year dawned: the leopard would appear in the spring in Osaka and the

[1] The following is based on Saitō Gesshin, *Zōtei Bukō Nenpyō* (Heibonsha, 1968), vol. 2, 187–92, except where otherwise noted.

[2] Ōhama Tetsuya and Kenichirō Yoshihara, *Edo Tōkyō Nenpyō* (Shōgakkan, 1993), 149.

[3] Isono Naohide, 'Umi o Koete Kita Chōjū-tachi', in Yamada Keiji (ed.), *Mono no Imēji: Honzō to Hakubutsugaku e no Shōtai* (Asahi Shinbunsha, 1994), 76–79.

tiger in the winter at a temple near Nagoya. The urban sophisticates of Edo needed something new, however, and a bet on a Bactrian must have seemed like a sure thing, trumping both dromedary and felines. But, sniffs the chronicler, 'it was not a real camel'. Given the ingenuity of the capital's showmen, it probably looked good enough. But Edoites took pride in their discrimination, the obverse of which was an ability not to be gulled. The fraud was made good a year later, with the real two-humped article, even if it was 'very small'.[4] There was no need to wait long, however. Other distractions were on the way.

In the second month, temples and shrines in Edo began to hold 'unveilings' (*kaichō*) of their own and others' treasures, as a way to raise funds. Asakusa-dera, the popular temple near the Sumida River in the northeast of the city, was the most experienced at the game and first out of the blocks on the eighteenth, with an image of Kishimojin, the goddess of the many aspects of parenting, who had been dispatched from Hokekyōji, a Nichiren temple at the north end of Edo Bay. It was followed a week later by Yushima Tenmangū, a shrine lying a mile or so northeast of the castle, and on the twenty-eighth by Kaianji, a temple in Shinagawa, the first of the official post towns on the Tōkaido, the main road from the shogun's capital to the emperor's residence in Kyoto. The shrine was noted for plum viewing in the spring and the temple for maple viewing in the autumn, but there were always more income streams to tap. In all three locations, as was customary, the show lasted for sixty days, with sacred revelation only one option among a teeming mass of secular attractions. Yushima Tenmangū included 'some remarkable examples of craftsmanship', most likely the automata that had long enchanted Edo audiences. Kaianji complemented numerous images, including some from nearby shrines, with plays and other entertainments.

There were plenty of things to look at, therefore, as Edoites whiled away the winter and welcomed the spring. Not all of them were staged. The lack of rain since the New Year, combined with strong wind, provided the right conditions for frequent fires, the 'flowers of Edo', which always prompted excited comment. There were also British diplomats, whose official residence at Tōzenji Temple in Takanawa had been attacked repeatedly over the previous year, who had therefore retreated to Yokohama and were now planning to move back into the city. There was a difference between fear of an unexpected conflagration, antipathy

[4] Asakura Kamezō, *Misemono Kenkyū* (Heibonsha, 1992), 74.

towards recently arrived foreigners, and anticipation of a scheduled show; but they were all fit subjects for the media gaze of the time and subsequently immortalized by its chroniclers.

There were also more didactic diversions to be had. The first 'product gathering' (*bussankai*) of the year opened in Kyoto on the tenth of the second month. The imperial capital was not well-known for such shows, however, and the exhibition only managed to attract a small number of participants and exhibits. The centre of gravity for the genre had long been Nagoya, where interest in traditional and Western medicine had taken institutional form as the Shōhyakusha.[5] The group held a gathering on the twenty-fifth of the third month. Its erstwhile leader, Itō Keisuke, had left Nagoya for Edo the previous year to work for the shogunate, and noted an event at its medical facility on the nineteenth of the fourth month. Kyoto was also seeing more interest. A product exhibition in the fifth month, held this time over two days, attracted 70 exhibitors and 388 exhibits. It included Itō and other Edo-based scholars, as well as some twenty specimens – of peonies, snapdragons, and the like – which had been grown from seeds brought back from the official mission to the US two years previously. Nor did the natural world lack for more demotic enthusiasm. The popular craze for goldfish had already lasted over two decades, above all in Osaka, where the obsession focused on *ranchū* (a small variety, bred with particular enthusiasm in the city). A show opened on the fifteenth of the same month in a pond in the Tenman district, spawning a catalogue of some sixty exhibits.[6]

Meanwhile, at the other end of Eurasia, Japan itself was being put on display. The London International Exhibition opened on 1 May (the third day of the fourth month, in the lunar calendar). A second British attempt to maintain its place in the increasingly competitive world of industry and empire, it devoted one small corner, sandwiched between Siam and Music, to things Japanese. The 614 exhibits ranged from lacquer, basketry, ceramics, metalwork, and textiles, to tempt the connoisseur and consumer, to more singular curiosities, to excite the ethnographer and/or the prurient: a lithochrome of some 200 illustrations of the 'manners, costume, and architecture' of the Japanese; maps of the 66 provinces, the shogunal capital of Edo, the main highway linking it to Kyoto and Osaka,

[5] Nagoya-shi Hakubutsukan, *Nagoya no Hakurankai* (Nagoya-shi Hakubutsukan, 1982), 12–21, and Maki Fukuoka, *The Premise of Fidelity: Science, Visuality, and Representing the Real in Nineteenth-Century Japan* (Stanford University Press, 2012), 166–68.

[6] Isono Naohide, *Nihon Hakubutsushi Nenpyō* (Heibonsha, 2002), 715.

and Mount Fuji; rain cowls made of bark and of reeds; a cable of human hair; and the 'short sword of one of the party of assassins which attacked the British Legation [in Tōzenji] on the 5th July 1861 (with a leather purse containing cash, and seal attached, belonging to the same)'.[7]

The potential for spectacle became a coup de théâtre when a shogunal mission to Europe happened to arrive in London on the first day of the exhibition. The compound excitement was enough to produce repeated comment on both officials and exhibits in the pages of the *Illustrated London News*, including a half-page illustration of the ambassadors themselves later in May, and three full pages in September.[8]

DISCRIMINATING EXHIBITIONS

To append an international exhibition in the capital of the nineteenth-century world to a chronicle of exotic animals, revealed deities, and animated puppets is not to argue that there are no salient differences between them. It was clear to the Japanese officials in London that they were seeing something new. The scale of the undertaking and of the architecture needed to house it created a world and a Japan that they had not seen before. Their reaction, which ran the gamut from shock to awe, may even suggest that the difference in degree adds up to a difference in kind. But we should be wary of jumping to the conclusion that the two worlds were incommensurable. There is always a temptation to start the story with a categorical difference between the ancient and the modern, the West and the rest. It is more honest and useful to admit that there were similarities, too. Great international exhibitions may have been a new species, but the genetic resemblance was clear. In Japan, this implied that awe could lead to assimilation, that the imported form could be grafted onto native stock. Comparing the shows of Edo with the one in London allows us to understand the common environment that fostered the practice of exhibition, the mutation that led to international versions, and the problems and possibilities these offered to a place such as Japan.

The early modern world proved fertile ground for the culture of display. The civil wars of the sixteenth century in the Japanese

[7] Rutherford Alcock, *Catalogue of Works of Industry and Art, Sent from Japan* (Clowes & Sons, 1862), 2, 8, and passim. See also *International Exhibition 1862: Official Catalogue of the Industrial Department* (Truscott & Simmons, 1862), 338–39.

[8] *Illustrated London News*, 1862/5/3, 432; 1862/5/10, 470 and 480; 1862/5/24, 535 and 537; and 1862/9/20, 318–20.

archipelago (the seventeenth in the British Isles) had given way to a Great Peace (Glorious Revolution), which spurred economic growth, accelerated urbanization, and spurred a culture of curiosity, both among would-be elites and the population at large. The early modern city was a complicated, sophisticated place, spawning a stream of entrepreneurs and crowds of consumers, who provided a ready and growing demand for and therefore prompted the supply of revelation, entertainment, and spectacle.[9] By 1700, Edo had over a million inhabitants and was likely the largest in the world, a position it seems to have maintained for the following century. The point here is not precedence or pre-eminence, however. What is clear is the extent to which the practices of display and exhibition became a part of the everyday urban landscape, routinized, semi-professionalized, and highly diversified.[10]

Some shows in Japan were sacred, at least in terms of site and sales. Images of deities were usually secluded from worshippers, hidden away in badly lit recesses if not behind closed doors. But this also made them amenable for recruitment, at home or away, in the prosaic business of raising temple funds. And the latter could only be enhanced by providing space for other, more worldly attractions, generically designated as 'shown things' (*misemono*).[11] There were many things that might prompt curiosity and wonder among the crowds, thereby qualifying them to be put on display: natural curiosities such as camels, peonies, and goldfish; the physical dexterity of acrobats; the technical ingenuity of automata. Nor were such shows, trading in novelty rather than eternity, confined to temples.[12] Priests needed showmen more than the other way round, perhaps. But the logic was complementary and together they could stage a catholic spectacle wherein the audience could wander at will, petitioning the gods, admiring the devices, applauding the performers, and horripilating at the latest curiosity to make its way from the provinces or overseas.

[9] James McClain, 'Japan's Pre-Modern Urbanism', in Peter Clark (ed.), *The Oxford Handbook of Cities in World History* (Oxford University Press, 2013), 328–45, for an overview.

[10] For London, Richard D. Altick, *The Shows of London* (Belknap Press, 1978). For Japan, Tōkyō Kokuritsu Hakubutsukan, *Tōkyō Kokuritsu Hakubutsukan Hyakunenshi* (Tokyo Kokuritsu Hakubutsukan, 1973), vol. 1, 2–8.

[11] Nam-lin Hur, 'Invitation to the Secret Buddha of Zenkōji: Kaichō and Religious Culture in Early Modern Japan', *Japanese Journal of Religious Studies* 36.1 (2009), 45–63.

[12] Asakura, *Misemono kenkyū*, and Andrew Markus, 'The Carnival of Edo: Misemono Spectacles from Contemporary Accounts', *Harvard Journal of Asiatic Studies* 45.2 (1995), 499–541.

FIGURE 1.1 Early modern Japan had a thriving culture of public display. In Owari, medicinal products were a particular interest.
Source: *Owari Meisho Zue* (1844), National Diet Library.

Other exhibitions were more scholarly, almost secular. The study of *materia medica* had developed on the basis of Chinese texts from the beginning of the Edo period. From the mid eighteenth century, groups of scholars began meeting to show each other their collections and discuss their significance. The remit and scale of such gatherings had quickly expanded to include anything, animal, vegetable, or mineral, which might be useful in augmenting the country's wealth and productivity.[13] The key figure here was Hiraga Gennai, who had encouraged his teacher to organize the first exhibition of medical specimens in 1757 and quickly took over the organization of such events himself. In 1762, he set up collection agencies throughout the archipelago, producing an exhibition of over 1,300 products. A year later, some 360 of the exhibits were described and illustrated in *Butsurui Hinshitsu*, the first multi-volume exhibition catalogue in Japan.[14] One hundred years later, the tradition was still going strong under the tutelage of Itō (Figure 1.1). In principle,

[13] Isono Naohide, 'Yakuhinkai, Bussankai Nenpyō (Zōteiban)', *Keiō Gijuku Daigaku Hiyoshi Kiyō, Shizen Kagaku* 29 (2001), 55–65.

[14] Fukuoka, *The Premise of Fidelity*, 84–86, and Federico Marcon, *The Knowledge of Nature and the Nature of Knowledge in Early Modern Japan* (University of Chicago Press, 2015), 213–27.

product exhibitions were not open to the general public, although Hiraga's catalogue and later publications suggest that a public was being sought. Other gatherings were more explicitly commercial. Calligraphy and painting gatherings also took off around the turn of the nineteenth century, providing an opportunity for a paying public to see artists at work and in their cups.[15]

By the second half of the nineteenth century, then, exhibition, entertainment, and spectacle were as familiar in urban Japan as they were in metropolitan Europe, even if their proto-national potential had not yet been scaled up to the international stage. In Edo, the shows continued through the remainder of 1862, punctuated by the disturbances provided by the foreigners.[16] The summer was a quiet one, with not that much to see. Authorities and subjects were struggling to cope with an outbreak of cholera and measles, thought to have arrived on a foreign ship in Nagasaki. By the autumn, the epidemic had killed 100,000 in Edo alone.[17] In the circumstances, there was not much enthusiasm for putting things on display. In the seventh month, Chōkoku-ji, a temple adjacent to the Yoshiwara pleasure quarter, tried to get things going by unveiling its main image of a bodhisattva perched on top of an eagle (*washi*). This was ahead of schedule, though, which usually saw it put on show early in the winter, to celebrate the day of the rooster. Given the epidemic, few people came.

Things began to improve as summer turned to autumn. The eighth month began with more goldfish in Osaka and a series of performances by Hayatake Torakichi, the reigning superstar of acrobatics, in Edo. It also saw a major media event with the assassination of a British merchant outside Yokohama by samurai from Satsuma, indignant at his refusal to pay his respects as their lord and his retinue passed. The Namamugi incident became grist both for the publishing mill, commemorated in a series of woodblock prints, and for imperial posturing, with British demands for reparation and the bombardment of Kagoshima by its navy the following autumn. Even in the midst of national crisis, however, the year continued on its familiar course. The intercalary eighth and ninth months brought temple festivals, the tenth chrysanthemum dolls in Komagome and Senba, and the eleventh a fire in the Yoshiwara

[15] Andrew Markus, 'Shogakai: Celebrity Banquets of the Late Edo Period', *Harvard Journal of Asiatic Studies* 53.1 (1993), 135–67, and Furuta Ryō, 'Nihon no Bijutsu Tenrankai: Sono Kigen to Hattatsu', *Museum* 545 (1996), 29–56.

[16] Saitō, *Zōtei Bukō Nenpyō*, vol. 3, 143–45.

[17] *New York Times*, 1863/1/12, 2.

brothel district. The chronicle ended the year with automata in Asakusa, more fires, and more attacks on foreigners, as well as memoranda from Itō about the need for regular product exhibitions.[18] There was never a shortage of things to see and always more that could be put on display.

Nonetheless, the difference between the attractions of Edo Japan and the spectacle that greeted the Japanese officials in London was clear enough. International exhibitions also had their precursors: older stories squabble over English and French claims of precedence.[19] More important is the extent to which the new initiatives were prompted by the development of industry and empire. New technologies of production, transportation, communication, and destruction had begun to enable access to new resources and markets overseas, increasing the capacity for and probability of violence, like that which punctuated the year in Edo. In East Asia, the changing rules of the game had been apparent to all in China's prostration following the First Opium War. In Europe too, the new calculus of technology and economy was accelerating the pace of change – transforming the urban landscape, disturbing social hierarchies, and exacerbating regional rivalries. By the mid nineteenth century, royals and others were therefore demanding comprehensive overviews and stricter accounts: surveying the state of the world and displaying its lessons to one's people at exhibitions, themselves made possible by new technologies of construction and illumination, became a concern of increasingly anxious states.[20] The ways in which international exhibitions therefore differed from their subnational predecessors were immediately apparent to even a sophomoric observer.

DEFINING EXHIBITIONS

Fukuzawa Yūkichi was twenty-seven years old when he arrived in Paris as one of the members of the Japanese mission to Europe in 1862. He had already visited the United States in 1860. He would go on to become the paradigmatic public intellectual of the late nineteenth century, trimming his sails neatly to steer the Japanese public from an embrace of Western ideas and institutions in the 1870s to shouldering the burdens of empire

[18] Isono, *Nihon Hakubutsushi Nenpyō*, 715.

[19] For example, Kenneth Luckhurst, *The Story of Exhibitions* (Studio Publications, 1951), 63.

[20] Emily S. Rosenberg, *A World Connecting, 1870–1945* (Belknap Press of Harvard University Press, 2012), for an overview. Jeffrey Auerbach, *The Great Exhibition of 1851: A Nation on Display* (Yale University Press, 1999), for anxiety.

from the 1880s. He was also careful to tend to the institutional basis for his intellectual endeavours, founding Keiō University, which remains a leading private university, and *Jiji Shinpō*, one of the most influential newspapers for the half century leading to World War II.[21] His place on today's 10,000-yen banknote, though, is principally underpinned by a stream of publications. Fukuzawa was too junior a member of the delegation to participate in the opening ceremonies of the London International Exhibition, but he visited it two weeks into its run. Five years later, he included exhibitions in a three-volume encyclopaedia, *Seiyō Jijō*, which enumerated the 'conditions in the West' that accounted, in his view, for its wealth and power. The book established both Fukuzawa's reputation and *hakurankai* as the authorized translation for 'exhibition'. Its use in what quickly became a bestseller ensured that it beat out other contenders as the neologism with which to identify what was clearly a key feature of the modern West.[22]

Fukuzawa lists exhibitions, neatly and suggestively, between museums (*hakubutsukan*), with which they shared a Chinese character, and steam engines (*jōki kikan*).[23] His entry not only notes their resemblance to the former but also explains why things such as steam engines make exhibitions necessary. The pace of change is such that what was rare yesterday is commonplace today. Exhibitions make sense of and further encourage this parade of invention. They bring immediate commercial benefits, but their main function is to allow the viewer to step back from the flurry of the market: to exchange knowledge and, crucially, to compare places. Seeing the products of each country enables one to evaluate its stage of moral and intellectual development (*enkaku fūzoku, jinbutsu no chigu*). Seeing where one stands in the parade serves to motivate the foolish, keep the intelligent on their toes, and further civilization (*bunmei*) throughout the world. Fukuzawa's account thus serves as a justification of the Japanese term for such events, with which to distinguish them from their early modern forbears and their contemporary kin. Museums provide palatial surroundings (*kan*) within which to spread out and contemplate a variety of 'products, antiques, and rarities' from throughout the world (*haku*); but an exhibition provides a place to gather (*kai*), not only to ponder but also to survey (*ran*).

21 Fukuzawa Yukichi, *The Autobiography of Yukichi Fukuzawa* (Columbia University Press, 2007).

22 Yoshida, *Bankoku Hakurankai*, 7.

23 Fukuzawa Yukichi, *Seiyō Jijō* (Keio Gijuku Daigaku Shuppankai, 2002), 49–50.

What Fukuzawa did not do in his encyclopaedia was communicate the experience of an exhibition, not least the anxiety that could strike visitors, overwhelmed by the magnitude of the undertaking, and alarmed by discovering their own place in the universal scheme of things. It was the scale of the Great Exhibition of 1851 that had marked it off from its domestic predecessors. In 1862, too, it was the sheer size of the exhibition that first struck Fukuzawa. His brief diary entry on what he then called a *tenkanjo* (literally, 'show and see place') and glossed in the phonetic syllabary as an *ekusubishon* (exhibition), begins by describing the 'vast brick structure ... covered with glass', which enables the interior to be flooded with light, despite the relative paucity of windows. It ends with the floods of visitors. 'Everyone comes. There is no one who has not seen it'.[24]

But, crucially, different people saw different things. Few visitors pondered the lessons of the whole, like Fukuzawa. Most lingered on particular exhibits. And even when similar visitors looked at the same exhibit, the lessons they drew could differ. Fuchibe Tokuzō and Takashima Yūkei were also members of the delegation in 1862 and were appalled by what they saw. The objects that comprised the Japanese exhibit – used women's clothes and straw raincoats, paper lanterns and wooden clogs – were ones that in Japan a samurai elite could hardly be expected to purchase, much less display (Figure 1.2). In London, they were even harder to contemplate. The Japanese section was clearly inferior to those of other countries. There was no way to trade the exhibits as goods 'anywhere in the world'. For Fukuzawa, however, the point was not the quality but the value. Small as it might have been, the exhibits as a whole still totalled 'some 20,000 ryō'.[25] An exhibition served to reveal the value of things. Even in 1862, Japanese things had value at the exhibition. Fukuzawa, in his late twenties, could see a future that his slightly older, certainly crustier companions could not. Both views of the archipelago at the exhibition – trapped in the past, ready for the future – were also echoed by their English hosts.

[24] Yamaguchi Kazuo, *Fukuzawa Yukichi no Saikō Junreki* (Fukuzawa Yukichi Kyōkai, 1980), 145.

[25] Ibid., 145–47, and Miyanaga Takashi, *Bunkyū Ninen no Yōroppa Hōkoku* (Shinchosha, 1989), 77–78. For the diaristic context, Susanna Fessler, *Musashino in Tuscany: Japanese Overseas Travel Literature, 1860–1912* (Center for Japanese Studies, University of Michigan, 2004), 97–106.

FIGURE 1.2 The Japanese Court in London in 1862 appalled some Japanese observers, but intrigued others.
Source: *Illustrated London News* (1862/9/2), New York Public Library.

DISCOVERING JAPAN

European markets and consumers had long been familiar with Japanese products, given more than two centuries of trade with the archipelago, facilitated by the Dutch East India Company. The shogunate had imposed restrictions on maritime travel and trade in the seventeenth century, which precluded any relationship with practising Catholics, along with foreign travel by Japanese, but left Nagasaki open for trade with both China, the dominant economy in the region, and the Dutch. The latter were eager to profit from the European demand for raw materials, metals, and luxury goods from eastern Asia, not least ceramics, whose supply was threatened given the disruption then roiling China. The demand would persist through the next two centuries.[26] 1862 was therefore not the first

[26] Robert Hellyer, *Defining Engagement: Japan and Global Contexts 1640–1868* (Harvard University Asia Center, 2009), for the system of trade and diplomacy; Adam Clulow, *The Company and the Shogun: The Dutch Encounter with Tokugawa Japan* (Columbia University Press, 2016), for the Dutch.

time Japan had appeared at an exhibition in the West, though previous exhibits had relied mainly on Dutch collections. The Great Exhibition of 1851 included some 'Japanese wares', but subsumed them under the China section in the official catalogue.[27] Two years later, exhibitions in Dublin and New York capitalized on the curiosity generated by the US expedition then under way to 'open' what the Irish promoters of the former claimed to be a 'forbidden land'.[28] The exhibit in New York was fragmentary, supplementing a small Dutch loan with some objects recovered from a shipwrecked junk.[29] Dublin was more comprehensive, ranging from architecture to armour and costumed mannequins to 'manufactured monsters', most likely mermaids.[30] The interest was sustained with a small exhibit at the Paris Exposition Universelle in 1855.[31]

It was in 1858, however, that the shogunate succumbed to American demands, soon followed by the British, French, Russians, and Dutch, for treaties that were soon understood in Japan to be unfair: opening four new ports to trade, lowering the duties that could be charged on both imports and exports, and allowing foreigners to live in these settlements, practise their religion, and be subject to their own rather than Japanese law. The following year saw Yokohama, the first treaty port, opened for foreign residence and trade. It was therefore London in 1862 that saw the first exhibit of 'Japan', somewhat on its own terms, catalysing the first in a series of waves of enthusiasm for things Japanese.

It is easy enough to find comments consigning Japan to a past that the West had left behind. William Burges, architect and organizer of a Medieval Court at the exhibition, suggested that someone who wanted to 'transport oneself into' the Middle Ages should visit the Japanese Court.[32] Expressions such as 'quaint picturesqueness' studded the accounts of the popular press, allowing for both curiosity and condescension.[33] The latter was tempered, however, with insistent admiration for

[27] *Official Catalogue of the Great Exhibition of the Works of Industry of All Nations, 1851* (Spicer Brothers, 1851), 214–16.

[28] *The Exhibition Expositor and Advertiser* 14 (Gunn & Cameron, 1853), 4–5.

[29] Holt, *The Art of All Nations*, 361–62.

[30] T. D. Jones, *Record of the Great Industrial Exhibition, 1853* (Dublin, 1853), 68–70.

[31] Ellen Conant, 'Refractions of the Rising Sun: Japan's Participation in International Exhibitions, 1862–1910', in Toshio Watanabe and Tomoko Sato (eds.), *Japan and Britain: An Aesthetic Dialogue, 1850–1930* (Lund Humphries, 1991), 79.

[32] William Burges, 'The Japanese Court in the International Exhibition', in Clive Edwards (ed.), *Nineteenth-Century Design: Volume IV, Networks Mediators and Design* (Routledge, 2021), 370–80.

[33] *Illustrated London News*, 1862/9/20, 318.

the quality of Japanese craftsmanship. An illustrated guided tour of the exhibition noted not only the ceramics and lacquerware, already well-known in Europe, but also the astonishing variety 'of beautiful, rich, and well-executed works with which this little Court abounded'.[34] The International Juries, charged with evaluating the comparative quality of the exhibits and awarding prizes on this basis, concluded that the bronzes were 'in advance of anything old or new that has been produced in Europe'.[35] The *Illustrated London News* allowed that the silks were remarkable examples of 'softness ... and durability', and even the mortifying straw raincoats were 'manufactured with great neatness'.[36] Some suggested that the exhibit therefore bore witness to indigenous Japanese capabilities, which marked them out from their Asian neighbours. The artist John Leighton argued that whereas the Chinese 'lack inventive power, working from instinct rather than from the dictates of reason – a fault with all Asiatics', the Japanese 'appear much more free in thought'.[37] It was only a short step to the enthusiastic embrace of Japanese art and design by the aesthetic avant-garde and social elite.[38]

For the organizer of the Japanese exhibit, this had been in part the point. Rutherford Alcock had been in China since 1844 and had taken the lead in arranging its representation at the Great Exhibition in 1851. He moved to Yokohama in 1859 as the first British Minister, where he extended an invitation for the 1862 exhibition to the Tokugawa shogunate, which promptly appointed him commissioner and asked him to organize their exhibit. For Alcock, the goal of doing this was to identify potential items of trade.[39] In the event, the exhibit was more diverse than this, including a miscellany of reference materials, necessary given the lack of European familiarity with the archipelago. These included the objects that had outraged some of the Japanese delegation, but Alcock's

[34] Taliaferro P. Shaffner and William Owen, *The Illustrated Record of the International Exhibition of All Nations, in 1862* (Ward and Lock, 1862), 112.

[35] *International Exhibition, 1862: Reports by the Juries on the Subjects in the Thirty-Six Classes into Which the Exhibition Was Divided* (Clowes & Sons, 1863), class XXXI.

[36] *Illustrated London News*, 1862/9/20, 318–19.

[37] Holt, *The Art of All Nations*, 366.

[38] For example, Christopher Dresser, *Development of Ornamental Art in the International Exhibition* (Dan and Son, 1862), 148. For the impact, Holt, *The Art of All Nations*, 360–77, and Yuko Kikuchi and Toshio Watanabe, 'The British Discovery of Japanese Art', in Gordon Daniels and Chushichi Tsuzuki (eds.), *The History of Anglo-Japanese Relations 1600–2000. Volume V Social and Cultural Perspectives* (Palgrave Macmillan, 2002), 151–53.

[39] Sano Mayuko, 'Banpaku no Hito, Razafōdo Ōrukokku', in Sano Mayuko (ed.), *Bankoku Hakurankai to Ningen no Rekishi* (Shibunkaku Shuppan, 2015), 21–35.

catalogue suggests that they too demonstrated the 'delicacy of workmanship and the perfection of material'. In the preface, he underlined the broader point. There were products in Japan 'that either cannot be produced in [European] workshops or ateliers, or only at a cost that would make them practically unattainable, and wholly unprofitable'.[40] An international exhibition worked by disaggregating a country in order to reveal the resources and products that might make their way to the international market, even if that was largely due, as here, to the cost of labour. The jury charged with evaluating silk and velvet concurred: Japan's strength lay in its industrial arts, but raw silk also had export potential. It was 'gratifying' to find that a 'supply of better qualities ... could, under a more liberal government ... be indefinitely increased'. They looked forward to the moment 'when Daimios no longer oppose the best kind of silk being exported'.[41]

London in 1862 thus pointed to the possibilities and the perils that faced Japan at the exhibition and beyond. True, it revealed that Japanese exhibits were at odds with the world on display in London. For Fuchibe and Takashima, preoccupied with status, this was a cause for embarrassment. For Alcock, concerned with trade, it meant opportunity. He admitted that some objects – old lacquer, bronzes, swords, and armour – are only to be had at prices 'which in Europe would probably be thought far disproportionate to their value', meaning that their interest was mainly historical and ethnographic. But others – ceramics and lacquer, already familiar in European markets, as well as metalwork, silk, and textiles, which were less so – are 'procurable ... at moderate prices; and some even at very little costs', piquing the interest of both the industry-minded juries and aesthetically inclined cognoscenti.[42] Fukuzawa was attuned both to political economy and to the institutions, culture, and values by which it was informed. London demonstrated the utility of exhibitions. They not only exposed comparative advantage and deficiency. They also revealed how to use the former and address the latter. As Japan began to engage with its new terms of trade, it could already find a place at the exhibition and in the global market. Making the most of both might turn its world upside down. The ability of exhibitions to do these things, however, would soon be seized on by the authorities as a way to capitalize on Japanese assets. Their first attempt

[40] Alcock, *Catalogue*, 2.
[41] *International Exhibition, 1862: Reports by the Juries*, class XX, 5.
[42] Alcock, *Catalogue*, 2.

to do so came five years later in Paris, where the lessons of London and the distance still to go were both underlined.

PRETENDING TO BE UNIVERSAL

By 1867, the point of international exhibitions was clear. The organizers of the second Exposition Universelle in Paris were determined to produce an epitome of the genre, displaying 'the resources which industry can produce to satisfy man's needs', grouping together 'the products which correspond to each of these needs', and so drawing 'a complete picture of human industry in the most true and compelling form'. They would meet the demands of the public, that exhibitions should be 'complete, that is, include all the nations of the world and all that is interesting to know about them', and who would be unhappy if 'one of these solemn occasions remained within the limits of those which had preceded it'. The advance of industry and the expectations of visitors combined to require a site covering four times as much ground as any previous exposition.[43] The ambition of the organizers was also conceptual. An exhibition should provide both a 'complete picture' of industrial production and a demonstration of how progress might guarantee social harmony. The system of classification therefore swelled to incorporate not only art and agriculture but also exhibits aiming at 'the amelioration of the moral and physical condition of the population'. The commissioners also promised to recognize establishments that 'guaranteed to the workers their material, intellectual, and moral well-being'.[44]

Architecturally, the official aspirations were given form by Gustave Eiffel, who would later erect his eponymous tower for the 1889 exposition. This time, he designed a massive oval building, divided into concentric galleries. Each gallery was dedicated to a particular class of objects. Each participating nation was allocated a particular slice of the cake.

[43] Commission Impériale, *Rapport sur l'Exposition Universelle de 1867 à Paris* (Imprimerie Impériale, 1869), 5, 16, 265–66; Exposition Universelle de 1867 à Paris, *Guide Officiel à l'Exposition Universelle de 1867* (Dentu, 1867), 5–6.

[44] Commission Impériale, *Rapport*, 23 and, for the full classification, appendix 5. Exposition universelle de 1867, *Guide Officiel*, 7. For overviews, Kashima Shigeru, *Zekkei Pari Bankoku Hakurankai: San Shimon no Tetsu no Yume* (Kawade shobo shinsha, 1992); Pieter Van Wesemael, *Architecture of Instruction and Delight: A Socio-Historical Analysis of World Exhibitions as a Didactic Phenomenon (1798, 1851, 1970)* (Uitgeverij 010, 2001), 221–330; and Ichikawa Fumihiko, 'Kindai Pari Banbutsuhaku no Kiseki: Sono "Ban'yū Rinen" ga Arawasu mono', in Sano Mayuko (ed.), *Bankoku Hakurankai to Ningen no Rekishi* (Shibunkaku Shuppan, 2015), 533–60.

The design was meant to promote flexibility for the exhibitor and visibility for the visitor, who could either dedicate themself to the industry in which they were most interested, strolling through the galleries in order and seeing 'the complete panorama of universal production unfold in succession'; or explore a particular country, moving from fine art, through 'materials and their applications in the liberal arts', to furniture, clothing, and 'extractive industries'. The innermost circle of the palace was occupied by the gallery devoted to the history of labour, intended to demonstrate the main 'epochs of each people'; the outermost by a Galerie des Machines, which incorporated a platform, allowing the visitor to surveil the whole, and was ringed by cafes and restaurants.[45] Beyond the palace spread the park, where the authorities had provided participating countries with enclosures in which to exhibit their characteristic architecture, and where the visitor might also find some respite.[46]

The official rhetoric was all scientific method and universal truths. On the ground, however, the exhibition was a coliseum of international rivalry and national distinction. It sought, above all, to place France, and particularly Napoleon III, at the centre of a world putting itself on show. Eighty years after the revolution, having gone through two republics and coming to the end of its second empire, the state and its monarch needed some help.[47] This was nowhere clearer than in the main exhibition building, where 'France and her colonies' occupied almost half the available space. The aspirations of the organizers were given their fullest expression, however, by the pen of Victor Hugo, in a vatic introduction to an encyclopaedic guide to the city. In 'the twentieth century', he prophesied, the influence of 'an extraordinary nation' would make the 'fallow lands of America, Asia, Africa, and Australia ... available to civilizing emigration'. Human needs would be satisfied, property 'democratized and universalized'. This nation 'will have Paris as its capital, and

[45] Commission Impériale, *Rapport*, 5–7, 16, 21.

[46] *Guide Officiel*, 111; Commission Impériale, *Rapport*, 9.

[47] For the importance of ritual for monarchy and empire during the Second Empire, Matthew Truesdell, *Spectacular Politics: Louis-Napoléon Bonaparte and the Fete Impériale, 1849–1870* (Oxford University Press, 1997). For its prehistory, Philip Dwyer, '"Citizen Emperor": Political Ritual, Popular Sovereignty and the Coronation of Napoleon I', *History* 100.339 (2015), 40–57. For spectacle in everyday life, Vanessa Schwartz, *Spectacular Realities: Early Mass Culture in Fin-de-Siecle Paris* (University of California Press, 1998).

will no longer be called France; she will be called Europe … In the centuries following, transformed again, she will be called Humanity'.[48]

These were fond hopes, but it did not take much to see things differently. Sir Henry Cole, the impresario of the Great Exhibition of 1851 and the British commissioner in Paris, pointed out that the exhibition itself stymied the vision of progress and harmony. The architecture, designed to reveal both, in fact prevented 'any lengthened views'. The longed-for panorama was always disappearing around the corner, over the horizon. Not only that. The 'specious and attractive' classification promised that each nation might be seen in its proper place, but the profusion of 'national varieties' rendered it 'impossible to fix any standard of excellence'.[49] Given such chaos, it was perhaps only from a distance that the system of the exhibition could reveal itself. One possibility was the official bird's eye view, such views having begun to take off during the decade.[50] Another was a small hill, some distance from the Champ de Mars, whose summit had been lowered and profile tidied during the run-up to the exhibition by Baron Haussmann, as part of his ongoing municipal improvements. The finished product provided not only an inviting space for the leisured strolling of the emerging middle class but also a place from which Edouard Manet was able to compare the pretensions of the spectacle with the fragmentation of modern life.[51] Unlike Manet, however, the Japanese participants did not have the luxury of distance. They were part of the show. As in London, so in Paris, the exhibition served to reveal both the problems and the possibilities of being Japan.

MULTIPLYING JAPANS

Some of the problems were homemade. By the second half of the 1860s, the need for Japan to engage with the challenge posed by the West was urgent. The shogunate had embarked on political reforms, but its efforts were dogged by conservative factions and sclerotic decision-making. It was also being challenged from above and below. Already in 1863,

[48] Victor Hugo, 'Introduction', in A. Lacroix, Verboeckhoven & Cie, *Paris Guide, par les Principaux Écrivains et Artistes de la France,* vol. 1 (Libraire internationale, 1867), i–v.

[49] *Reports on the Paris Universal Exhibition, 1867, Volume 1, Containing the Report by the Executive Commissioner, and Appendices* (1868), ix–xxv.

[50] For example, Eugène Ciceri and Philippe Benoist, Vue Officielle à Vol d'oiseau de l'exposition Universelle de 1867 (1867), www.loc.gov/pictures/item/00652015/.

[51] T. J. Clark, *The Painting of Modern Life: Paris in the Art of Manet and His Followers* (Knopf, 1984), 60–66.

the Chōshū and Satsuma domains had taken matters into their own hands, in response to an imperial order to 'expel the barbarians', prompting violent reaction by the foreign powers and punitive expeditions by the shogunate. By the middle of the decade, shogunate and domains were seeking their own connections to and support in the West to bolster their domestic cause. Satsuma went the furthest. It had long used its suzerainty over the Ryūkyū Islands to escape shogunal oversight and boost its own economy. Now it inflated this into a claim of independence from the Tokugawa state.[52] The year 1865 saw missions to France by the shogunate and Satsuma, as well as the French extending an invitation to the exhibition through their envoy in Edo. The shogunate slowly overcame in-house protests to assemble an exhibit and a mission, headed by the thirteen-year-old brother of the shogun, Tokugawa Akitake. They were outpaced, however, by the delegation from Satsuma, which signed a contract while they were in France with the Comte Cantons des Montblanc, the Paris-born son of a minor Belgian aristocrat, who became their commercial agent in France, with various responsibilities, including their representation at the exhibition.[53]

The shogunal party thus arrived in Paris in April 1867 to find a fait accompli. The Satsuma delegation had arrived before Akitake and his entourage had even left Yokohama. They had participated in the opening ceremonies. More galling still, the Satsuma exhibits, only half-vetted by the shogunal authorities, were on display in the palace and listed in the catalogue as those of 'His Majesty, King of the Ryukyu Islands', asserting a sovereignty all its own. A long and emotional meeting between the two sides was swiftly arranged and a diplomatic compromise produced, with some tinkering to the French translations of their respective titles. But the solution did little either to clarify matters on the ground or to derail the tireless Montblanc, who produced a stream of articles in the Paris press,

52 W. G. Beasley, *The Meiji Restoration* (Stanford University Press, 1972) and Conrad Totman, *The Collapse of the Tokugawa Bakufu, 1862–1868* (University of California Press, 1980), for the standard accounts. Hellyer, *Defining Engagement*, for Satsuma.

53 Tōkyō Kokuritsu Hakubutsukan, *Hyakunenshi*, vol. 1, 13–19, and vol. 2, 553–61, for an overview and relevant documents. Sumi Yutaka, *Tokugawa Akitake: Banpaku Tonosama Ichidaiki* (Chūō Kōronsha, 1984), for the shogunate. Inuzuka Takaaki, 'Pari Bankoku Hakurankai to Satsuma gaikō: "puropaganda" no gaikō senryaku', in Kagoshima Junshin Joshi Daigaku Kokusai Bunka Kenkyū Senta, *Shin Satsuma Gaku, Vol. 1, Sekai No Naka No 'Satsuma'* (Nanpō Shinsha, 2002), 13–52, for Satsuma. Willy Vande Walle, 'Le Comte Des Cantons Charles de Montblanc (1833–1894), Agent for the Lord of Satsuma', in Ian Neary and Stephanie Rogers (eds.), *Leaders and Leadership in Japan* (Routledge, 1996), 39–55, for Montblanc.

as well as a medal to be pinned on the chests of worthy Europeans, continuing to argue Satsuma's case for independence. In the event, the debate was only a minor distraction in a busy French summer. By early August, it was clear that Montblanc had failed to convince the French public.[54] He was not the only problem facing Japan, however.

As in London, five years previously, the exhibition served to show up the extent to which the archipelago found itself at odds with the ways of the world. Part of this was simply a question of space. In the palace, this had been allocated to countries on the basis of various criteria – space occupied in previous expositions, number of exhibitors, 'productive forces', political importance, and nearness to Paris – none of which favoured Japan, as the official shogunal account pointed out. France occupied nearly half of the total area, England 1/6, Prussia 1/16, and the US 1/32, while Japan together with China and Siam had only been given 1/128.[55] Japan took up half the available sliver of space, but the problem was compounded by the way in which it had been framed. A Monseiur Chapon had designed the cases by mashing up a series of Moorish, Indian, and Chinese motifs. The arrangements appalled the more discriminating observer, who noted how the Japanese exhibits were 'badly housed, piled up along an obscure corridor'.[56] But there was also a problem with Japan itself.

The various Japanese parties had sent over a huge amount of material. The shogunate's 189 cases and Satsuma's 506 included a bit of everything, from ceramics, lacquerware, metalwork, and weaponry to various primary products, including quartz, sulphur, timber, tobacco, and sugar – the last a proof of Satsuma's claim to the Ryukyus. The southwestern domain of Saga, meanwhile, had sent 520 cases of Arita ceramics, long well-known and much-appreciated in Europe. The three governments had also been joined by an assortment of Edo merchants, who sent 157 cases, also mainly of arts and crafts.[57] By the standards of the exhibition,

54 Angus Lockyer, 'The Problem of Sovereignty in an Age of Empire: Representing Japan in Paris in 1867', *Critical Asian Studies* 45.4 (2013), 615–42. See also Mark Ericson, 'Sovereignty on Display: The Tokugawa Bakufu and the Paris Universal Exposition of 1867', *The Journal of Japanese Studies* 47.2 (2021), 277–311.

55 Commission Impériale, *Rapport*, 5, 53, and 440; Shibusawa Eiichi, *Shibusawa Eiichi Taifutsu Nikki* (Nihon Shiseki Kyōkai, 1928), 81.

56 David Bromfield, 'Japanese Representation at the 1867 Paris International Exhibition and the European Response to It', in Alan Rix and Ross E. Mouer (eds.), *Japan's Impact on the World* (Canberra: Japanese Studies Association of Australia, 1984), 135–37.

57 Sumi, *Tokugawa Akitake*, 52; Inuzuka, 'Pari Bankoku Hakurankai', 23–24.

however, all this still fell considerably short. Japan had nothing to show in thirty-eight of the ninety-five classes into which the exhibits were officially divided, and in many of the others it could only produce one or two pieces.[58] Most galling, it was out of place in the outermost and innermost galleries of the palace. Interrupting the parade of machinery around the edge of the palace was 'le Kiosque Japonais', with Islamic decoration evoking the riverside arbour of a domainal lord. Inside could be found some 'magnificent porcelains' sent by the shogunate. Outside, 'all-powerful' Satsuma stood guard, with a warrior in a 'uniform unchanged for centuries'. 'Why … this motionless, terrible sentinel … in the peaceful precincts of the Exposition?'[59] Meanwhile, in the centre of the palace, the Orient confounded the possibility of a history of labour, by presenting the 'monuments of the past' among 'the products of … modern industry'.[60] Japanese prints, together with Chinese vases, Egyptian costumes, and 'the white elephant of the King of Siam' were to be understood rather 'from an ethnographic point of view than that of science and industry'.[61]

The possibility and threat of ethnography was further exacerbated by the Japan on display in the park, where the official aim of considered pedagogy was again subverted by the unexpected and picturesque. 'At the bend of a green path one suddenly finds oneself in front of a small state … "It's the empire of the Mikado! … Isn't it charming?"'[62] Satsuma had built a hunting pavilion, which drew admiring comments from the jury for its delicate craftsmanship.[63] But it was overshadowed by two buildings, installed by Shimizu Usaburō, an Edo merchant: a teahouse and 'l'appartement des femmes', which housed three 'authentic daughters of Yeddo' (Figure 1.3). The shogunal diarist noted the sensation provoked by 'the first Oriental women who have ever come to Europe. The crowd, forbidden to enter the room, jostled with one another, and peered at them intently through spyglasses'.[64] A French reporter noted their skin complexion, 'exactly that of a dark-haired Andalusian', but bemoaned the practice of 'lightening [it] with rice powder in such a quantity as to prove

58 Commission Impériale, *Rapport*, 442–46.

59 *L'Exposition Universelle de 1867 Illustrée*, no. 12 (1867/6/8), 186, and no. 15 (1867/6/21), 234–35.

60 Michel Chevalier, *Rapports du Jury International* (Paul Dupont, 1868), vol. 1, 240.

61 Exposition universelle de 1867, *Guide Officiel*, 102–3.

62 *L'Exposition Universelle de 1867 Illustrée*, no. 23 (1867/7/22), 363.

63 Chevalier, *Rapports*, vol. 1, 329.

64 Shibusawa, *Shibusawa Eiichi Taifutsu Nikki*, 95.

FIGURE 1.3 An Edo merchant brought three geisha to Paris in 1867. They caused a sensation.
Source: *Le Monde Illustré* (1867), Fine Art Images/Heritage Images via Getty Images.

that rice is a Japanese product'. He also admitted he found it difficult to distinguish some mannequins, 'exact' copies of Japanese workmen, from the 'living, breathing' Japanese selling products on the farm.[65]

So far, so Orientalist, perhaps. Japan was not alone in the late nineteenth century in being subjected to the prurient gaze of uninformed observers. By the same token, though, Japan could turn that attention to useful account. In years to come, the features on display in the park – manicured nature, wooden buildings, intricate handicraft, green tea, and kimono-clad women – would become staples of Japanese participation in international exhibitions and thereby its self-definition on the global stage. Auto-exoticism has its uses as a way of defining a cultural heritage, thereby providing an indigenous foundation for modern achievement.[66] Even in the 1860s, it was clear that this might be possible. As in London, so in Paris, when the reviews came in, it became clear that the exhibition may have disadvantaged, but it also disaggregated Japan to reveal the

[65] *L'Exposition Universelle de 1867 Illustrée*, no. 23 (1867/7/22), 364.
[66] For auto-exoticism, Tim Winter, 'Auto-Exoticism: Cultural Display at the Shanghai Expo', *Journal of Material Culture* 18.1 (2013), 69–90.

value of the archipelago. Some of this was as a source of agricultural products and raw materials. The shogunate was awarded a silver medal for its quartz hyalin and, alongside Satsuma, bronze for its tobacco. Even more promising was Japanese silk. Satsuma received a silver medal for its silk thread and fabrics, but more telling was the *grand prix* awarded to the shogunate for sericulture. Given the blight then infesting the European industry, the resistance of Japanese silkworms was noteworthy, even if it was still unclear just how much raw silk the archipelago could provide.[67]

But there was more. The objects on display in Paris only deepened the abiding fascination in Japanese art and craft, evident in the comments of the juries. The *grand prix* for the shogunate also encompassed its exhibits of paper, 'industrial arts', and lacquerware. The appreciation of Japanese stationery extended to Satsuma and the various Edo merchants, with a scattering of silver and bronze medals for paper, inkwells, and the like. The enchantment with Japanese lacquer was underlined, with Satsuma acknowledged as *hors concours* for items of 'incomprehensible dexterity'. The jury's judgement was echoed by the press, the art world, and the market. Ernest Chesneau, art historian and critic, rhapsodized about the 'remarkable aesthetic power' and 'inimitable perfection of design' of Japanese art. The following year, Manet was including Japanese objects in his own painting as a way into modern art. Soon enough the gold and white Satsuma ware at the exhibition had prompted a boom in European demand and the construction of kilns to produce it as far away as Edo, soon to become Tokyo.[68]

More than this, reassembled in the aggregate, the objects suggested a moral about the country as a whole. For the official illustrated newspaper, the Japanese exhibit was 'the most complete and brilliant ... of all the

[67] Exposition Universelle de 1867 à Paris, *Catalogue Officiel des Exposants Recompensés par le Jury International* (Dentu, 1867), and Chevalier, *Rapports*, vol. 4, 174. For the story of silk, Yasuhiro Makimura, *Yokohama and the Silk Trade: How Eastern Japan Became the Primary Economic Region of Japan, 1843–1893* (Lexington, 2017), and Simon Partner, *The Merchant's Tale: Yokohama and the Transformation of Japan* (Columbia University Press, 2018).

[68] For the prizes, Exposition Universelle de 1867, *Catalogue Officiel des Exposants*, and Chevalier, *Rapports*, vol. 3, 494. For the contemporary press, *Revue des Deux Mondes*, vol. 70 (1867), 730–31, and Gabriel P. Weisberg, 'Japonisme: Early Sources and the French Printmaker 1854–1882', in Weisberg (ed.), *Japonisme: Japanese Influence on French Art, 1854–1910* (Cleveland Museum of Art, 1975), 1–19. For the retrospective revelation, Martin Eidelberg and William R. Johnston, 'Japonisme and French Decorative Arts', in Weisberg (ed.), *Japonisme*, 18. For the boom in Satsuma ware, Sarah Teasley, *Designing Modern Japan* (Reaktion Books, 2022), 47–48. For Manet, Aiko Okamoto MacPhail, 'Imagining Modernity: European Japonism and Japanese Westernism' (PhD thesis, Indiana University, 2001).

Asian states'. For the jury responsible for textiles and silk, where China 'has lost its creative spirit', in Japan 'the movement of progress has not been halted'. To the extent that it thereby conformed to the general logic of the exhibition, its exhibits suggested that Japan might even 'take its place among the civilized nations'.[69]

LEARNING FROM THE EXHIBITIONS

Before it did, there was work to do, which was evident to the Japanese observers in Paris, as they had been to Fukuzawa in London. In 1867, his insights were widely shared by members of all the Japanese parties, but most clearly in the official shogunal diary, which was produced by Shibusawa Eiichi. Then in his late twenties, Shibusawa would go on to become the quintessential entrepreneur of modern Japan, in which capacity he returned again and again to exhibitions as a key to the development of the modern nation. Like Fukuzawa, Shibusawa's first impressions were of the size of the building and the quantity of the exhibits, but he soon turned to the way in which the exhibition served to illustrate and further promote industrial competition. In general, he noted, 'the European countries compete in the delicacy of their craftsmanship and the novelty of their knowledge'. At the exhibition, 'they carry ingenuity to extremes, striving to outdo themselves in enhancing their reputation in the world'. The exhibition facilitated this by enabling a threefold comparison, among fields of endeavour, peoples in space, and development over time. The exhibits thereby allowed an observer to 'ascertain the customs of each country and the intelligence of its people'. The 'rough and remote' nature of countries such as Arabia was apparent, as was 'the extent to which civilization has developed in a country like harmonious Norway'. Difference and deficiency could be directly 'inferred from [their] products'. Given the frame, inevitably, Japan sometimes stood out. 'Amidst the round currencies of the European state and all the other countries, it is the only one with a square shape'. In general, though, Shibusawa noted, the Japanese exhibits were well-regarded, ranking with Britain and the US in being recognized with the highest award.[70]

69 *L'Exposition Universelle de 1867 Illustrée*, no. 21 (1867/7/15), 331; Chevalier, *Rapports*, vol. 4, 206.

70 Shibusawa, *Shibusawa Eiichi Taifutsu Nikki*, 82–84, and Sekine Hitoshi, 'Shibusawa Eiichi to Bankoku Hakurankai: Pari Banpaku (1867 nen) kara Panama Taiheiyō Banpaku (1915 nen) made', in Sano Mayuko (ed.), *Banpakugaku: Bankoku Hakurankai toiu, Sekai o Haakusuru Hōhō* (Shibunkaku shuppan, 2020), 118 and

More than this, the exhibition offered an opportunity to examine the technology through which Euro-America had begun to transform the world. As another member of the shogunal party noted, 'there are many things which could be studied with profit'.[71] For Shibusawa, it was clear that the governing principle was the 'investigation of things', through empirical observation and experimentation. This was not a new idea. Hiraga, Itō, and their colleagues had been doing much the same in Japan for at least a century.[72] But there was a difference between materia medica and metal machinery. Shibusawa did his best. 'The steam engine . . . is a marvel of design and technical achievement. One should be able to grasp its working principles, but to my regret, I lacked sufficient knowledge and, uncomprehending, merely gazed at it as one might at a passing cloud or wisp of smoke.'[73] Part of the problem was language. 'Water thunder fire' (*suiraihi*) involved attaching three cords to 'something like a twenty-gallon barrel made out of steel' and attaching the cords to a stand on the ocean floor so that the barrel floated in the water. When a ship hit what would later become known as a mine (*shokuhatsu suirai*), the 'gunpowder in it explodes and the ship breaks up'. Searchlights were 'like an earthenware mortar', with a surface 'like a mirror', placed on a 'high stand like a firetower'. The search for similes signalled, clearly, an order of things not yet represented in Japanese. It also suggested, however, the desire and possibility of assimilating these products, making the knowledge embodied in them available to and productive in Japan, even if this would require considerable labour. 'It is hard to understand them even if one looks at them for a long time, and they are surely even harder to understand written down like this'.[74]

The exhibition exposed Japan, therefore, revealing the ways in which it fell short in linguistic capacity, technical understanding, and industrial development. But it also showed that the archipelago had resources, with which to remedy the deficiencies, and provided access to the technology, with which to exploit the resources. The problem, in the first instance, was the Japanese state. By 1867, the shogunate was doing everything it could at home to reorganize the government and modernize the military, the latter with French assistance. Its decision to participate in the

124. See also Shibusawa Eiichi, *The Autobiography of Shibusawa Eiichi: From Peasant to Entrepeneur* (University of Tokyo Press, 1994), 162.

[71] Sumi, *Tokugawa Akitake*, 55–56.

[72] Fukuoka, *The Premise of Fidelity*, and Marcon, *The Knowledge of Nature.*

[73] Shibusawa, *Shibusawa Eiichi Taifutsu Nikki*, 81–82.

[74] Sumi, *Tokugawa Akitake*, 55–56.

exhibition had been one demonstration of intent. But its nominally subordinate domains were already calling for power to be returned to the emperor, in part to pre-empt the growing power of Chōshū and Satsuma. The shogunate's performance in Paris had also revealed the extent to which its anachronistic constitution made it incapable of conforming to newly emerging, rapidly evolving international norms.

The general lesson was underlined late in the year by Godai Tomoatsu, a samurai from Satsuma. Godai would eventually go on to do for Osaka what Shibusawa would do for Tokyo, based in large part on his experience in the 1860s. In 1863, he had fought against and been captured by the British. Come 1865, he was studying at University College London and involved in the negotiations with Montblanc.[75] Now, in a conversation with the head of the delegation on the latter's return from France, he was scathing. The shogunate's peremptory demands were 'contrary to human nature'. They demonstrated the extent to which existing Japanese practice was bedevilled by 'evil customs', 'old abuse[s]' that demanded 'rectification'. There was a solution to hand, however. Japan should 'return to a unified country' (*tenka issei*) and 'through every exertion make up what was deficient'. With the emperor as the head of state, Japan would not be 'the only example in the world of a country with two lords, but rather no exception to the law of nations' (*bankoku kōhō*).[76] His wish would soon be granted. One week after the close of the exhibition, the last shogun would admit his time was up, submitting his resignation and returning his authority to the newly installed emperor.

So much for domestic politics. But Godai went on to underline the difference between the old world and the new that had been evident at the exhibition. Until now, he noted, Japan's whole world had been Japan, but from now on, everything would be related to public opinion in the treaty countries. This is why it was impossible to practise the 'old customs' (*kyūhō shikitari*) of 'ancient Japan' (*korai Nihon*). Nowadays, it was the 'trends of the times' that drove events, which arose beyond the domainal and archipelagan borders that had previously secured the political order. That order had been regulated by the 'principles of public debate', but once its borders had been thrown open, the imperatives that drove action

[75] For Godai and Europe, John H. Sagers, *Origins of Japanese Wealth and Power: Reconciling Confucianism and Capitalism, 1830–1885* (Palgrave Macmillan, 2006), 82–90, and Hellyer, *Defining Engagement*, 200–4.

[76] Kagoshima-ken Shiryō Kankōkai, *Kagoshima-Ken Shiryōshū* (Kagoshima Kenritsu Toshokan, 1962), vol. 4, 481–83.

'originated outside' such principles. This was why, Godai suggested, Japan had to unite if it was going to withstand the corrosive trends of the times.[77] This was why, as Fukuzawa and Shibusawa had noted, the industrial, imperial powers had turned to exhibitions as a way of surveying and keeping up with such trends. This was why, when the various Japanese parties made their way back from Paris, they included a number of officials who were convinced that exhibitions should be a priority for the modern state.

77 Ibid., 484–86.

2

Industrial Promotion

If you find yourself in Nagoya, it will not be long before you meet the *shachi*. Two mythical beasts, one male, one female, with the bodies of carp and the faces of tigers, they sit at either end of the ridgepole crowning the tower of the castle. But they do not stay there. You can find them on manhole covers and as bean-paste-stuffed confectionery souvenirs. They have lent their name to a J-pop group, a language resource database, various local sports teams, and an army division based in the city. Sometimes, they even come down to the ground. In 2005, when a world expo came to the suburbs, they were the centrepiece of a concurrent exhibition in town, celebrating the castle and the new century. They descended again in spring 2021 to help fight Covid.[1] But their celebrity – their apotheosis as a symbol of Nagoya and as the model of the breed – is due mainly to their adventures in the 1870s.

In the early seventeenth century, you could find *shachi* on castles throughout the archipelago. The ones in Nagoya were solid gold, and had slimmed down somewhat over the years, providing relief when the local Owari domain ran into fiscal problems. Then, in 1871, with the domain scheduled for dissolution and the castle for demolition, they were donated to the newly prominent Imperial Household. They soon found themselves on the road. The first stop for the male was a specially built case at the centre of an exhibition in spring 1872 at Yushima Seidō in Tokyo (Figure 2.1). With his celebrity canonized in woodblock prints, he continued on a tour of the further reaches of the archipelago. But his

[1] Nagoyajō Kinshachi Tokubetsu Tenran, 'Nagoyajō Kinshachi Tokubetsu Tenran' (2022), https://kinshachi-2021.jp.

FIGURE 2.1 The *shachi* belonged on the roof of Nagoya castle. Their appearance at ground level in Tokyo provided an opportunity to make fun.
Source: Shōsai Ikkei, *Moto Shōheizaka Hakurankai* (1872), Keio University Library.

partner was travelling much further afield. By May 1873, she stood on a pedestal, guarding the entrance to the Japanese gallery at the Vienna *Weltausstellung* (World Exhibition), which was stuffed with ceramics and other export crafts. When she finally made it back to Japan, she was reunited with her partner. They starred in an exhibition in Nagoya in 1878 and reassuming their customary positions on the roof of the castle, by then saved from demolition, the following year.[2]

Putting the *shachi* on display in Tokyo and Vienna was part of a deliberate strategy by the new Japanese government, which embraced exhibitions as a catalyst for industrial promotion and economic development. This strategy was formulated, in the first instance, by a number of men who had attended the Exposition Universelle in Paris in 1867. They had been on different sides then. They continued to differ in interest, vision, and prescription, but they were united in their belief that putting objects on display would benefit the modern state. The exhibitions in which they participated and which they organized – Vienna 1873, Philadelphia 1876, Paris 1878 and 1889; and Tokyo 1877, 1881, and 1890 – have been well-chronicled, for good reason.[3] For their contemporaries, they were among the most visible harbingers of what was initially called 'civilization and enlightenment', that is, what it meant to live in a world of industry and empire.[4] In retrospect, they have proved a spawning ground for numerous subsequent histories, of art and industry, museums and technology. At the time, they also reinforced the first wave of Euro-American interest in Japan, which went under the name of Japonisme.[5]

[2] Tōkyō Kokuritsu Hakubutsukan, *Hyakunenshi*, vol. 2, 622–27. Once reinstalled, the *shachi* were surrounded by wire cages, to prevent theft, not always successfully, and destroyed in American air raids in 1945. Their replacements use less than half of their ancestors' weight in gold.

[3] For an overview, Kuni Takeyuki, *Hakurankai to Meiji no Nihon* (Yoshikawa Kōbunkan, 2010), for domestic exhibitions, Kuni Takeyuki, *Hakurankai no Jidai: Meiji Seifu no Hakurankai Seisaku* (Iwata Shoin, 2005), and for international exhibitions, Itō Mamiko, *Meiji Nihon to Bankoku Hakurankai* (Yoshikawa Kōbunkan, 2008).

[4] For their prevalence in the woodblock print record, Konishi Shiro, *Nishikie Bakumatsu Meiji no Rekishi*, vol. 6 (Kōdansha, 1977).

[5] For art, Satō Dōshin, *'Nihon Bijutsu' Tanjō: Kindai Nihon no 'Kotoba' to Senryaku* (Kōdansha, 1996). For industry, Kuni, *Hakurankai no Jidai*. For museums, Seki Hideo, *Hakubutsukan no Tanjō: Machida Hisanari to Tōkyō Teishitsu Hakubutsukan* (Iwanami Shoten, 2005), and Alice Tseng, *The Imperial Museums of Meiji Japan: Architecture and the Art of the Nation* (University of Washington Press, 2008). For technology, Yoshida, *Bankoku Hakurankai*. For Japonism, Gabriel P. Weisberg (ed.), *Japonisme: Japanese Influence on French Art, 1854–1910* (Cleveland Museum of Art, 1975), and Teramoto

The story is worth revisiting, though, to highlight the multiple, sometimes incompatible threads of which it was comprised. Exhibition bureaucrats and their masters were clear about what they wanted exhibitions to do. There was an urgent need to promote exports abroad and to improve production at home if Japan was to avoid a semi-colonial fate. And exhibitions did help do this. But it was hard to translate the bureaucratic vision into social practice. It was not easy to reconcile differences among the elite in Tokyo, to get the word out to the provinces, or to convince farmers, craftsmen, and the rest to put their goods on show. It was not easy to design exhibitions so that the desired lessons were clearly apparent. It was not easy to get the audience to look at things in the way they were meant to, as the dutiful students the government wanted, rather than through the more familiar lens of distraction and discernment, as the sophisticated consumers they already were. And it was almost impossible to prescribe the moral of the story for foreign audiences, or to sustain their interest in Japan when it was no longer the next new thing. Exhibitions did their job. The story does not end in 1890. But both they and Japan looked different by the time they got there.

PROMOTING EXPORTS, TRAINING EYES

By the time the various parties made their way back from Paris, Japan itself was changing. The last shogun had resigned his position at the end of November 1867. The newish emperor was installed as head of state in January 1868. By the end of the month, the country was at war, as the new imperial forces tried to mop up the pro-Tokugawa resistance. It took a while for the confusion to clear. The new imperial era was only inaugurated in the autumn, and the war dragged on until the following summer. It was only in 1872 that the government began to construct the foundations of the modern state, adopting the Gregorian calendar, instituting centrally accountable prefectures, and holding the first domestic exhibition to be designated as a *hakurankai*. The official initiative was catalysed by foreign exposure. A handful of bureaucrats, who had been in Paris in 1867, were spurred into action early in 1871 by an invitation to participate in the Vienna *Weltausstellung* in 1873. They each had their own interests, however, rooted in earlier experience and personal networks. They were also not alone in seeing the promise of exhibitions.

Noriko, *Pari Bankoku Hakurankai to Japonisumu no Tanjō* (Shibunkaku Shuppan, 2017).

Other authorities were also getting in on the act. The resulting expo boom would persist until the end of the decade.

The year 1871 saw three exhibitions. The first, in the fifth month, was an official undertaking, prompted by the invitation to Vienna, but undercut by the initial reluctance of the Foreign Ministry to accept it. The proposal by the Product Bureau (*Bussan Kyoku*) was for a month-long *hakurankai*, which would collect the 'products of the world', identify them accurately (literally, 'rectify their names'), and clarify how they might be used. In the event, however, the exhibition was reclassified as a product exhibition (*bussankai*) and ran for only seven days, in an Army Ministry building next to the Shōkonsha shrine, which had appeared two years earlier to 'summon the souls' of those who had given their lives for the emperor in the recent civil war, and would be renamed at the end of decade as Yasukuni. In the meantime, it proved a convenient stage for the exhibition, which bore witness to the enthusiasms of its organizers. Tanaka Yoshio, a youthful botanist turned shogunal servitor, had been employed in the Product Bureau since the 1860s, had been in charge of the shogunate's exhibits in Paris in 1867, and had arrived back from France with quite a haul: seeds and bulbs, cacti, some guinea fowl. Machida Hisanari had been sent to Europe by Satsuma in 1865, had attended the opening ceremonies in Paris, and had joined the Bureau in 1870, intent on 'collecting old things', investigating 'times ancient and modern' (*kokon jisei*), and thereby preserving Japanese material culture from the depredations of acquisitive foreigners and anti-Buddhist zealots. Sure enough, the exhibition included not only some ancient Chinese bronzes but also more recent imports: scientific instruments, oil paintings, and a camera obscura. It also built on existing practice. One-third of the 2,347 exhibits were mineralogical, with another third devoted to the animal kingdom. The first exhibition in modern Japan combined a well-established scholarly interest in the natural world together with the imports and novelties, which had long attracted crowds. It also piggybacked on two newly invented shrine festivals, which celebrated key imperial victories in the recent civil war, thereby rehearsing a familiar blend of secular and sacred.[6]

[6] Asakura Haruhiko, *Meiji Kansei Jiten* (Tōkyōdō shuppan, 1969), 550–51, and Tōkyō Kokuritsu Hakubutsukan, *Hyakunenshi*, vol. 1, 28–37 and vol. 2, 572–605. For Tanaka, *Hyakunenshi*, vol. 2, 561–70, Murasawa Takeo, *Kindai Nihon o Kizuita Tanaka Yoshio to Yoshikado* (Tanaka Yoshio Yoshikado Kenshōkai, 1978), and Kutsuna Takehiko, 'Kannō Kaibutsu Ō no Bakumatsu, Meiji: Tanaka Yoshio to Hakurankai, Hakubutsukan', in Sano Mayuko (ed.), *Banpakugaku: Bankoku Hakurankai toiu, Sekai*

Come the autumn, moreover, the central government was being gazumped by others, who were just as eager to exploit the new-fangled word. A newspaper company in Nagoya held a five-day 'mini-exhibition' (*hakuran shōkai*) in a temple, but this was clearly another throwback, with exhibits of antiques and other curiosities.[7] Kyoto quickly followed with something more substantial. The city had recently lost the emperor and so status, which was already causing concern among the local elite. A month-long *hakurankai* at Nishi-Honganji, the huge temple in the south of the city, would 'open the way to improving knowledge and skill in crafts and [provide] a means of restoring the capital's fallen splendour'. The exhibits included a few imports, for example, a model steam engine and a coffee maker, but were heavily biased towards antiques and curios. The combination proved popular, attracting well over 10,000 visitors, and encouraged further efforts. The Kyoto Exhibition Company was established the following year and held the first in what would become over half a century of exhibitions, attracting 40,000 visitors to exhibits spread over three temple sites. From 1873, the company moved its operations to the former imperial palace.[8] It was only one of a number of such initiatives during the decade, however. Local authorities and boosters held a stream of exhibitions in shrines and temples, as well as previously inaccessible castle grounds, trying to improve the region's fortunes in the present – and thereby carry it into the future – while preserving the distinctive, local past. Meanwhile, the state was gradually getting its own act together.[9]

o *Haakusuru Hōhō* (Shibunkaku shuppan, 2020), 80–96. For Machida, Seki, *Hakubutsukan no Tanjō*, and Hiroko McDermott, 'The Hōryūji Treasures and Early Meiji Cultural Policy', *Monumenta Nipponica* 61.3 (2006), 339–74.

[7] Nagoya-shi Hakubutsukan, *Nagoya no Hakurankai*, 28, and Peter Kornicki, 'Public Display and Changing Values: Early Meiji Exhibitions and Their Precursors', *Monumenta Nipponica* 49.2 (1994), 191.

[8] Maruyama Hiroshi, 'Meiji Shoki no Kyōto Hakurankai', in Yoshida Mitsukuni (ed.), *Bankoku hakurankai no kenkyū* (Shibunkaku Shuppan, 1986), 221–48. See also Kornicki, 'Public Display', 189–90, Hiroko McDermott, 'Meiji Kyoto Textile Art and Takashimaya', *Monumenta Nipponica* 65.1 (2010), 41–43, and Alice Tseng, *Modern Kyoto: Building for Ceremony and Commemoration, 1868–1940* (University of Hawai'i Press, 2018), 37–38 and 111–13. For the company's own account, Kyōto Hakuran Kyōkai, *Kyōto Hakurankai Enkakushi* (Fujimi Shobo, 1997).

[9] Maruyama, 'Meiji Shoki no Kyōto Hakurankai', 222–23. Kornicki, 'Public Display', for an example, and Nonaka Katsutoshi, 'Meiji Shoki ni Shiroato de Kaisai Sareta Hakurankai ni Kansuru Kenkyū', *Toshi Keikaku Ronbunshū* 41 (2006), 911–16, for castles.

The first *hakurankai* to be named as such by the new government was the one that opened in 1872 in Yushima Seidō. This had originally been built in the seventeenth century as a Confucian temple, then combined with the Product Bureau in 1870 to form the 'Southern College' of a new 'University' (*Daigaku Nankō*). The announcement of the exhibition reiterated the determination to collect everything under the sun, 'without distinguishing between [products] created by nature and those crafted by human hand', but it also emphasized the usefulness of antiquities in tracing change over time and institutional development. The visual record reveals the resulting mix: shells, skulls, and stuffed animals, but also ancient comma-shaped jewels (*magatama*) and old coins, arms and armour, kimono and musical instruments, as well as what was beginning to be identified as fine art (*bijutsu*).[10] But the new location, one-and-a-half-month duration, and enlarged scale – as well as the presence of the gold, male *shachi* – also suggest the growing importance of exhibitions for the state, which had realized that it had to not only confront but also engage the Western world.

The previous autumn the government had dispatched over a hundred ministers, bureaucrats, academics, and students on what would become a twenty-one-month mission to 'seek knowledge throughout the world', headed by and subsequently named after Iwakura Tomomi. The Austrian Minister had quickly seized the opportunity to renew the invitation to Vienna, pointing out that the exhibition would provide as much benefit as the mission at much less cost. Within a couple of months, the government had formed an official commission and asked would-be exhibitors to send products to a newly formed Exhibition Bureau. The Vice-President, Sano Tsunetami, had also been in Paris in 1867, as part of the delegation from Saga. He wasted no time formulating a rationale. By participating in the Vienna exhibition, Japan would both enhance its international prestige, by exhibiting its resources and its crafts, and improve its industrial know-how, by studying the exhibits of the West. Learning about overseas markets would lead to better Japanese product quality, increased exports, and improved terms of trade. The virtuous cycle could be underwritten, he argued, by setting up a museum and holding an exhibition in Japan, 'indispensable for the advancement of knowledge and learning'.[11]

10 Tōkyō Kokuritsu Hakubutsukan, *Hyakunenshi*, vol. 1, 49–60; Mitsukuni, *Bankoku Hakurankai*, 66–67; McDermott, 'The Hōryūji Treasures', 343–45.

11 Tanaka Yoshio and Hirayama Narinobu, *Ōkoku Hakurankai Sandō Kiyō* (Meiji Bunken Shiryō Kankōkai, 1964), part 1, 7–14; Commission Impériale Japonaise, *Notice sur*

Sano's syllogisms were persuasive, but they failed to stipulate either what the government should put on show or to predict how hard it would be to assemble the Japanese exhibit.

For advice, the Bureau turned to foreign government employees, notably Gottfried Wagner, a German chemist who had arrived in Saga in 1870 to provide instruction in Western ceramic techniques. Wagner was clear that the Bureau should focus on the decorative art that had already proved its worth as an export good: lacquer from Tokyo, textiles and ceramics from Kyoto, porcelain from Saga, and Satsuma ware from Kagoshima, among others. Another adviser, Alexander von Siebold, suggested that it might be useful to supplement these with one or two much larger objects, so that European visitors would stop 'marvelling at Oriental customs' and focus on the exhibits themselves. Hence the *shachi*. But local authorities needed help in assembling what was required. The call for exhibits produced an 'exhibition craze', but the early results were disappointing. The bureaucrats quickly fanned out around the country to explain, cajole, and coordinate. In Kyoto, the silk weavers were told to alter the dimensions of their product, from the broad bolts they produced for the Japanese elite, to a narrower bolt that might appeal to European consumers. Local authorities were ordered to send their ceramics to Tokyo unglazed, to be finished by a factory attached to the Bureau, which had a decent grasp of how to 'improve the old style of decoration and make it similar to painting, which would open up the path to exports, so benefiting the country'. Finally, a Tokyo merchant, Wakai Kensaburō, was commissioned to complete the exhibit and to sell the goods when they arrived in Vienna. The final haul arrived at the Bureau in the eleventh month, where it was displayed to the Emperor, then the general public, and dispatched to Vienna in what was now, following the adoption of the Gregorian calendar, January 1873.[12]

l'Empire de Japon et sur Sa Participation à l'Exposition Universelle de Vienne, 1873 (Lévy, 1873), 58; and Tōkyō Kokuritsu Hakubutsukan, *Hyakunenshi*, vol. 2, 161. For Sano and exhibitions, Matsumiya Hideharu, 'Bankoku Hakurankai to Myūjiamu', in Nishikawa Nagao and Matsumiya Hideharu (eds.), *'Beiō Kairan Jikki' o Yomu: 1870-nendai no Sekai to Nihon* (Hōritsu Bunkasha, 1995), 237–44, and 'Meiji Zenki no Hakubutsukan Seisaku', in Nishikawa Nagao and Matsumiya Hideharu (eds.), *Bakumatsu Meijiki no Kokumin Kokka Keisei to Bunka Hen'yō* (Shin'yōsha, 1995), 264–71; as well as Tsunoyama Yukihiro, *Uīn Bankokuhaku no Kenkyū* (Kansai Daigaku Keizai Seiji Kenkyūjo, 2000), 89–96 and 106–14. For his life, Kuni Takeyuki, *Sano Tsunetami 1822–1902* (Fukuhaku Insatsu, 2013).

12 Tanaka and Hirayama, *Ōkoku Hakurankai Sandō Kiyō*, part 1, 14–24; Tōkyō Kokuritsu Hakubutsukan, *Hyakunenshi*, vol. 1, 65–73; Tsunoyama, *Uīn*

When they arrived, the commission faced a similar situation and audience as in Paris six years earlier. Situated in the Prater, the public park in the east of the city, the main building – a massive central rotunda flanked by two naves, each with seven galleries on each side – looked to the diarist of the Iwakura mission like 'a comb'. Japan had been given a shared space, with China and Siam, in the northern half of the easternmost gallery, but took over the whole thing and more. The *shachi* took its place at the entrance to the Japanese gallery, which was stuffed with export goods – porcelain, lacquer, and bronzes; textiles, bamboo, and woodwork. At the back were the other large objects, including a Buddhist altar, a model of the pagoda from a temple in Tokyo, and a papier mâché reproduction of the head of the great Buddha at Kamakura. (The rest of him had gone up in flames.) A pamphlet in French acknowledged that Japan had been 'isolated from progress, especially in the mechanical arts', but noted that articles such as those on display in the gallery had recently generated considerable exports. They were also available for purchase in the park, where the Austrian authorities had provided space for countries to erect their own structures, and where the commission had created 'Little Japan'. Passing under a torii gate, visitors could buy examples of Japanese craft, before proceeding into a garden, with a kagura stage on the right and a shrine across a pond. 'The purity and simplicity of style', noted the pamphlet, 'is the characteristic trait of our religious construction'. Back in Japan, Buddhist priests, sculptures, and temples were emerging from a wave of destruction, following the government's decree that Buddhism and Shinto should be separated. In Vienna, the distinction between the two was elided. There was no explanation how the large Buddhist objects in the palace related to the Shinto elements in the park.[13]

No matter. Most Europeans were indifferent to the niceties of Asian belief but captivated by the charms of Japanese products. As in Paris, the exhibits were a resounding hit. Already in May, Sano could report back

Bankokuhaku, 21–38, 45–49; and Kuni, *Hakurankai no Jidai*, 33. For an example of local cluelessness, Kornicki, 'Public Display', 184–85.

[13] Tanaka and Hirayama, *Ōkoku Hakurankai Sandō Kiyō*, part 1, 26–34 for a list of exhibits, 36–37 for their display, as well as 'Fuzu' (illustrations), np, at the rear of the volume; Commission Impériale Japonaise, *Notice*, 65–74. See also Tsunoyama, *Uīn Bankokuhaku*, 33–38; Kuni, *Hakurankai no Jidai*, 33–34; Itō, *Meiji Nihon*, 20–21; and Teasley, *Designing Modern Japan*, 60–62. For the anti-Buddhist storm, James Ketelaar, *Of Heretics and Martyrs in Meiji Japan: Buddhism and Its Persecution* (Princeton University Press, 1990), 43–86, and John Breen, 'Ideologues, Bureaucrats and Priests: On "Shinto" and "Buddhism" in Early Meiji Japan', in John Breen and Mark Teeuwen (eds.), *Shinto in History: Ways of the Kami* (Routledge, 2000), 230–51.

that the press was extremely favourable. The international juries endorsed the media judgement, giving some 200 awards to Japanese exhibits and making Japan the highest-ranked non-Western country in the medal table. Official recognition in the palace was matched by popular enthusiasm in the park. The Austrian emperor praised the 'skilled craftsmanship' of the Japanese workmen constructing the buildings. The shop was 'constantly thronged'. 'Those who ... left without buying something Japanese felt as if they had missed the opportunity to acquire a valuable rarity'.[14] The commercial potential of Japanese goods was evident, for example, to the Alexandra Park company in London, which petitioned the Japanese commission to sell them the garden and all its fittings. The result of all the interest was the creation of the Kiryū Kōshō Kaisha ('pioneering craft and commerce company'), by Wakai, the Tokyo merchant, together with a tea merchant from Saga. It would go on to coordinate the production and export of Japanese craft for European and American exhibitions and markets for the next two decades.[15]

Their experience of Vienna confirmed to the official delegation that Japan had things to sell overseas and that more work would be needed at home in order to meet and sustain the demand. Again, as in Paris, the exhibition provided some starting points. First, it allowed the government to get its hands on samples of European technology and industry. The effort was derailed somewhat when one of the ships carrying the haul sank on the way back to Japan, with the loss of 193 boxes of goods, including a Jacquard loom, though 68 boxes were later recovered. The materials that did make it back were put on display in the Exhibition Bureau's 'museum', which had been set up following Sano's recommendation in 1871 and which continued to prove highly popular. They also eventually got their hands on a replacement loom, installing it in a 'testing running place', so that the Japanese public could see machines in action.[16]

[14] Tanaka and Hirayama, *Ōkoku Hakurankai Sandō Kiyō*, part 1, 39–45; Kunitake Kume, *The Iwakura Embassy, 1871–73: A True Account of the Ambassador Extraordinary & Plenipotentiary's Journeys of Observation through the United States and Europe* (The Japan Documents, 2002), vol. 5, 40; and Kuni, *Hakurankai no Jidai*, 35–36.

[15] Tsunoyama, *Uīn Bankokuhaku*, 119–68. See also Hidetoshi Miyachi, 'Successes and Failures among Porcelain Trading Firms in the Early Meiji Period: The Cases of Morimura Bro., Inc. and Kiryū Kōshō & Co.', *Japanese Research in Business History* 24 (2007), 99–105, and Teasley, *Designing Modern Japan*, 62–64.

[16] Tanaka and Hirayama, *Ōkoku Hakurankai Sandō Kiyō*, part 1, 60–64 for a list of exhibits. Tsunoyama, *Uīn Bankokuhaku*, 207–35 for the larger story, and 241–55 for looms.

For Japan, as for all the rest, textiles would be the gateway through which industrialization would proceed.

Second, the exhibitions in Vienna and back home thus allowed their Japanese observers to discern the secrets of technology, to compare the products of industry, to understand where Japan stood, and so to prescribe how it might improve. Sano had included a bevy of craftsmen in the official delegation to the exhibition, who were predictably shocked at the gap between their own efforts and what was on display, providing an epiphany for several who went on to become industrial leaders back in Japan.[17] The lesson was also clear to the Iwakura mission, whose diarist acknowledged the Japanese achievement at the exhibition, but emphasized that it was contingent. The Japanese exhibits had 'won particular acclaim' in Vienna because they were 'different from European ones' and so 'had the charm of exoticism'. None of these advantages were permanent, however. In some fields – lacquerware, cloisonné, and damascene; leather, paper, and linen – Japanese technical superiority was evident, but absent novelty, it was not clear that Japanese supply could compete effectively for European demand. In others, Japan clearly lagged behind. Japanese porcelain was renowned, but the 'study of firing temperatures, mixing pigments and decorative techniques has hardly begun'. Japanese weaving was 'for the most part uneven', the joins in marquetry 'less than perfect', straw and raffia work 'cheap and crudely made', and Japanese portraits 'frankly embarrassing'.[18]

Third, the exhibition also suggested a strategy through which such deficiencies might be made up. In a memorandum on his return, Sano underlined how exhibitions and museums could serve as a starting point for a wholesale transformation of Japanese economy and society.[19] He elaborated the point in the prefaces he wrote for each volume of the Commission's official report. The archipelago was rich in resources and manufactures. But its agricultural and industrial techniques were stagnant, since they were only ever revealed in private (*ikko no ken ni idezuru ni sugizu*). Given the ignorance of practices and products elsewhere, it was inevitable that Japanese industry had fallen behind. But, he insisted, this could be fixed by 'training the eye' (*ganmoku no kyō*). Looking at an object allowed an observer to discern the general principles behind its

[17] Tanaka and Hirayama, *Ōkoku Hakurankai Sandō Kiyō*, part 1, 48–50; Tsunoyama, *Uīn Bankokuhaku*, 40–45.

[18] Kume, *Iwakura Embassy*, 31–32.

[19] Tanaka and Hirayama, *Ōkoku Hakurankai Sandō Kiyō*, part 1, 52–60.

manufacture, understand in particular how it had been made, and so evaluate its overall quality. This 'training' could happen anywhere, but its chances of success would be immeasurably increased by establishing a 'school for everyday enlightenment', such as the one that was emerging in London, in the wake of the first Great Exhibition of 1851.

By the 1870s, this comprised the South Kensington Museum, incorporating what would later become the separate Victoria and Albert and Science Museums, as well as the germ of what would become Imperial College. Japan, too, Sano argued, should have a universal survey museum, making it easy for visitors to inspect particular groups of objects; a technical school with sections for arts and crafts and for industrial technology; and a park surrounding the museum, with zoological and botanical gardens, so that visitors might have the horizons of their knowledge imperceptibly broadened (*shikirazu shirazu kaichi no iki ni susumu*). Above all, Sano argued, the government should hold exhibitions, which were both '*fons* of enriching the state and *origo* of enlightening the people'. He reiterated the by-now familiar advantages that exhibitions had over museums, in terms of gathering products, enabling comparison, and surveying development. He also made explicit the importance of competition among producers, who would 'exert themselves furiously to broadcast their reputation and win prizes'. 'They will strive to discard their shortcomings and build on their strengths, change the old and move towards the new, leave behind the ugly and approach the beautiful' and so 'contribute to the wealth of the nation'.[20] To cut a long story short, the gap separating Japan from the West, evident at the exhibition in Vienna, could be made up by adopting the practice of exhibition in Japan. There was some work to do before the government would be ready to stage such a show, but Sano suggested 1880 as a suitable date.

In fact, the first such exhibition took place in 1877, thanks to an imperious statesman and an eager United States. Raised in Satsuma, Ōkubo Toshimichi's epiphany had arrived during the Iwakura mission, by mashing up the lessons he learned about English industrial productivity, French political suppression, and German power politics. He returned

[20] Tanaka and Hirayama, *Ōkoku Hakurankai Sandō Kiyō*, part 2, 4–13. For the subsequent emergence of the museum system, Tseng, *Imperial Museums*, and Noriko Aso, *Public Properties: Museums in Imperial Japan* (Duke University Press, 2014). For the park, Thomas Havens, *Parkscapes: Green Spaces in Modern Japan* (University of Hawai'i Press, 2011), 19–32. For the zoo, Ian Jared Miller, *The Nature of the Beasts: Empire and Exhibition at the Tokyo Imperial Zoo* (University of California Press, 2013).

home determined to centralize the authority he would need to ensure civil order, modernize industry, and empower Japan. His vehicle was the Home Ministry, established in 1873, and in 1874 he formulated the mantra that would underpin its work: 'a country's strength depends on its people's wealth, and a people's wealth on the amount of production'. Without industrial capital and economic development, educational and military institutions were 'simply superficial display'. Within a couple of years, the annual budget for the Home Ministry had multiplied seven times and that of the Ministry of Industry fifteen. But there was also cultural work to be done, to disseminate the lessons. Imperial tours, museums, and exhibitions were the carrots with which to complement the stick of the newly formed police.[21] The Ministry began to draft regulations for a 'domestic industrial exhibition' (*naikoku kangyō hakurankai*) in 1874 but had to put the project on the back burner for a while. It was busy dealing with the Saga Rebellion, the first in a three-year series of uprisings by disgruntled samurai in the southwest of the country. But it was also getting ready to appear in Philadelphia in 1876, which would provide Ōkubo the opening he needed.

Japan's participation in the Centennial International Exhibition followed a familiar pattern. Again, it took a while to accept the invitation, initially extended in 1873, given the absence of half the government on the Iwakura mission, and a brewing row among those left behind as to whether or not to invade Korea, in response to the latter's failure to acknowledge the Japanese monarch as an emperor (*kō*). Again, it required a determined hustle to round up the exhibits, with Machida in charge, Tanaka and Wagner along for the ride, and the Kiryū Kōshō Kaisha helping out in New York. The focus remained on export craft, though this time the government provided subsidies for production, transport, and travel, as well as a pattern book, *Onchi Zuroku*, to help them embrace good design. Again, the efforts produced a mountain of 1,966

[21] Ōkubo Toshimichi, *Ōkubo Toshimichi Monjo* (Tokyo Daigaku Shuppankai, 2014), 561–65, for the mantra. Sasaki Suguru, 'Bunmei Kaika no Seiji Shidō: Ōkubo Toshimichi o Chūshin ni', in Hayashiya Tatsusaburō (ed.), *Bunmei Kaika no Kenkyū* (Iwanami Shoten, 1979), 78–102, for his thinking. To place the budgets in context, during the same period the Council of State's funding had decreased by a third, the Foreign Ministry's had stayed roughly the same, Finance had doubled, Education increased by a quarter, the Army decreased by a quarter, and the Navy doubled. For imperial tours, Takashi Fujitani, *Splendid Monarchy: Power and Pageantry in Modern Japan* (University of California Press, 1996), 42–55. For the links between early Meiji cultural politics and the emperor system, Makihara Norio, 'Bunmei Kaika Ron', in Asao Naohiro (ed.), *Iwanami Kōza Nihon Tsūshi*, vol. 16 (Iwanami Shoten, 1994), 285–86.

exhibits, which were crammed into the assigned space – a tenth of that allocated to the US, but slightly less than Austro-Hungary and slightly more than both Russia and Spain – but again supplemented in the park by a 'Japanese dwelling' for the commissioners and a Japanese Bazaar surrounded by a garden, with attendants in 'native costume'. And, again, the American public was impressed, in the now familiar mix of racist condescension and aesthetic enchantment. The 'leathery old Oriental' at the Bazaar had 'a face like that of a wrinkled old monkey', but he was selling objects of 'delicate and intricate workmanship', which 'cannot be reproduced by the most skilful artificer in either Europe or America'. In retrospect, noted Edward Sylvester Morse, zoologist and archaeologist, it was in 1876, that 'the Japanese craze took firm hold of us'. For Japanese observers, too, the experience reprised those of London, Paris, and Vienna. In light of industrial civilization, the 'skill and craft' of the Japanese exhibit could only make one blush.[22]

Ōkubo did not need to wait for the reports. Before the exhibition opened, he seized the moment. The situation was becoming urgent: potters in Satsuma persisted in making products at odds with the market; coppersmiths in Kaga were happy making inferior products in pursuit of short-term gains; weavers in Kyoto were simply behind the technological curve. It was the government's 'obligation to develop and enlighten', he protested. It took his colleagues a little while to come around, but in July 1876 the Council of State announced that the first domestic

[22] For the official accounts, Imperial Japanese Commission to the International Exhibition at Philadelphia, 1876, *Official Catalogue of the Japanese Section, and Descriptive Notes on the Industry and Agriculture of Japan* (The Japanese Commission, 1876) and *The Empire of Japan: Brief Sketch of the Geography, History and Constitution* (Imperial Commission for the Philadelphia International Exhibition, 1876), and Japanese Department of Education, *An Outline History of Japanese Education* (Appleton, 1876). For Japanese intentions, Kuni, *Hakurankai no Jidai*, 41–47; Itō, *Meiji Nihon*, 22–27; and Daniel Hedinger, 'Fighting a Peaceful War: Japan at World Exhibitions in the 1860s and 1870s', in Volker Barth (ed.), *Innovation and Education at International Exhibitions* (Bureau international des expositions, 2007), 88–91. For the Japanese exhibit and the importance of the *Onchi Zuroku*, Tōkyō Kokuritsu Hakubutsukan, *Meiji Dezain no Tanjō: Chōsa Kenkyū Hōkokusho 'Onchi zuroku'* (Kokusho Kankōkai, 1997), and Teasley, *Designing Modern Japan*, 65–67. For 'leathery old Oriental', J. S. Ingram, *The Centennial Exposition Described and Illustrated* (Hubbard Bros., 1876), 733 for 'leathery old Oriental'. For craze, Edward S. Morse, *Japanese Homes and Their Surroundings* (Charles E. Tuttle, 1972), xxxvii. For the subsequent impact, Clay Lancaster, *The Japanese Influence in America* (Walton H. Rawls, 1963), 46–49, and Neil Harris, 'All the World a Melting Pot? Japan at American Fairs, 1876–1904', in Akira Iriye (ed.), *Mutual Images: Essays in American-Japanese Relations* (Harvard University Press, 1975), 26–36.

exhibition would take place in a year's time in Tokyo. Ueno had flourished as temple town and cherry-blossom viewing site under the Tokugawa, been destroyed by the anti-Tokugawa forces in the civil war, and then turned into a park four years earlier.[23] The exhibition went ahead, but it was touch and go. The following year would turn out to be a watershed for the new Meiji regime.

BROCADE OF ENLIGHTENMENT

On or around Valentine's Day 1877, war broke out in the southwest. Saigō Takamori had initially been one of a triumvirate, including Ōkubo, his compatriot from Satsuma, which led the overthrow of the shogunate and assumed leading roles in the new regime. But Saigō had left the government in a fit of pique in 1873, given its failure to endorse the war with Korea, and had retreated home to Kagoshima, accompanied by similarly proud local boys. Come 1877 they prevailed on Saigō to come out of retirement and lead a last hurrah. His former colleagues, however, were fed up with petulant samurai. In mid-February, they decided to send the new Imperial Army into the field. The outcome was never really in doubt. By early April, Saigō and his remaining troops were in retreat, though the mopping up dragged on into the summer. Some suggested postponing the exhibition, but Ōkubo would hear none of it. On 19 August, following a penultimate stand on the west coast of Kyushu, Saigō retreated to Kagoshima with his remaining troops. On 21 August, as scheduled, the Emperor opened the exhibition in Ueno Park. On 24 September, with only forty rebels still alive, Saigō committed ritual suicide (or so the legend went). And on 30 November, the exhibition came to a triumphant close. Ōkubo was delighted. A few weeks later, he was boasting to the British minister about its 'financial success' and petitioning the Chancellor for a repeat. By the end of the year, he had got his way. The next one would be held in Tokyo 1881, supplemented in the meantime with smaller, regional exhibitions to sustain the momentum.[24] Ōkubo was not around to see it, however.

[23] Ōkubo, *Ōkubo Toshimichi Monjo*, vol. 7, 45–48, and Sasaki, 'Bunmei kaika no seiji shidō', 103. For Ueno, Julia Rasche, 'Ueno Park during Meiji Times – A Mirror of Its Time: Discursive Space and Symbolic Representation of Modernity', *Vienna Graduate Journal of East Asian Studies* 1 (2011), 91–122.

[24] For the apotheosis of Saigō, Mark Ravina, 'The Apocryphal Suicide of Saigo Takamori: Samurai, Seppuku, and the Politics of Legend', *The Journal of Asian Studies* 69.3 (2010), 691–721. For the determination of Ōkubo, Kuni, *Hakurankai no Jidai*, 75 and 82–83.

He was assassinated in May 1878 by disgruntled samurai, albeit not from Satsuma.

From Ōkubo's elevated perspective, it may indeed have seemed as if his vision had won the day, but a closer look reveals that the picture is more mixed. A year ahead of time, still, most people thought that an exhibition was little more than an 'old curiosity shop' (*furu dōgu ya*).[25] The bureaucracy had done its best to set them right, issuing regulations about how exhibitors should apply, how exhibits should be displayed, and how visitors should comport themselves.[26] An 'Advice to Exhibitors' underlined what an exhibition was for: 'to make enterprises prosper in the years to come', by enabling comparison between products, 'determining their ranking', and displaying the possibility and desirability of 'daily improvement in agricultural and industrial technique'. They should also pay attention to demand. 'Exhibits are to be things that are needed by people, things that can be seen to lead to prosperity, things that will find a market within the country, or things that can be sold abroad'. This needed further specification, though, if the exhibition were to do its work. 'Plants, flowers, and minerals' could be displayed, but not strange fauna, antiques, and old paintings. This did not mean, however, that 'we should discard old products ... and only seek out the new'. Different regions were at different stages of development. Products from one place might prove to be highly profitable when exhibited elsewhere.[27]

The work of demarcation did not end there. One problem was how to square the circle that led ineluctably from competitive products to immediate sales. 'This exhibition is about profit', the potential exhibitors were reassured.[28] But the exhibition itself was not a place for commercial exchange. Exhibits could be marked as sold, but they should not change hands.[29] This in turn posed its own dilemma, however: exhibitors-as-sellers would have to wait to get their money; visitors-as-buyers might 'lose their desire' (*nozomi o ushinahi*). The solution was to provide exhibitors with rent-free space, 'outside the exhibition but inside the park', where they could build stalls at their own expense and sell the products on display.[30] A similar problem and solution emerged when the bureaucrats

[25] *Yomiuri Shimbun* 1876/1/25, 3, cited in Kuni, *Hakurankai no Jidai*, 54.
[26] Ōkubo Toshimichi, *Naikoku Kangyō Hakurankai Shokisoku* (Tsuruda Shin'ya, 1876).
[27] 'Meiji Jūnen Naikoku Kangyō Hakurankai Shuppinsha Kokoroe', in ibid. [28] Ibid.
[29] 'Meiji Jūnen Naikoku Kangyō Hakurankai Shuppin Kisoku,' in Ōkubo, *Naikoku Kangyō Hakurankai Shokisoku*.
[30] 'Meiji Jūnen Naikoku Kangyō Hakurankai Fuzoku Urimono Mise Kisoku,' in Ōkubo, *Naikoku Kangyō Hakurankai Shokisoku*.

turned to the foreign products, on which domestic producers were expected to model their own efforts, but which would obviously outcompete Japanese offerings, given free rein. 'Imported foreign goods are not to be exhibited by private citizens. Foreign goods that can serve as samples for comparison shall be exhibited by officials. (Things produced inside the country that copy foreign goods are not subject to this restriction.)'[31] The three-step regulatory waltz was an attempt to ensure that the exhibition might incubate domestic industrial transformation. Officials would provide examples of industry to which Japanese manufacturers should aspire, protect them from potentially devastating foreign competition, and thereby encourage the baby steps required for them to become competitive themselves.

Regulations alone were not enough, however. Potential exhibitors also needed encouragement. This time, at least, the bureaucrats could rely on an emerging network of local officials, who had been charged with promoting local industry. In August 1876, the prefectural functionaries gathered in Tokyo for a week, to be told what kinds of exhibits should be encouraged and what kinds of incentives might be offered, including both support with shipping and the promise that outstanding exhibits would be taken to Paris for the Exposition Universelle of 1878. Back in their regions, they got to work. In Nagano, for example, they chose nine district and village heads to act as local agents, who prepared their own materials to drum up interest, toured their local region to tout for exhibits, and finally took them to Tokyo, where they took charge of the resulting exhibit. The various efforts bore considerable fruit. Some 84,523 exhibits eventually made their way to Ueno, from almost every part of the archipelago, including both the Colonial Office (*Kaitakushi*), which was then busy developing Hokkaido, and the Ryukyu Islands, which had been annexed in 1872 but would only become a prefecture in 1879. The only unsurprising no-show was Kagoshima, busy with its rebellion. The authorities were forced to expand the size of the halls. Inside them, the exhibits were organized by prefecture, somewhat compromising the possibility of comparing competing products, which had been Sano's goal, but ready for evaluation by the juries and the crowds.[32]

The official view of the results was mixed. In the aggregate, the official report insisted, the exhibition had been a success. There had not been 'one

[31] 'Meiji Jūnen Naikoku Kangyō Hakurankai Shuppin Kisoku,' in Ōkubo, *Naikoku Kangyō Hakurankai Shokisoku*.

[32] Kuni, *Hakurankai no Jidai*, 55–58 and 62.

word of criticism' from the press. One in three exhibitors, including a number of craftsmen who had been to Vienna, received some kind of award, which seems to have increased the value of the products on which they were bestowed. As intended, however, the exhibition had revealed as many weaknesses as it had strengths. Agriculture was insufficiently attuned to the needs of the market. Most farmers rarely exhibited the independence required for commercial success (*dokuritsu yūi no kishō*). When they did, it was apparent that the 'hand of the bureaucrat' was at work. They were also slow to adopt Western technology, which had been available for at least a decade. The same was true more generally. There was the occasional bright spot. Gaun Tatchi had invented a low-cost spinning machine in 1873, which received an award, sold well during the exhibition, and quickly took off thereafter. But, overall, modern technology was still a novelty in an economy still dominated by primary industry and manual labour. Of the 211 machinery exhibits, roughly a quarter were intended for agriculture, just over a quarter for textiles, and another quarter had been supplied by the government. Tokyo had supplied a fifth of them, but twelve prefectures had nothing to show. There was clearly work to do and things to learn. The work started in October, which saw a series of meetings for the local bureaucrats in charge of industrial promotion. The reports were then compiled, disseminating the exhibition's survey of the economic situation nationwide.[33] Thus, the exhibition did enhance the emerging network of actors, information, and activity dedicated to industrial development, drawing the regions into national and international circuits of exchange. On site, though, this was not the only lesson on display.

The problem was that the exhibition was never going to be a simple survey of industry. As with the exhibitors, the organizers were aware that they had to train the eyes of the visitors. The official guidebook began with some 'advice for the visitor' (*kansha chūi*), glossed as 'rules for looking' (*mite no kokoroe*). 'The aim of this domestic industrial exhibition is to aid in the progress of craft (*kōgei*) and to open the wellsprings of production and trade. It is not to construct a place for entertainment or to provide an excuse for sightseeing.' It went on to provide five rules: to

[33] Naikoku Kangyō Hakurankai Jimukyoku, *Meiji Jūnen Naikoku Kangyo Hakurankai Shuppin Hōkokusho* (Meiji Bunken Shiryō Kankōkai, 1964); and Kuni, *Hakurankai no Jidai*, 64–72 and 206–10. For Gaun, Murase 1965 and Yukihiko Kiyokawa, 'Entrepreneurship and Innovations in Japan: An Implication of the Experience of Technological Development in the Textile Industry', *The Developing Economies* 22.3 (1984), 211–36.

look past surface appearances; to evaluate differences in form, colour, decoration, elegance, and proportion, especially among goods destined for export; and to consider whether the product was well designed for use; whether it matched the needs of the market; and whether it had been priced appropriately. A separate publication, a 'commentary on the exhibits', provided some preliminary suggestions about the conclusions a visitor might reach, noting that some fields have made 'remarkable advances', but others continued to cling to 'old ways'.[34] The visitor was also instructed when they arrived at the imposing main gate, which had yet more rules posted on either side. Lunatics and drunks, large packages and alcohol would not be admitted. Lighting fires and smoking, touching the exhibits without permission, taking photographs of the grounds, or making reproductions of the exhibits were all forbidden. A visitor who damaged an exhibit or a case would have to reimburse the cost.[35] The exhibition was not the place for the kind of activity that might have accompanied the cherry viewing of yore.

Despite their best efforts, though, the exhibition remained a transitional, compound thing, which incorporated both the newer logic of modern pedagogy and the older attractions of spectacle. Like its early modern predecessors, the exhibition traded in novelty. The government exhibits themselves provided not only an opportunity to study but also a chance to gawp at the technology of the future: an American windmill (Figure 2.2); a steam-powered spinning machine, operated by kimono-clad young women; a steam-powered pump, which brought water up from Shinobazu pond, to gush in the fountain in the centre of the courtyard, to put out fires, and to water the animals. The technology may have been intended to demonstrate how production would have to change, but they were remarkable due to the distance they had travelled, the charms of their operation and, perhaps, the incomprehensibility of their effects. Exhibits of machinery in 1877 satisfied a similar demand as exotic animals and mechanical puppets in earlier years.

For the private exhibitor, too, the point was not to allow a viewer's attention to wander back and forth between their exhibit and the next, but rather to attract that attention to the exclusion of everything else. Sure

[34] Naikoku Kangyō Hakurankai Jimukyoku, *Meiji Jūnen Naikoku Kangyō Hakurankaijō Annai* (1877), 1–6; and Naikoku Kangyō Hakurankai Jimukyoku, *Meiji Jūnen Naikoku Kangyō Hakurankai Shuppin Kaisetsu* (Meiji Bunken Shiryō Kankōkai, 1962). See also Kuni, *Hakurankai no Jidai*, 63–64.

[35] Nakamura Juntarō, *Naikoku Kangyō Hakurankai Michishirube* (Itō Shōnosuke, 1877), 4, for the rules on the gate. See also Naikoku Kangyō Hakurankai Jimukyoku, *Annai*, 67.

(a) (b) (c)

FIGURE 2.2 Kyōsai captured both official pretension and popular distraction at the exhibition in this triptych, setting them in the enduring landscape of Ueno. *Source*: Kawanabe Kyōsai, *Meiji Jūnen Ueno Kōen jinai Naikoku Kangyō Hakurankai Kaijō no zu* (1877), Tokyo Metropolitan Library.

enough, the official report noted that farmers were not exhibiting their everyday products, but crowd-pleasing one-offs. A reporter was disconcerted to find a nine-foot-long treasure ship, modelled in cake, not too far – maybe not too different – from the oil paintings in the art gallery. The continuity with earlier practice was even clearer outside the main gate. The signboards hanging outside the stalls lining the approach to the exhibition, together with the calls of the salespeople, reminded one reporter of the atmosphere of the unveiling of an icon at a temple (*kaichō*), where sacred enlightenment blurred with secular distraction. The extent to which the new world would struggle to emerge from the old was underlined in at least one private guide, which provided what these days is known as a 'model course' through what had long been a famous place (*meisho*). It started by celebrating the regeneration of Ueno after years of destruction and neglect. It provided an extensive description of the various sights along the way: the temples and shrines that dotted the hillside; the lacquerware, dried seaweed, and endless variety of candy available for purchase. Finally, it suggested, you may, or may not, want to enter the exhibition.[36] Modern industry was subsumed within a much older landscape and way of doing things.

[36] Kuni, *Hakurankai no Jidai*, 62–65, for the machines. Yoshimi, *Hakurankai No Seijigaku*, 133–34, for the cake and the stalls. Nakamura, *Michishirube*, for the guide. For pictures of *meisho*, Mary Elizabeth Berry, *Japan in Print: Information and Nation in the Early Modern Period* (University of California Press, 2006), 149–52, and Robert Dale Goree,

The media, too, bore witness to the manifold nature of the exhibition. Like polity and economy, the publishing industry was also in transition, as would-be journalists and illustrators navigated the passage from woodblock print to modern press, trying to work out whether or not it was possible to speak one's mind. The newspapers, for the most part, parroted the party line. The *Chōya Shimbun* echoed Sano's characterization of the exhibition as a place to discover profitable resources and promising technologies, such as spinning. The *Yomiuri* provided verbatim quotation of official announcements, dry recitations of exhibits and ceremonies, and concerned comment when the principles of exposition seemed to be violated. But the dailies did not have a monopoly. They had emerged in tandem with weekly lampoons, above all the *Marumaru Chinbun*. In its first issue, earlier in the year, it had poked fun at the rage for exhibitions, modifying the characters for domestic exhibition so as to suggest the need for a public cleaning of the Japanese house (*naikoku takurankai*). For the next ten months, the column provided a regular rogue's gallery of fads and fools (Figure 2.3). When the actual exhibition rolled around in August, the paper provided tongue-in-cheek commentary on the opening day, soon followed by a take-down of the exhibits.[37] Woodblock print artists also made their own considerable hay. Even in 1877, the fourth estate was an unruly beast. Conversely, the official inability to control the spectacle was at least part of the reason for its success.

But it also may not have mattered, too much. There was no shortage of visitors, with 454,168 making their way through the gates by the time they closed.[38] It is hard to believe that everyone was there to have their eye trained. But the exhibition did inaugurate a framework, some incentives, and a network through which producers could be encouraged to up their game. And it did enough, galvanizing local officials and animating

'Fantasies of the Real: Meisho Zue in Early Modern Japan', PhD dissertation (Yale University, 2010).

[37] Kuni, *Hakurankai no Jidai*, 68, for the *Chōya Shinbun*. *Yomiuri* 1877/8/18, for an example of stodge. For 'naikoku takurankai', *Marumaru Chinbun*, nos. 1–41 (1877/3–12), and for the exhibition, nos. 25, 27, and 28. See also Peter Duus, 'Weapons of the Weak, Weapons of the Strong: The Development of the Japanese Political Cartoon', *Journal of Asian Studies* 60.4 (2001), 977–81, and '"Punch Pictures": Localising Punch in Meiji Japan', in Hans Harder and Barbara Mittler (eds.), *Asian Punches: A Transcultural Affair* (Springer, 2013), 324–33; and Kenneth Masaki Shima, 'The Laughing Masses: Comedy and Visual Media in Imperial Japan' (PhD thesis, University of California Los Angeles, 2020), 98–176. For the media industry in general, James L. Huffman, *Creating a Public: People and Press in Meiji Japan* (University of Hawai'i Press, 1997) and Kinoshita Naoyuki and Yoshimi Shun'ya, *Nyūsu no Tanjō: Kawaraban to Shinbun Nishikie no Jōhō Sekai* (1999).

[38] Kuni, *Hakurankai no Jidai*, 59–61.

FIGURE 2.3 Rebel samurai idols on display. Saigō Takamori, characterized as monkish doer of evil deeds, on the left, with three deceased rebel leaders, characterized as bodhisattvas, on the right.
Source: *Marumaru Chinbun*, no. 2 (1877/3/31), Stanford University Libraries.

existing producers, to suggest that the experiment should be repeated. Takamura Kōun, a well-known wood carver, had been asked to exhibit something at the exhibition. He did not really understand what the bureaucrat was on about, but he reluctantly agreed. His white-robed Kannon went on show, but he still thought the exhibition was much ado about nothing. He won a grand prize, but he did not know whether or not it was a good thing. But then he saw his name and the award mentioned at a local newsstand and began to realize that something might come of it. By the turn of the century, Takamura was a leading figure in the national arts establishment. His reputation was underpinned by his prominence at exhibitions at home and abroad, both as an exhibitor and a juror, though he was not above more populist forms of urban spectacle.[39] In his interwar reminiscence, he protested too much, perhaps. There were other examples,

39 Takamura Kōun, *Bakumatsu Ishin Kaikōdan* (Iwanami Shoten, 1995), 122–23. For his biography, Hashizume Shinya, *Jinsei wa Hakurankai: Nihon Rankaiya Retsuden* (Shōbunsha, 2001), 17–34.

though, of the way in which national exposure bolstered brand value. Prizes given in 1877 would soon become proud boasts in product advertising at the second domestic industrial exhibition in 1881.

ON WITH THE SHOW AND TELL

Before they did, some products made their way to France. The Exposition Universelle 1878 proved the crest of the wave of European enthusiasm for things Japanese, even if this only became clear in retrospect. On the Japanese side, the rationale for participating in international exhibitions remained compelling. Matsukata Masayoshi had been appointed as vice-president of the commission in March 1877, while his former compatriots from Satsuma were besieging Kumamoto castle, in the opening salvo of their doomed rebellion. In 1878, following their slaughter, he noted that there was still considerable ignorance about Japan in the West. This was due, he insisted, to the still immature economy, which exhibiting in Paris would again help improve. His concerns were given concrete form by another former Satsuma samurai, Maeda Masana, recently returned from Paris, who lobbied for Japan's participation, was put in charge of the bureau, and doubled down on export craft; and by the Kiryū Kōshō Kaisha, which assembled much of the exhibit. In Paris, it made an impressive show. The official exhibit could be found on the left bank, on the Rue des Nations, between Italy and China. The entrance was dominated by a massive, elaborate gate, provided by the Mitsui trading company, with a map of Japan to the right and Tokyo to the left, the former including things such as population statistics, the latter with details of the imperial palace and the other institutions that made the city a modern capital. There was still no Japanese machinery to display in the Galerie des Machines, but the Ministry of Education had produced some materials which suggested that the necessary infrastructure was being put in place.[40]

On the French side, the reception remained enthusiastic, but discriminating. There was still an unabated demand for the baroque extravagance of Japanese export craft. But there was also a search for a simpler and

[40] Itō, *Meiji Nihon*, 27–30, for an overview. Teramoto Atsuko, '1967 nen Pari Bankou Hakurankai ni okeru Maeda Masana no Yakuwari', in Sano Mayuko (ed.), *Bankoku Hakurankai to Ningen no Rekishi* (Shibunkaku Shuppan, 2015), 73–102, for Maeda's role. He would subsequently suggest a new direction for industrial policy in the 1880s. See Crawcour, '*Kōgyō Iken*'.

aesthetic aspect, which, it was believed, truly represented Japan. The Mitsui gate provided a surprising example, combining 'extreme simplicity' and 'some small refinement, which one would not expect to find in such a rudimentary construction'. But it was easier to discover authenticity back on the right bank, where colonial and foreign exotica sheltered between the curving wings of the Palais du Trocadero. A Japanese farm offered 'light bamboo fences ... a charming little wooden door ... light and rustic construction ... under an ingenious shelter of bamboo', providing an ideal frame for decorative art and the 'thousand nothings, fanciful fripperies and children's playthings', in which the Japanese excelled. Given this native genius, any attempt by Japanese producers to mimic European practice was clearly a mistake. There was general opprobrium for contemporary Japanese oil painting in the Galerie des Beaux-Arts. The deal and the dilemma were clear. 'Japan borrows our mechanical and military arts and our sciences, we take its decorative arts.'[41] The Japanese government was trying to assimilate the former, but Euro-American observers would continue to demand the latter. Development, for Japan, was degeneration. The archipelago should remain the pastoral, pre-industrial idyll first observed in London a decade and a half earlier.

For now, though, for the government, it was on with the shows, which would enable the economy to transition from decorative to mechanical arts. Back in Japan, the Second Domestic Industrial Exhibition went ahead in 1881, with a royal in nominal charge of the exhibition office, Matsukata as number two, and Sano and Tanaka among its seven members. A 50 per cent increase in budget translated into a concomitant increase in the size of the site, but two-and-a-half times the amount of exhibit space, again on top of Ueno Hill. This time, there were two courtyards: four large exhibition halls in the first, machinery and art in

[41] Ernest Chesneau, 'Le Japon à Paris', *Gazette Des Beaux-Arts* 18.2 (1878), 2 for European madness, and 386 for the exchange of mechanical for decorative arts. Philippe Burty, 'Le Japon Ancien et Le Japon Moderne', *L'Art* 4 (1878), 243, for the Mitsui gate. *Gazette Des Beaux-Arts*, 1878/11, 915, for the farm. See also Geneviève Lacambre, 'La Découverte de l'art Japonais à travers les Expositions Universelles de Paris, 1867, 1878, 1889, 1900', in Terukazu Akiyama and Tōru Haga (eds.), *L' Age du Japonisme. Le IIème Colloque Franco-Japonais* (Société Franco-Japonaise d'Art et d'Archéologie, 1983), 38–39; Deborah Levitt-Pasturel, 'Critical Response to Japan at the Paris 1878 Exposition Universelle', *Gazette Des Beaux-Arts* 119 (1992), 68–80; Jeehyun Lee, 'Resisting Boundaries: Japonisme and Western-Style Art in Meiji Japan' (PhD thesis, University of Pennsylvania, 2011), 16–18 and 110–30; Kimi Coaldrake, 'Fine Arts versus Decorative Arts: The Categorization of Japanese Arts at the International Expositions in Vienna (1873), Paris (1878) and Chicago (1893)', *Japan Forum* 25 (2013), 179–82.

the second, with the latter installed in the almost-completed Museum at the apex of the site. There were four times as many exhibits, which were this time arranged by class rather than prefecture, making them easier to compare. Again, however, Tokyo dominated the show, providing almost a quarter of the exhibits, and again the exhibition revealed the still-transitional nature of the economy. Machinery, in general, was overwhelmingly dedicated to textile production, copying foreign models rather than inventing new solutions. Again, though, the crowds proved enthusiastic, with twice the number of visitors. The exhibition also continued to provide grist to the mill for both the burgeoning press and woodblock prints.[42]

And again, the exhibition did its job. It made its numbers and documented development. It revealed the progress achieved and the work that remained, sustaining official enthusiasm for exhibitions, as a way of doing it, at least for a while. In its immediate wake, Sano proposed that the next in the series, scheduled for 1885, should be expanded to become an all-Asia Exhibition, which would not only help revive the Asian economy, he suggested, but also address familiar concerns about Japanese trade. It would also, not incidentally, be a useful step towards Sano's long-held dream of an international exhibition in Japan.[43] Already in the early 1880s, however, the times were changing, and as the decade went on, so the function and form of exhibitions began to shift, their urgency to diminish.

Some of the changes were testament to official achievement. The utility of putting things on display was widely acknowledged as Japan entered the 1880s. The early regional experiments in the early 1870s had become a boom following 1877, with an average of ten a year until 1885. Exhibitions were also being supplemented by other opportunities to compare progress, attract customers, and shift product. Following the exhibition in 1877, Tokyo had opened a *kankōba* – a place to 'encourage industry' – providing space to sell exhibits left over from Ueno. It was an immediate hit, attracting over 1,500 exhibitors, 350,000 exhibits, and over 150,000 visitors in its first year of operation, and such bazaars would flourish into the new century. The most important innovation,

42 Kuni, *Hakurankai no Jidai*, 88–94 for planning, numbers, and evaluation, 213–16, for machinery. Tseng, *Imperial Museums*, 59–65, for the museum. Hashizume Shinya, *Nihon no Hakurankai: Terashita Tsuyoshi Korekushon* (Heibonsha, 2005), 18–23, for plans and prints.

43 Kuni, *Hakurankai no Jidai*, 103–4.

however, was the result of a discovery by Matsukata in Paris, where he had noticed the agricultural *concours*, running alongside the exhibition, which allowed a professional attention to detail that could easily be lost in the welter of an exhibition. Translated as *kyōshinkai* ('meeting to progress together'), the first two Japanese examples were held in Yokohama in 1879, devoted to tea and silk thread, both major export goods. Osaka followed in 1880, with the same two commodities, as well as a *hinpyōkai* ('meeting to evaluate goods'), for the two great imports, raw cotton and sugar, which attracted exhibits from throughout the country. By the mid-1880s, the local authorities were also getting in on the act, turning to *kyōshinkai* rather than *hakurankai* as the preferred way to catalyse regional development. Between 1885 and 1900, the country as a whole saw on average 400 of these more focused exhibitions a year. Most were not large: 6 days long, with 380 exhibitors and 3,000 visitors. But they were also not one-offs. Local *hinpyōkai* would precede and produce prize winners for regional *kyōshinkai*, which would do the same for national exhibitions, which would do the same for international events, which would send the prize-winners back down the ladder, to encourage the others. They were joined by product exhibition halls, permanent facilities wherein local producers might seek instruction and inspiration. By the mid-1880s, a network of institutions had emerged, catalysed by but no longer entirely subject to the national bureaucracy, refining the practice of exhibition and calibrating its effects.[44]

By the middle of the decade, though, the proliferation of exhibitions, broadly conceived, was unfolding against a very different backdrop. Already, in the wake of the first exhibition in 1877, the government was beginning to face the fiscal consequences of samurai rebellion, which had morphed into a movement for popular rights, fuelled in part by the disruption and distress caused by a new land tax system. Persistent inflation did not help. The year 1881 saw not only the second exhibition but also a political crisis, with a promise by the government that it would establish a parliament within a decade, but also a turn to violent repression of the rights movement. Matsukata, now Finance Minister, sought to

44 Kiyokawa, Yukihiko, 'Shokusan Kōgyō Toshite No Hakurankai, Kyōshinkai No Igi', *Keizai Kenkyū* 39.4 (1988), 341–48, for the boom and diversification of exhibitions, and Tom Nicholas, 'The Origins of Japanese Technological Modernization', *Explorations in Economic History* 48.2 (2011), 284–88 for the way in which they boosted domestic innovation and technological modernization. Suzuki Hideo, *Kankōba no Kenkyū: Meiji Bunka to no Kakawari* (Sōeisha, 2001), for *kankōba*. Teasley, *Designing Modern Japan*, 81–84, for product exhibition halls.

stabilize the economy through fiscal restraint but thereby exacerbated the distress in the countryside.[45] At the same time, it was clear that the industrial sectors of the economy were beginning to make their own way. The need to catalyse domestic industry had therefore passed. The need to mitigate its effects was now the priority. But the need to promote exports and to demonstrate Japanese progress overseas was ongoing.

The place of exhibitions in government policy therefore began to shift. In 1883, Saigō Tsugumichi, a younger surviving brother of the deceased Takamori, then Minister of Agriculture and Commerce, announced that the third industrial exhibition, scheduled for 1885, would be postponed until 1889. A cheaper, more cost-effective *kyōshinkai* would take its place. Two years later, he revived Sano's earlier idea for a pan-Asian exhibition, pushed back a further year to 1890, to make an even bigger splash, commemorating what the government had earlier determined to be the 2,550th anniversary of the enthronement of the first, mythical emperor and coinciding with the opening of the new parliament. But when a detailed plan emerged the following year, it quickly attracted a veto from Matsukata's Ministry of Finance. The Third Domestic Industrial Exhibition would indeed take place in 1890, but within the template established by its predecessors.[46] By the time it did, its significance was also in question, a consequence of changes both at home and overseas.

The year before, 1889, had seen Japan again back in Paris. There too, the attraction was beginning to pall, both for Japanese exhibitors, who were battling to navigate the rapids of the domestic economy, and for European visitors, jaded with the same old Japanese exhibits. The official exhibit of raw materials and manufactures suggested that 'the day is coming when the Empire of the Rising Sun will occupy an important place as a producer'. A survey of art showed how 'the Japanese are quick to engage, invent, create and copy'. A Japanese garden on the Trocadero provided the expected calm. But the novelty was wearing off for

[45] For the rights movement, Daikichi Irokawa, *The Culture of the Meiji Period* (Princeton University Press, 1985), and Mara Patessio, *Women and Public Life in Early Meiji Japan: The Development of the Feminist Movement* (University of Michigan Press, 2011), 141–60. For rural transformation, Kären Wigen, *The Making of a Japanese Periphery, 1750–1920* (University of California Press, 1995). For the reforms, Steven J. Ericson, *Financial Stabilization in Meiji Japan: The Impact of the Matsukata Reform* (Cornell University Press, 2020).

[46] Suzuki Hiromune, 'Dai Sankai Naikoku Kangyō Hakurankai to Ajia Dai Hakurankai Keikaku', *Hōsei Shiron* 20 (1993), 44–52, and Kuni, *Hakurankai no Jidai*, 104–10.

Europeans, who were distracted by the Eiffel Tower, electricity, and a full-blown human zoo, all making their first appearance. From Japan, 'we are waiting for something else'. This left Japanese exhibitors with a dilemma. On the one hand, their charms were paling. 'The Japanese neither understand nor seek the grand, the monumental; in their hands, everything shrinks, even nature itself.' On the other, their attempts to cater to European demand were marked as degenerate. Japan seemed to be happy to make 'japonerie', which was 'born in a factory . . . and ends up in a bazaar'. The official Japanese report took note. In future, it suggested, exhibits at overseas exhibitions should be divided into trade goods and art objects, the latter to be housed in a building 'in pure Japanese style', which might stimulate demand for the goods available elsewhere on site. They would also begin to deploy a neologism, *bijutsu kōgei* (artistic crafts), to emphasize the distinction of Japanese handicraft. The problem, though, was more general. 'I don't like these little Japanese strapped into our European clothes.'[47] Japan had to continue to modernize; but, if it did, it would no longer be Japanese.

Nor was it clear that government-led exhibitions at home were still necessary, to make Japanese agriculture and industry fit for purpose. The Third Domestic Industrial Exhibition of 1890, back in Ueno, showed that the economy was humming, despite the privations of the previous decade. There was a third more exhibition space and well over three times as many exhibitors, double the number of exhibits, and quadruple the number of prizes. There was a two-storey 'Reference Building' for foreign products, with 1,726 specimens brought back from Paris (of which 344 were wallpaper). True, the economy remained Tokyo-centric and in transition. But development was under way. More than 100 of the machinery exhibits were still textile-related, but more than 200 had something to do with steam. Private enterprise was on the march and

[47] *L'Exposition de Paris (1889)*, 2.44, 31, for approval, and 1.21, 167, for contempt. Lucien Falize, 'Les Industries d'Art à l'Exposition Universelle 1889: L'Orfèvrerie', *Gazette Des Beaux-Arts* (1889/8), 222, for wanting something new. Lucien Huard, *Livre d'Or de l'Exposition* (Boulanger, 1889), 610–11, for japonerie and Japanese strapped into European clothes. Nōshōmushō, *Fukoku Pari Bankoku Dai Hakurankai Hōkokusho* (Fujimi Shobō, 1890), 100–1, for the official response; Hino Eiichi, 'Bankoku Hakurankai to Nihon no "Bijutsu Kōgei"', in Mitsukuni Yoshida (ed.), *Bankoku Hakurankai no Kenkyū* (Shibunkaku Shuppan, 1986), 21–43, for artistic craft. See also Itō, *Meiji Nihon*, 33–37, for Japanese intention, planning, and reaction, and Lacambre, 'La Découverte', 40–41, for the European response. For human zoos, Pascal Blanchard, Nicolas Bancel, Gilles Boetsch, Eric Deroo, and Sandrine Lemaire, *Human Zoos: Science and Spectacle in the Age of Colonial Empires* (University of Chicago Press, 2009).

eager for recognition, which might be turned to market advantage. The organizers were deluged by complaints and even lawsuits from those who felt their efforts were insufficiently acknowledged by the juries. Not that they needed the validation. There were other places where people could learn a trade and access information about what the market might want; where producers could congregate, compare notes, and scope out rivals. There were also other attractions for the press and the public. The exhibition provided useful copy, for print artists, guidebook publishers, and the press, but its representation had become formulaic, even dry. It attracted more visitors than its predecessors, but interest waned over time. The crowds were thick in the first few weeks, buoyed by the cherry blossom that had long drawn crowds to Ueno. They picked up again in May, drawn by an electric tram that circled the site, as well as three days of horse racing around Shinobazu Pond. By June, however, they were thinning out, put off by the rainy season, an outbreak of influenza, and the inadequacy of the exhibition itself. Visitors wanted attractions, not instruction, as the press was quick to point out. The last two months saw less than half the numbers of the opening month. The diminishing enthusiasm was underlined by the fact that half the goods for sale were left unsold.[48]

By 1890, then, exhibitions had changed. Overseas, the enthusiasm for the débutante had receded, leaving Japan exposed to the double bind of the newly racist, imperialistic world: damned as inauthentic, if it tried to prove that it belonged among the industrial powers; damned as irrelevant, if it accepted its relegation to pastoral fairyland and purveyor of overly familiar, insufficiently exotic goods. In the years to come, the government would try to use international exhibitions to square the circle, insisting on both distinction and relevance. The shift was foreshadowed, however, with the closing of the Kiryū Kōshō Kaisha in 1891, its subsidies cut off, accounts in arrears, and customers drying up. Its work – building a first bridge between traditional Japanese manufacturers and modern Western consumers – was done.[49] Back home, too, the landscape looked different.

[48] Kuni, *Hakurankai no Jidai*, 110–20, and Suzuki, 'Dai Sankai Naikoku Kangyō Hakurankai', 52–54, for an overview. Kuni, *Hakurankai no Jidai*, 217–23, for a breakdown of exhibits. For two of many guidebooks, Itō Ken'ichi, *Dai Sankai Naikoku Hakurankai Kaijō Annai* (Meikōsha, 1890), and Ōkubo Tsunekichi, *Dai Sankai Naikoku Kangyō Hakurankai Kaijō Michishirube* (Kōgyokudō, 1890). For desiccated reporting, *Fūzoku Gahō*, no. 15 (1890/4/15).

[49] Tsunoyama, *Uīn Bankokuhaku*, 167–69, and Miyachi, 'Successes and Failures', 101–5, for decline and dissolution.

Exhibitions were an established feature of mid-Meiji time and space. They were a recurrent fixture in local, regional, and national calendars, an addition to the existing round of religious festivals and secular attractions. They were not as distinct from these as officials might have wished. But they had also begun to diversify, spawning offshoots in the early forms of trade fairs, retail emporia, and the occasional museum. To some extent, then, at least within Japan, the early official work had been done. The state could afford to step away, though the Ministry of Agriculture and Commerce, among other agencies, continued to provide a repository of regulations, data, samples, standards, and expertise for any and all exhibitions.[50] This did not mean an end to government-led exhibitions, much less to the flotilla of smaller events that bobbed along in their wake. It did mean, however, that exhibitions in the 1890s and after were about more than promoting industry.

[50] Kiyokawa, 'Shokusan Kōgyō', 342–43.

3

Municipal Branding

White City, in the west of London, is best known today for a sprawling shopping mall and the former BBC Television Centre, a 'cultural icon' from the 1950s, now converted into 'inspirational offices' and 'premium apartments'. Just to the west of these, however, is a manicured Japanese garden, and to the north, a 1930s housing estate, spread out along South Africa and Australia Roads, Canada, India, and New Zealand Ways. Both garden and estate are testament to the area's original incarnation as the Great White City, which opened in 1908 with a Franco-British Exhibition on what had been fifty-seven hectares of farmland. Its mastermind was Imre Kiralfy, who had spent half a century honing his trade as a showman, with theatrical extravaganzas and eventually full-blown exhibitions in Europe, the US, and finally London. The empty land and excellent transport links provided a suitable canvas on which to plagiarize the original White City, which he had seen in Chicago in 1893. 'Anglo-Mughal' buildings in the southern half of the site provided space for stuffier displays. A stadium on the east of the site accommodated an Olympics, following the model of Paris in 1900 and St Louis in 1904. But it was the northern half of the site, where the housing estate is now situated, which held the native villages and other attractions that, he knew, would draw the crowds. It was not only European empires that Kiralfy turned to profitable account.[1]

1 'Television Centre', https://televisioncentre.com/, for 'cultural icon'. David Littlefield, 'White City: The Art of Erasure and Forgetting the Olympic Games', *Architectural Design* 82.1 (2012), 70–77, for the history of White City, and Nathaniel Télémaque, 'Annotating Black Joy on the White City Estate', *Transactions of the Institute of British*

Following an Imperial International Exhibition in 1909, Kiralfy turned in 1910 to Japan, which had been a British ally since 1902 and was still a reliable source of crowd-pleasing attractions. 'The buildings', noted the press, 'which are after Indian models, ought to be easily adapted to the purposes of another Eastern nation.' He found a willing partner in the Japanese government, which was eager both to shore up the alliance and to show off its own recently acquired imperial possessions. The White City promised to satisfy what a Japanese commissioner described as a 'desire to be known to you thoroughly'. Japan did not have an easy time of it in London, however. The Japanese commissioners did their best, in the southern half of the site, to convince the British public both that they belonged among the great powers and that industry and empire rested on indigenous foundations. The evidence for the latter included a Garden of Peace, which was surrounded by 'high and extensive canvas screens', covered with 'scenic pictures', to obscure the 'many ugly objects in the vicinity' and provide the illusion of a 'vast expanse' far from London. They thereby echoed, however, Kiralfy's own strategy to maximize his gate receipts, by depicting Japan as a place apart from the modern world. In the northern half of the site, Kiralfy supplied what was advertised as 'Japan in Essence' and reported as 'a Far East Fairyland for Londoners'. Not for the first time, it was native bodies – of craftsmen and wrestlers, Taiwanese aborigines, and Hokkaido Ainu – that would captivate the crowds.[2]

London in 1910 provided a fitting coda for Japan's experience and practice of exhibitions over the previous half century. The modern state had initially seized on them in its attempt to become an equal of the great white powers. By 1890 or thereabouts, the economy had been placed on a

Geographers 46 (2021), 810–14, for contemporary black experience on the estate. Breandan Gregory, 'The Spectacle Plays and Exhibitions of Imre Kiralfy, 1887–1914', PhD dissertation (University of Manchester, 1988), for Kiralfy.

[2] Imre Kiralfy, 'Prospectus', item 4, box 7, Kiralfy Collection, Museum of London, for his vision. 'Japan-British Exhibition 1910', www.bl.uk/onlinegallery/onlineex/maps/uk/004880579.html, for a map. *Morning Advertiser* 1909/8/13, for Indian models. *Official Report of the Japan British Exhibition 1910 at the Great White City, Shepherd's Bush, London* (Unwin Brothers, 1911), 509, for desire to be known, and 247, for screens. *Japan-British Exhibition, 1910, Shepherd's Bush, London: Official Guide* (Bemrose & Sons, 1910), 2 and passim, for 'Japan in Essence'. *The Graphic* 1910/8/23, for 'Far East Fairyland'. *Illustrated London News* 1910/7/9, for native bodies. Ian Nish, *The Anglo-Japanese Alliance: The Diplomacy of Two Island Empires, 1894–1907* (Athlone Press, 1966) and *Alliance in Decline: A Study in Anglo-Japanese Relations, 1908–23* (Athlone Press, 1972) for the alliance.

sound, repressive footing and was beginning to exhibit signs of modern industrial life; the polity had achieved institutional form with an imperial constitution and a propertied parliament; and the foundations had thereby been laid for imperial crusades, which distinguished the geopolitical great and good and whose booty could be unveiled at international exhibitions. Japanese efforts to be 'known thoroughly' proved Sisyphean, however. They were stymied in part by the burgeoning belief in immutable racial hierarchies. When Japan tried to be modern, it was accused of mimicry. When it tried to be authentic, it was confined to the native enclosure.[3] More generally, the officials were frustrated by the logic and course of exhibitions themselves, which were evolving to accommodate the increasingly diverse interests of organizers, exhibitors, and visitors. Official attempts to police the representation of Japan overseas were repeatedly undermined by Japanese showmen, brokers, and other compradors, who were determined to profit from the ongoing demand for a less nuanced, more familiar Japan.

At home, conversely, the state was beginning to step back as impresario and exhibitions to reveal the municipal diversity and commercial energy that were beginning to germinate in the newly ploughed economic soil. The early industrial exhibitions of the 1870s and 1880s had done their job, catalysing enthusiasm among central bureaucrats, local authorities, and would-be entrepreneurs. The first to seize the opportunity for grander statements were the three major cities, Kyoto, Osaka, and Tokyo, who were not only eager to exploit the potential of an exhibition to promote local industry, but also to assert their own place in the emerging national landscape, by exhibiting their industry, but with other attractions too. Soon enough, other players were getting in on the act. By 1910, therefore, exhibitions were moving on from their earlier, didactic concerns. They allowed countries and cities to claim their place on the map and to parade their difference. They were opportunities for entrepreneurs to engage in modern marketing or to turn a quick buck. They were a familiar form, which afforded space for a changing world, providing something for everyone and so pulling in the crowds.

After 1910, back in London, the exhibition buildings slowly disappeared over time. A four-fifths-sized replica of a ceremonial gate from Kyoto was carted off to Kew Gardens. The stadium was knocked down in 1985 to make space for another building for the BBC. By the end of the

[3] For mimicry, Homi Bhabha, 'Of Mimicry and Man: The Ambivalence of Colonial Discourse', *October* 28 (1984), 125–33.

century, there was little left: cavernous overhead halls, which had served to link a couple of stations to the main entrance; the grubby, inconspicuous remnants of the Garden of Peace. Come the new millennium, though, the siren call of profit transformed the south end of the original expo site: the overhead halls gave way to the shopping centre; the television centre was converted into apartments; and the garden was restored. Six-foot-tall lanterns and cherry trees now line the path to a new, traditional gateway, beyond which water, rocks, and vegetation provide a Japanese time out from everyday London life.[4] The demand for spectacle may have moved on from exhibitions to smaller, more individualized screens. The demand for Japan as a place apart remains.

INDIGENOUS ARTISTRY

The modern state's early enthusiasm for exhibitions had been driven by the need to foster trade and promote industry, which would allow Japan to compete on the international stage. However, its officials came back from Paris in 1889 somewhat chastened. It looked as if European enthusiasm for export craft, the staple of Japanese exhibits at international exhibitions for over two decades, was beginning to collapse. Still, overseas exhibitions remained a priority. First, pragmatically, they remained a place to promote trade. The World's Columbian Exhibition in 1893 was an opportunity to sell to Japan's biggest foreign market by far, responsible for over one-third of Japanese exports. More than this, exhibitions were becoming an opportunity to show that Japan belonged. Chicago provided a chance for 'the world [to] gain a more accurate and comprehensive knowledge of our country, its history, its progress and its aspirations', to show that Japan was 'worthy of full fellowship in the family of nations', and to argue that the time had come to abolish the unequal treaties, whose 'necessity ... has entirely passed away'. Sure enough, in 1894, Japan would sign a new treaty with Britain, which agreed that extraterritoriality should come to an end five years later. It also went to war with China over Korea, gaining Taiwan as a colony, in 1895. By the time the bureaucrats started preparing for Paris in 1900, participation in the exhibition had become a 'necessary duty', incumbent on its newfound position in the world. The Japanese exhibit as a whole would express 'the external form of the nation' (*kokka no taigai-teki keitai*)

[4] The Japanese Garden Centre, 'About This Project', www.buildajapanesegarden.com/blog/portfolio/hammersmith-park-london/, for the restoration.

and its 'degree of enlightenment' (*kaimei no shindo*), not least its success in improving imported mechanical and industrial arts.[5]

Fond hopes and inflated expectations made for swelling costs and swollen exhibits. Chicago 1893 was lavished with an official budget nearly five times that for Paris 1889, and 16,500 Japanese exhibits; Paris 1900 with a budget that had doubled again, and 26,640. Still, the majority remained the agricultural products, arts, and handicraft familiar from previous exhibitions overseas. The pre-industrial effect was exacerbated in Chicago by using bamboo to frame the exhibits, to create 'the appearance of Japanese environment', and in Paris by the continued emphasis on horticulture, not least chrysanthemums. In both, however, the bureaucrats tried to make the familiar artefacts tell a new story. First, they insisted, the exhibits came from a place that was modernizing fast. In Chicago, the story was told in a *History of the Empire of Japan*, and supported by photographs of railroad lines and telegraph systems; samples of school textbooks; plans and photographs of the new Tokyo Central Prison; and a demonstration of how Hokkaido had progressed from a barren wilderness to a 'rich source of national wealth'. Even the aristocratic 'Japanese Lady's Boudoir' in the Women's Building was accompanied by a booklet including 'statistics of female education and employment'. In Paris, too, the commissioners tried their best to argue that the exhibits of forestry, fishery, and foodstuffs, and crafts, textiles, and furnishings demonstrated ongoing progress, despite the continuing absence of machinery.[6]

Second, they rooted the story of industry and empire not simply in aboriginal imperial history, but in indigenous Japanese culture. They were building on more than a decade of efforts within Japan to identify a distinctive Japanese aesthetics. These had been coordinated in large part

5 Yoshida, *Bankoku Hakurankai*, 90, for the export opportunity, and Gozo Tateno, 'Foreign Nations at the World's Fair, I: Japan', *The North American Review* 156.434 (1893), 34–43, for the argument about fellowship. Yoshida, *Bankoku Hakurankai*, 102–3, and Itō, *Meiji Nihon*, 64–70, for the aspirations for Paris 1900.

6 Yoshida, *Bankoku Hakurankai*, 90–93, for an overview of Chicago, and 96–104, for Paris; Itō, *Meiji Nihon*, 38–42 and 71–79, similarly. Lisa Kaye Langlois, 'Exhibiting Japan: Gender and National Identity at the World's Columbian Exposition of 1893' (PhD thesis, University of Michigan, 2004), for a comprehensive analysis of Chicago, and 238–320 for the boudoir. Tateno, 'Foreign Nations', 40, for 'Japanese environment'. Imperial Japanese Commission of the World's Columbian Exposition, *History of the Empire of Japan* (Dai Nippon Tosho, 1893), 411, for Hokkaido. Shirahata Yōzaburō, 'Kiku to Bankokuhaku', in Yoshida Mitsukuni (ed.), *Bankoku Hakurankai no Kenkyū* (Shibunkaku Shuppan, 1986), 111–32, for chrysanthemums.

by Kuki Ryūichi, who had been on staff for the Paris exhibition in 1878, served as ambassador in Washington DC in the early 1880s, director of the new imperial museum from 1888, and chair of the jury for the Third Domestic Industrial Exhibition in 1890, where his nativist predilections had infuriated artists working in Western oil rather than indigenous ink.[7] Now he made sure that the Japanese exhibit in Chicago bent to his will. Oil painting was excluded, while the authorities were petitioned to allow decorative art into the Palace of Fine Arts. More visibly, his vision took form in an exhibit of 'historical architecture' on a Wooded Island at the centre of the lagoon, designed as pastoral refuge amidst the gleaming white, neoclassical stucco of the site as a whole. The Japanese commissioners modelled the Phoenix Hall on the main building of Byōdō-in, a thousand-year-old temple south of Kyoto, thereby creating the stage for a textbook of Japanese art, with three period rooms: one wing dedicated to the Fujiwara period (ca. 1200), modelled on the Kyoto Imperial Palace, one to the Ashikaga (ca. 1400), on Ginkaku-ji, also in Kyoto, and the central hall to the Tokugawa (ca. 1700), on Edo castle, in what was now Tokyo.[8]

Chicago in 1893 thus formalized the use of traditional architecture to frame Japanese culture, which had first been deployed in Paris in 1867 and would continue for the next half century. Seven years later, Paris got a Japanese pavilion modelled on the Kondō of Hōryūji, the oldest extant wooden building in the world, which provided an appropriate frame for the more than 750 treasures on display, with the imperial collections joined by religious institutions in opening their vaults. The French were also gifted with the first-ever History of the Art of Japan, with a preface by Kuki, telling the reader what it all meant. Japan had absorbed

[7] For the state's invention of Japanese art, Satō Dōshin, *Meiji Kokka to Kindai Bijutsu: Bi no Seijigaku* (Yoshikawa Kōbunkan, 1999). See also Mimi Hall Yiengpruksawan, 'Japanese Art History 2001: The State and Stakes of Research', *The Art Bulletin* 83.1 (2001), 105–22. Ellen Conant, *Nihonga: Transcending the Past: Japanese-Style Painting, 1868–1968* (The Saint Louis Art Museum, 1995), 31, for Kuki's alienation of contemporary artists in 1890.

[8] Conant, 'Refractions', 31, for the art exhibit. Harris, 'All the World', 39–40, for the Wooded Island. Hannah L. Sigur, 'A Neoclassical Translation: The Hōōden at the 1893 World's Columbian Exposition', in David Raizman and Ethan Robey (eds.), *Expanding Nationalisms at World's Fairs* (Routledge, 2017), 92–108, Langlois, 'Exhibiting Japan', 165–237, and Lancaster, *The Japanese Influence*, 76–83, for the Phoenix Hall. 'Souvenir Map of the World's Columbian Exposition at Jackson Park and Midway Plaisance, Chicago, Ill, U.S.A. 1893', www.loc.gov/item/2010587004/, for a map. Christopher Vernon, 'Daniel Hudson Burnham and the American City Imperial', *Thesis Eleven* 123.1 (2014), 80–105, for its governing vision.

influences from India and China but had refined these as those countries had declined. Japanese art was thus a distillation of Asian values, but had to be understood on its own aesthetic terms.[9] By 1900, Kuki's star was on the wane back home, but the corollary of his message, that Japan's leadership in Asia was the natural historical order of things, was widely shared.

The double-barrelled message, however, about Japan ancient and modern, was lost in translation. For foreign observers, the mounting evidence of industry and empire were hard to reconcile with the Japan they had grown to expect. In Chicago in 1893, the Americans were impressed. Japan was trying to join the Occident in its world historical mission. Perhaps it might want to send its 'missionaries over here'. The Wooded Island, however, served to set Japan off and apart. The Phoenix Hall showed Japan 'at her very best, and let ... the world see her in a native village', 'gorgeous' but also 'grotesque'. The impression was reinforced by the native craftsmen, and the Japanese Bazaar, sandwiched between Irish and Javanese villages in the entertainment zone, the Midway Plaisance. On the one hand, 'the Japans [sic] are neat, industrious, studious, and progressive, ardent in desirin' knowledge'. On the other, again and again, the Japanese are 'child-like' and Japanese products 'queer'.[10]

In Paris in 1900, the ambivalence was underlined. Kuki's efforts bore some fruit. The historical exhibit was 'a revelation'. Modern art was puzzling, however, and decorative export craft roundly condemned, at odds with the contemporary prescriptions of Art Nouveau, that form should follow function. The dilemma was more general, however. The commissioners had tried to resituate Japan among the Euro-American powers and to police its appearance in venues beyond the official exhibit. The French authorities had been adamant, however, that the Japanese pavilion should take its place among the European colonies and that attractions were useful in showing the public its 'beautiful customs'. The latter were certainly popular. Fashionable women used the Japanese pavilion as a backdrop for proto-selfies. A virtual round-the-world tour revealed 'charming creatures' when it stopped in Japan. The sensation of

[9] Itō, *Meiji Nihon*, 75–87, for the exhibit and the history.

[10] Marietta Holley, *Samantha at the World's Fair* (Funk & Wagnalls, 1893), for neat, gorgeous, grotesque, child-like, and queer; Harris, 'All the World', 41–43, Rydell, *All the World's a Fair*, 50–51, and Langlois, 'Exhibiting Japan', 224–35 for the general American reception; Itō, *Meiji Nihon*, 41–42, for the Bazaar.

the summer, though, was a 198-member dance troupe led by Kawakami Otojirō and Sadayakko, who provided 'a striking sensation of reality, all the while embodying the grace … of the figures in the scrolls'. The moral was clear. 'What pleases us most… is precisely everything that is, for them, traditional, the opposite of progress, the past.'[11] It would be rehearsed again in St Louis in 1904 and London in 1910. Meanwhile, the ability of exhibitions to provide more than one account of the world, to cater to more than one demand, was also becoming apparent at exhibitions in Japan, where it was put to quite different effect.

IMPERIAL HERITAGE

With economic infrastructure in place, come the 1890s, the Japanese state could step back from promoting industry at home as assiduously as it had in the previous two decades. There was no shortage of municipal demand, however, for the opportunity to host the next Industrial Promotion Exhibition blessed by the imperial government. Kansai was the obvious place, including as it did the former imperial capital (Kyoto), the emerging centre of Japanese industry (Osaka), and the second most important international port (Kobe). The year 1892 saw a flurry of attempts to secure the fourth in the series. It was Kyoto that came out on top. The city had got into the exhibition game early and had already tried to host both the second and the third. By the end of the 1880s, its situation was becoming dire. The new governor's pet projects, not least a canal, which would provide water, generate power, and prevent fire, were stalling for lack of funds. Luckily, though, it had an ace in the hole; 1894 was going to be the 1,100th anniversary of its founding. What better opportunity to combine industry and empire? Not everyone was convinced, including the minister initially in charge. But Kyoto had its supporters, and with his replacement by Sano Tsunetami as minister, the die was cast. Sano had

[11] Lacambre, 'La Découverte', 42, for revelation and puzzlement, Clare Pollard, *Master Potter of Meiji Japan: Makuzu Kōzan (1842–1916) and His Workshop* (Oxford University Press, 2002), 80–81 for condemnation. For Art Nouveau, Teasley, *Designing Modern Japan*, 113–14. Itō, *Meiji Nihon*, 72–75, for the doomed efforts of the commissioners. *Le Livre d'Or de l'Exposition de 1900* (Édouard Cornély, 1900), 135–36, for proto-selfies and 'what pleases us most'. *Le Panorama: Exposition Universelle 1900* (L. Baschet, 1900), no. 13, for the round-the-world tour, and no. 16 for Sadayakko. See also Ayako Kano, *Acting Like a Woman in Modern Japan: Theater, Gender, and Nationalism* (Palgrave, 2001), 85–95, and Lesley Downer, *Madame Sadayakko: The Geisha Who Seduced the West* (Review, 2004), 160–79.

provided the intellectual justification for exhibitions as industrial promotion in the 1870s and had presided over the efforts of Kuki and his cronies to conserve the historical patrimony in the 1880s. His decision was Solomonic. Kyoto would hold the fourth – in 1896, he suggested, due to the need to prepare for Chicago – Osaka the fifth and Tokyo the sixth. The city got the government to move the date to 1895, to align better with the municipal anniversary. Kobe was offered an aquarium as a consolation prize. And the planning took off.[12]

The Fourth Domestic Industrial Exhibition took place in a very different landscape from the first three. The 1877, 1881, and 1890 exhibitions in Tokyo had served to erase the past, in order to make space for the educational and recreational institutions of a modern imperial capital. The 1895 exhibition in Kyoto sought to restore it, so that a superseded imperial hometown might find its place in the modern world. There were two parts to the equation. The first was the millennium and more of imperial history, which provided the city's warrant as a repository of tradition, and which was embodied in the centrepiece of the anniversary commemorations. The newly built Heian Shrine was a five-eighths-sized recreation of the main building of the original capital. It provided a new home for the city's founder, Kammu, in a rough, undeveloped area to the east of the Kamo River, which bordered the new canal and would become the city's first modern park. The exhibition, meanwhile, just to the south of the shrine, would show how the ancient city still had a part to play in modern Japan. Together, they formed a multiplex, making space for Kyoto in the world of industry and empire by keeping one foot in the past. They were only the most visible of a series of initiatives, however. Land had to be acquired, sites prepared, roads improved, transport upgraded, lodgings inspected, vaccinations ordered, rubbish removed, and slums cleared. It would not be the last time that an exhibition would provide cover for municipal improvement, beneath which the poor could be displaced.[13]

[12] Kasahara Kazuto, 'Rekishi, Kankō, Hakurankai: Dai Yonkai Naikoku Kangyō Hakurankai to Heian Sento Senhyakunen Kinensai no Toshi Kūkan', *10+1* 36 (2004), 166–67; Kuni, *Hakurankai no Jidai*, 128–34. For Tokyo's slump, Henry D. Smith, 'The Edo-Tokyo Transition: In Search of Common Ground', in Marius B Jansen and Gilbert Rozman (eds.), *Japan in Transition: From Tokugawa to Meiji* (Princeton University Press, 2014), 347–74.

[13] Tseng, *Modern Kyoto*, 51–65, for the shrine. Kasahara, 'Rekishi, Kankō, Hakurankai', 167–69, for siting and squabbles. Allan G. Grapard, 'Institution, Ritual, and Ideology: The Twenty-Two Shrine-Temple Multiplexes of Heian Japan', *History of Religions* 27.3 (1988), 246–69, for multiplex. Kuni, *Hakurankai no Jidai*, 135–43, for public works.

When the exhibition opened, on April Fool's Day, it therefore faced two ways. Past the inevitable stalls, organized by prefecture, and an eye-catching fountain, the visitor passed through an elaborate towered gateway into the exhibition itself. The site was a third larger than its Tokyo predecessor, and industry, rather than agriculture, now comprised the majority of the 169,098 exhibits, occupying all four sides of the first, central courtyard. Osaka had provided a tenth of them, but Kyoto was not far behind. The city was still wedded to traditional crafts. A quarter of its 10,000 exhibits were textiles. More generally, machinery remained light on the ground and had been consigned, with agriculture, forestry, and marine products, to some smaller halls a little further north. The war with China was still wrapping up when the exhibition opened. The Qing government had sued for peace in February, but the Treaty of Shimonoseki would only be signed two-and-a-half weeks after the opening ceremonies, so leading firms such as Mitsubishi were absent.[14]

For Kyoto, however, the point of the exhibition was to show how the city could hold its own – and maintain its distinction – even as the world turned. The lesson was underlined in the art gallery, just to the east of the shrine. On the one hand, there was oil painting. It may have been condemned as inauthentic by Kuki and the like, but Kuroda Seiki's 'Morning Toilette', the first nude to be publicly exhibited in Japan, caused a media uproar, and usefully associated the city with the avant-garde. On the other, there were modern reinterpretations of older themes. Kano Hōgai's 'Merciful Mother Kannon' had been painted in the mid-1880s and copied a number of times since. Now it was replicated as a monumental woven tapestry by Kawashima Jinbei II, who with other Kyoto textile artists had participated in earlier exhibitions, including Paris in 1889. One of the few craft exhibits to garner approval by the press, it was quickly snapped up by the Imperial Household, and would make its way to London in 1910, as an authentic representative of both imperial Japan and Kyoto craft.[15]

More generally, too, reaction to the exhibition was mixed. By the time the numbers were tallied, visitor numbers had exceeded those of its Tokyo

[14] *Fūzoku Gahō*, no. 94 (1895/8/18), 6–7 for a map and 15–22 for a guide. Kuni, *Hakurankai no Jidai*, 146–48 for numbers.

[15] Alice Tseng, 'Kuroda Seiki's Morning Toilette on Exhibition in Modern Kyoto', *The Art Bulletin* 90.3 (2008), 417–40, for Kuroda; Chelsea Foxwell, 'Merciful Mother Kannon and Its Audiences', *The Art Bulletin* 92.4 (2010), 326–47, for Hōgai; and Hiroko McDermott, 'Meiji Kyoto Textile Art and Takashimaya', *Monumenta Nipponica* 65.1 (2010), 37–88, for Kawashima.

predecessor, which was quite an achievement, given that new capital had four times the population of the old. They had declined precipitously over the course of the run, however. The opening day and weeks were buoyed by various concurrent ceremonies – for victory in the war with China, for the fallen, and for the reconstruction of the largest wooden building in Kyoto, the Founder's Hall at Higashi Honganji, a huge temple a couple of miles to the southwest. In the absence of Buddhist celebration, though, and given adjustments to train schedules, to accommodate soldiers returning from the war in China, visitors became much thinner on the ground. Sales were disappointing, too. The metropolitan press was also predictably disparaging about the regional cousin's inability to stage a proper show.[16] At least one outlet, however, suggested how a nominally industrial exhibition was beginning to serve more local, idiosyncratic ends.

Fūzoku Gahō had started up six years earlier, the first graphic magazine in Japan, using copperplate to illustrate the major media events of the day, as well as the customs and mores of its title. There was too much going on in the spring – the war in China and the new colony in Taiwan; the doll's festival and plum viewing – for there to be more than a brief mention of the exhibition. Come the summer solstice, however, there was less competition and so space for a comprehensive special issue. It did justice to the show: illustrations of the kimono-clad crowds, the buildings, and the slew of exhibits; a piece by Kuki, as president of the jury, pontificating about its significance; dutiful accounts of the opening ceremony, the rules, the officials, the exhibits. The issue as a whole, however, served to situate the exhibition as one among many attractions in a modern tourist destination: shrine and streetcars; an imperial tomb and the local office of the Imperial General Headquarters (Figure 3.1). Kyoto was full of famous places (*meisho*) and famous products (*meibutsu*).[17]

If its predecessors in Tokyo had been attempts to promote industry, the exhibition in Kyoto was one in a battery of attempts to promote the city – to catalyse a transformation of the municipal topography, to reassert its importance to the region, and to resituate the result in the national landscape. The exhibition was supported by events in cities ranging from Nagoya to Hiroshima. The most successful of the 33 guides to exhibition

[16] Kuni, *Hakurankai no Jidai*, 144–49, for visitor numbers and sales. *Asahi Shimbun*, 1895/4/24, 4/30, 5/3-5, and 5/10 for sniffy press.

[17] *Fūzoku Gahō*, no. 88 (1895/3/25), 16–18, for the brief mention; no. 94 (1895/6/18) for the special issue.

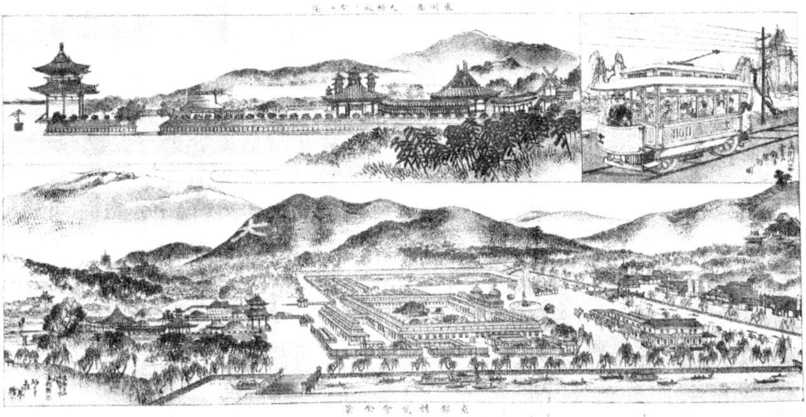

FIGURE 3.1 In Kyoto in 1895, shrine and exhibition helped to redefine the city as a modern tourist landscape.
Source: *Fūzoku Gahō*, no. 94 (1895/6/18), NOMURA Co., Ltd.

and city took in 150 sites in neighbouring prefectures, selling 300,000 copies in three editions in a single month.[18] Backed by the region, Kyoto was able to turn its imperial past to modern account. In doing so, however, it was turning exhibitions into a slightly different thing. The change would become even clearer as the circus moved on to Osaka for the Fifth Domestic Industrial Exhibition in 1903.

MUNICIPAL PROMOTION

Sano may have been Solomonic in his awarding of the Fourth Domestic Industrial Exhibition to Kyoto, but it took a while for Osaka to secure and plan its successor. Victory over China in 1895, the imminent prospect of treaty revision, and the growing enthusiasm for an international exhibition in Japan in 1907 prompted the government to push back the date for the Fifth Domestic Industrial Exhibition from 1899 to 1902, setting off a renewed bidding war between Osaka, Nagoya, Tokyo, and Sendai. It was only at the turn of the century that Osaka was confirmed as host, with the exhibition now scheduled for 1903.[19] The decision proved merely a starting point, however, from which different visions of the

[18] Kasahara, 'Rekishi, *Kankō, Hakurankai*', 172–76; Kuni, *Hakurankai no Jidai*, 153.
[19] For the complications, Matsuda Kyōko, *Teikoku no Shisen: Hakurankai to Ibunka Hyōshō* (Yoshikawa Kobunkan, 2003), 20; Kuni, *Hakurankai no Jidai*, 162–67, and Itō, *Meiji Nihon*, 97–101. See also Kuni, *Hakurankai no Jidai*, 245–54, and 'Naikoku

event diverged. Seen from Tokyo, the exhibition was still meant to promote industry (*kangyō*), although through export promotion rather than industrial pedagogy. It took a few words to specify what this meant. Exhibitors had to be directly involved with the 'extraction, production, or manufacture' of any proposed exhibits, which should not have been exhibited in a previous exhibition or produced before December 1897, unless they were animals and plants (though any cow born before December 1894, any horse born before December 1892, anything related to educational achievement, except in technical schools, and anything injurious to public morals was also beyond the pale). For the first time, too, there was a 'Reference Building' for foreign products, which would 'afford Japanese manufacturers an opportunity of studying the latest products of Western invention with a view to the way of industrial improvement'.[20]

Seen from Osaka, however, manufacturing was secondary to the city's municipal ambitions, which lay as much with trade as with industry. The exhibition would reveal Osaka to the world as 'a global marketplace, an entrepot for the cargoes of East and West, a gateway to the Pacific'. To look the part, though, upgrades were needed: a port, to compete with Kobe for ocean-going trade; meeting facilities and a modern hotel; and improvements to the transport infrastructure, which prompted a confrontation with 3,000 irate jinrikisha drivers, aware that their livelihood was under threat. Most pressing was the need to clean up the main street leading from the exhibition, in the south of the city, to the government and transport centre in the north. The thoroughfare had been a running sore in the municipal imagination for a quarter of a century, with unprepossessing row houses, designated as slums from the 1880s, and repeated outbreaks of cholera, dysentery, and smallpox, culminating in 1899–1900. For the emperor and foreign dignitaries to see the 'separate race' of slum dwellers in such a 'filthy place' (*fuketsu no tokoro*) 'would reflect badly not only on the honour of the city but on the face of the nation'. The landlords, who were profiting mightily from the poverty, protested. But the order came down to widen the roads, clear the 'poor

Kangyō Hakurankai no Kisōteki Kenkyū: Shokusan Kōgyō, Fubyōdō Jōyaku, 'Naikoku' no Imi', *Nihonshi Kenkyū* 375 (1993), 54–68, for the importance of the unequal treaties.

20 Hasegawa Nyogen, *Daigokai Naikoku Kangyō Hakurankai Shokisoku* (Nagoya Tsūshinsha, 1902), 9, and Yoshii Takao, 'Daigokai Naikoku Kangyōhaku no Disuprē', in Yoshida Mitsukuni (ed.), *Bankoku hakurankai no kenkyū* (Shibunkaku Shuppan, 1986), 288–91, for the rules. Kuni, *Hakurankai no Jidai*, 170, for the Reference Building.

neighbourhoods' (*saimin buraku*), and move the residents south, outside the city limits. *Buraku*, 'small village', was becoming a charged term by the turn of the century, and the people who lived in them the object of official and popular discrimination, thanks to their hereditary trades, their traditional designation as outcastes, and the exclusion of their villages from the new local government system. Their labour was still required, however, to build the exhibition, and they proved adept at exploiting the opportunity to demonstrate their own commercial prowess.[21]

By the time the exhibition opened, on the surface, in the aggregate, things looked pretty good. The economy was humming and Japan finding its place in the world. The overall area, the available display space, and the number of exhibitors had all doubled since Kyoto, and the number of exhibits by two-thirds. 'Industry' now comprised well over half the total, occupying a single massive building with four internal courtyards. Even so, the exhibition was bulging at the seams. The prefectures had proposed enough exhibits to fill each pavilion three to twelve times. Four colonies and fourteen foreign countries had also requested space. The Reference Building was quintupled in size, but it was not enough. A handful of foreign countries, Tokyo, and Aichi were allowed to build their own pavilions, the latter two flanking the main gate, with the last in the form of Nagoya castle, complete with gold *shachi* on top.[22] On closer inspection, however, the cracks began to show. There had been some progress in silkworm rearing, horse breeding, and food processing. The leather

[21] Daigokai Naikoku Kangyō Hakurankai Kyōsankai, *Ōsaka to Hakurankai* (Daigokai Naikoku Kangyō Hakurankai Kyōsankai, 1902), for the vision, together with Matsuda, *Teikoku no Shisen*, 35–46. For slum clearance, ibid., 22–35; Nagahashi Temesuke, 'Daigokai Naikoku Kangyō Hakurankai no Shikichi Settei ni okeru Suramu Curiaransu e no Eikyō', *Randosukēpu Kenkyū* 58.5 (1995), 21–24; and Katō Masahiro, *Osaka no Suramu to Sakariba: Kindai Toshi to Basho no Keifugaku* (Sōgensha, 2002), 102–7. For jinrikisha drivers, Hirai Seiji, 'Hakurankai kara Mita Kamagasaki no Rekishi: Kokusaku no Kage ni tsukai Sterareru Hiyatoi Rōdōsha', *Shin Nihon Bungaku* 43.1 (1988), 111–12. For *burakumin*, David L. Howell, *Geographies of Identity in Nineteenth-Century Japan* (University of California Press, 2005), 79–89 for early Meiji, and Ian Neary, 'Burakumin and Human Rights', in Laura Hein (ed.), *The New Cambridge History of Japan*, vol. 3 (Cambridge University Press, 2023), 622–26 for subsequent developments. For contemporaneous efforts to clear slums and beautify cities in other parts of the world, James Alfred Yelling, *Slums and Slum Clearance in Victorian London* (Allen & Unwin, 1986), and Vernon, 'Daniel Hudson Burnham'.

[22] Yoshii, 'Daigokai Naikoku Kangyōhaku', 287–92 for an overview and stats, 295–303 for the scramble to show off and the resulting space issues. See also Kuni, *Hakurankai no Jidai*, 181–83, and Itō, *Meiji Nihon*, 94–95. For pictures, *Daigokai Naikoku Kangyō Hakurankai Kinen Shashinchō* (Gyokumeikan, 1903).

workers of Nishihama, a Burakumin district in Osaka, were supplying both the military and white-collar markets, exhibited in numbers, and saw a further surge in demand as a result. While there was nine times as much machinery on display as in Kyoto, however, the sector still lagged behind foreign competitors and relied on foreign imports. Incentives were lacking. Innovation was piecemeal. Rickety wooden looms showed where the economy was actually at. Even agriculture and the livestock industry attracted journalistic ire.[23] The exhibition revealed a gap between aspiration and achievement, which suggested that Japan was still behind.

The same was true of empire. Taiwan had been a colony for some eight years, providing a few exhibits in Paris in 1900 and Hanoi in 1902, but remained foreign to 'our understanding of ourselves as a nation-state'. The exhibition was an opportunity to make Taiwan known, to promote its exports, and to enlighten visitors from the island itself. The official position took form as the Taiwankan, or the 'Formosa Building' in the English-language publicity, which could be found at the far end of the site. A Chinese-style compound, with space for exhibits, retail, and recreation, it included an overview of the new acquisition, which made careful distinction between 'native' customs, illustrated by mannequins, and the 'barbarian' ones of the eight, recently classified, indigenous tribes; 4,000 exhibits, with tea, sugar, and camphor most prominent, which prompted comment on both the immature state of the island's industry and its potential as a treasure house of empire; and the Tokkeido, a pavilion from Tainan, in the south, where a member of the imperial family had rested during his military exertions in 1894, before dying of malaria and being enshrined in Taipei, which was brought over and reconstructed on site.[24] The combination was a hit, but it was not the only colonial attraction on the ground.

[23] Yoshida, *Bankoku Hakurankai*, 130–31, and Kuni, *Hakurankai no Jidai*, 229–33, for the judges. Yoshimura Tomohiro, '1900 nendai Shotō no Nishihama Buraku to Hikakugyō: Daigokai Naikoku Kangyō Hakurankai ni miru Seihin o Chūshin ni', *Osaka Jinken Hakubutsukan Kiyō* 4 (2000), 61–73, for the leather workers. For dyspeptic metropolitan journalists, *Asahi Shimbun,* 1903/3/11, 2, and 7/3, 2, for machinery, 5/7, 3, 5/10, 5, and 5/16, 3, for agriculture, and 5/21, 7, for livestock.

[24] Matsuda, *Teikoku no Shisen*, 55–81; Kuni, *Hakurankai no Jidai*, 174–80; Itō, *Meiji Nihon*, 106–12; and Ha Sae-bong, 'Taiwan and Its Self-Images: The Case of the Osaka Exhibition in 1903', *Taiwan Shi Yanjiu* 14.7 (2007), 1–39, for details. Miura Ippei, *Daigokai Naikoku Kangyo Hakurankai Yōran* (Meiji Bunken Shiryo Kankokai, 1973), 264, for understanding ourselves. Kadota Masatsune, *Daigokai Naikoku Kangyō Saikai Benran* (Taiwan Kyōkai, 1903), for a guide for Taiwanese visitors. Matsuda, *Teikoku no Shisen*, 82–118, and Paul Barclay, 'Contending Centres of Calculation in Colonial

A group of local notables had decided that Osaka should also have a native village, by then de rigueur at international exhibitions. They subcontracted its creation to Tsuboi Shōgorō, Professor of Anthropology at the Imperial University in Tokyo, who had been smitten by such villages in Paris in 1889 and in charge of Japan's ethnology exhibit in 1900. Ten days after the exhibition opened, therefore, the *Jinruikan* (Human Pavilion), opened its doors, just outside the main gate, to reveal five Hokkaido Ainu, four Taiwanese aborigines, two Okinawans, two Koreans, three Chinese, three Indians, a Javanese, and a Bulgarian, as well as a large map of the world's races, using fifty pairs of cut-out paper dolls. Each of the participants had their own reasons for being there: some to advance their position as imperial subjects or to attract support for education back home; others simply to earn some cash. But the outrage was immediate, from Chinese students and diplomats, Korean intellectuals, and an Okinawan writer who protested the damage to 'our harmony with the nation'. The Foreign Ministry got involved. By April, the residents comprised seven Ainu, three Taiwanese aborigines, two Malays, one Javanese, one Turk, and a Zanzibar islander. Its critics were hardly opposed to the notion of racial hierarchy. Rather, they were outraged that their status should be reduced to that of 'dead countries' such as India, the 'barbarian Ainu', or the 'world's lowest races ... no different to deer or pigs'. Japanese commentators, meanwhile, used the uproar to their own ends, pointing out how similar exhibits overseas served to denigrate Japan, or lampooning Japanese fears of 'savage' others. A couple of months in, though, the excitement had died down and the media had moved on.[25] There was too much else to look at.

Taiwan: The Rhetorics of Vindicationism and Privation in Japan's "Aboriginal Policy"', *Humanities Research* 14.1 (2007), 67–84, for tribal classification. Teasley, *Designing Modern Japan*, 141, for the jury's criticism of Taiwanese weaving.

[25] Itō, *Meiji Nihon*, 113–18, for an overview, and Matsuda, *Teikoku no Shisen*, 119–41, for a detailed account. Kirsten Ziomek, *Lost Histories: Recovering the Lives of Japan's Colonial Peoples* (Harvard University Asia Center, 2019), 30–59, for the participants' motivations and reception. For Tsuboi, Tsuboi Shōgorō, 'Daigokai naikoku kangyō hakurankai ni okeru jinruigaku', *Tokyo Jinrui Gakkai Zasshi* 26 (1903), 299–303; Nishino Yoshiaki, 'Meiji 37-nen no Tsuboi Shōgorō: Jinruigaku Kyōshitsu Hyōron Tenrankai o megutte', in Tōkyō Daigaku (ed.), *Gakumon no arukeorojī* (Tōkyō Daigaku, 1997), and Yamaji Katsuhiko, *Kindai Nihon no Shokuminchi Hakurankai* (Fūkyōsha, 2008), 6. For the Korean response, Hyeokhui Kwon, 'An Analysis of Korean Intellectual Responses to the Exhibition of Koreans at Japanese Expositions: Nationalism and the Discourse on Northeast Asian Solidarity at the Turn of the Century', *Sungkyun Journal of East Asian Studies* 17.1 (2017), 27–31. For the Okinawan intellectual and 'barbarian Ainu', Ōta Chōfu, *Ōta Chōfu Senshū* (Daiichi

FIGURE 3.2 In Osaka in 1903, industrial exhibitions became things of fragments, ranging from a Hall of Wonders (top left) to Tokyo's bazaar (bottom right).
Source: *Daigokai Naikoku Kangyō Hakurankai*, poster, NOMURA Co., Ltd.

In Osaka, the Human Pavilion was only one among a welter of attractions, inside and outside the exhibition grounds, which featured in the guidebooks, attracted the crowds, and provided copy for the increasingly competitive press (Figure 3.2). Developing industry and emerging empire were subsumed within the entertainment afforded by the modern city. This included some familiar sights: a female headliner in the Hall of Wonders, whose performance concluded with 'real smoke and flames' and her 'disappearance in a mass of ashes'; a fifty-metre tower, complete with elevator and floodlight, which provided a bird's eye view of site and city, and an advertising platform for Ōbayashi Yoshigorō, whose company had built much of the site. But the attractions ran the gamut: from

Shobo, 1995), 210–16. For the other racial slurs, by Chinese students, Sakamoto Hiroko, 'Chūgoku Minzokushugi No Shinwa: Shinkaron, Jinruikan, Hakurankai Jiken', *Shisō* 849 (1995), 78. *Fūzoku Gahō*, no. 269 (1903), 37, and no. 275 (1903), 42, for passing mentions in contemporary media. Chinen Seishin put the Jinruikan again under the microscope in 1976: 'The Human Pavilion', in Davinder L. Bhowmik and Steve Rabson (eds.), *Islands of Protest: Japanese Literature from Okinawa* (University of Hawai'i Press, 2016). For a present-day Ainu perspective, Kaiho Yōko, *Kindai Hoppōshi: Ainu Minzoku to Josei to* (San'ichi Shobō, 1992), 155–62.

natural wonders (an aquarium in Sakai); to nutritional possibilities (a Canadian demonstration of bread making); to kinaesthetic enchantment (a merry-go-round and a water chute); to the technological sublime (a Refrigeration Building and the 11,000 electric bulbs that lit up the site at night). There was more. A massive naval review in Osaka Bay in April attracted the emperor, thirteen foreign ministers, a number of foreign vessels, and a vast flotilla of smaller ones, as well as huge crowds on land. The imperial couple subsequently stayed in Kyoto for a month, visiting the exhibition five times and providing useful copy for the press. The combination – sprawling exhibition, modernizing city, imperial state – proved irresistible. By the time the gates closed, the exhibition had seen 5,305,209 visitors, nearly five times as many as any previous domestic exhibition, of whom 700,000 had come at night.[26]

The numbers, together with the spectacle to which they flocked, made Osaka a turning point for exhibitions. The bureaucracy still busied itself with regulation, classification, and judgement. There would always be new things to learn about the state of industry and the development of empire. But, it was clear, exhibitions were hardly the place to stabilize the signifiers with which the state and other authorities sought to make claims about the world. Rather, they served, insistently, to reveal the gaps between rhetoric and reality, aspiration and achievement, which made modern development an uneven thing. Economic ambition would always be undercut by industrial achievement. Empire would never mean only one thing. Nor could either compete on the ground with more demotic, visceral attractions. In other words, exhibitions still had their uses, but they were becoming something else: an opportunity for a city to turn itself around and assert its status to its peers; an opportunity for folk from the countryside to see the modern city – and so begin to understand the country they were about to serve – as they made their way to war. Six months after the exhibition closed, the Russo-Japanese War broke out. The buildings remained, however, a continuing attraction for soldiers on

[26] Yoshimi, *Hakurankai No Seijigaku*, 104–6, for an overview. Kuni, *Hakurankai no Jidai*, 186–90, for detail and stats. *Fūzoku Gahō* 269 (1903), 35–38, for descriptions, and 275 (1903), passim, for illustrations. *Kinen Shashinchō*, n.p., for the female headliner. Hashizume, *Nihon no Hakurankai*, 58, for Ōbayashi, and Hashizume Shinya, 'Toshi no Denka to Hakurankai', in Sano Mayuko (ed.), *Bankoku Hakurankai to Ningen no Rekishi* (Shibunkaku shuppan, 2015), 454–58, for the light bulbs. Matsuda, *Teikoku no Shisen*, 49, and Itō, *Meiji Nihon*, 104–06, for the naval review and imperial visit. *Fūzoku gahō* 275 (1903), 13–24 for a blow-by-blow account and illustrations, *Taiyō* 9.4 (1903), 1–2 and 162–66 for photographs.

their way from their villages in the countryside to the Manchurian front. They featured in their letters home, once they got there, a way of connecting back to Japan and its imperial aspirations, in defence of which they were prepared to die on a foreign field. By then the site itself was in the process of becoming an amusement park, but the once-was exhibition still allowed them to find a way back home.[27]

SOCIAL SPECTACLE

Nearly 90,000 would die overseas. Their deaths, however, helped to secure Korea as a protectorate and, before long, a colony, its acquisition providing new opportunities for Japanese capital and new possibilities for exhibitions, both on the peninsula and back in the archipelago. The Korean government had observed exhibitions since the 1880s, participated in them since the 1890s, and established its own system for export promotion through exhibitions in 1902. In September 1907, however, the Japanese authorities opened the first exhibition in the Korean capital, a ten-week 'meeting to progress products together' (*bussan kyōshinkai*), echoing efforts in the archipelago thirty years earlier and revealing the work that lay ahead. The vast majority of exhibits had been made in Japan, officially due to the 'lack of development' of local industry and 'of the Korean people themselves'. In fact, the neophyte Japanese administration was busy trying to suppress armed resistance, which meant that Korean exhibits and visitors were in short supply.[28] The travails of the new possession not only attracted attention back in Japan but also had to compete for column inches with the flourishing exhibition calendar in the archipelago. *Kyōshinkai* had taken off in Japan in the 1880s. Prefectures had begun to combine their efforts in the mid-1890s. The ninth edition of the Kansai one had been held in the spring, attracting over twenty local authorities, many from well beyond the region, over 80,000 exhibitors, and over 750,000 visitors to the small city of Tsu, some fifty miles south of Nagoya.

The year 1907 was dominated, however, and Seoul and Tsu dwarfed, by Tokyo. Fifteen years earlier, the imperial capital had been promised

[27] Matsuda, *Teikoku no Shisen*, 191, and Naoko Shimazu, *Japanese Society at War: Death, Memory and the Russo-Japanese War* (Cambridge University Press, 2009), 75.

[28] Teasley, *Designing Modern Japan*, 145, for the Japanese takeover of Korean exhibition practice. Tae-woong Kim, 'Industrial Exhibitions ("Gongjinhoe") and the Political Propaganda of Japanese Imperialism in the 1910s', *International Journal of Korean History* 3 (2002), 181–83, for 1907.

the sixth in the series of domestic industrial exhibitions. The enthusiasm grew after the Russo-Japanese War, with businessmen, bureaucrats, and eventually politicians calling for an international rather than domestic exhibition, as a way of demonstrating Japan's new-found status to the world. It would take some time, however. Rather than a state-sponsored industrial exhibition, therefore, 1907 saw official plans for a Great Japan Exhibition in 1912, with three times the space and ten times the budget of Osaka. The announcement was met with considerable excitement, not least by landlords in Ueno over the prospect of urban improvement, and set off a wave of property speculation, prompting the government to move the prospective site to Aoyama, where it owned a military parade ground.[29] The municipal authorities, meanwhile, were not resting on their laurels. In March 1906, they renamed a craft exhibition, which had been planned for the following year, the Tokyo Industrial Promotion Exhibition. Given the short lead time, it was not going to be able to compete with its predecessor in terms of space or exhibits. Ueno, the site of the first three industrial exhibitions, was picturesque, however, and Tokyo was big. The exhibition that opened in March 1907 showed that cities could go it alone – and confirmed that exhibitions were changing as a result.

Some things stayed the same, not least grumbling about leaky roofs, inadequate signage, and poorly maintained paths, as the exhibition got under way. Still, industry was there. Mitsubishi steam turbines had contributed to the success of the navy in the recent war with Russia and were on display, but the omnipresence of foreign machinery and the low numbers of Japanese patents provided cause for ongoing concern. Empire was increasingly prominent. Inscriptions at the entrance to the Taiwanese building emphasized how grateful the island's inhabitants were to have been made imperial subjects. Korean products provided a welcome distraction from the overly familiar Japanese offerings, and the exhibition itself an education in commerce for six businessmen and fifty assemblymen from Korea, who visited in May. Art supplied its usual opportunity for discrimination and outrage. Western-style painters, 'sometimes criticized as traitors to the country', had upped their game, but the biased judging provoked a media storm. Entertainment was ever more evident, drawing on newly established hits. Two Ainu were

[29] Kuni, *Hakurankai no Jidai*, 263–67; Furukawa Takahisa, *Kōki, Banpaku, Orinpikku: Kōshitsu Burando to Keizai Hatten* (Yoshikawa Kōbunkan, 1998), 22–26; and Itō, *Meiji Nihon*, 162–67.

employed in the Ainu Building and two Koreans worked as touts in the Hall of Mirrors, exciting complaints from some Korean students, but failing to attract the crowds. Instead, there was a Ferris wheel, which produced some conservative grumbling, about the impropriety of viewing the city, especially the imperial palace, from above. There was a round-the-world tour, an aquarium, and a water chute. There were visiting dignitaries – Japanese, Southeast Asian, and Mongolian sovereigns – and a horse race. Like Osaka in 1903, the exhibition was lit up at night, but with three times the number of bulbs.[30]

The exhibition as a whole, however, presented a new and compelling sight. Previous exhibitions in Ueno had been confined to the top of the hill, which was now only the first of three sites. The second brought the exhibition down around the pond, which now boasted a moon-viewing bridge linking the island and its temple to the shore. Previous exhibitions had housed exhibits according to the official classification, with uniform architecture providing a blank canvas on which their comparative merits might be judged. This time, the authorities produced an architectural smorgasbord, and allowed exhibitors who could afford it free rein, providing visual novelty as much as industrial instruction. Up on the hill, the five buildings devoted to industry illustrated the diversity of European architecture, from Renaissance to various flavours of Gothic, and appealed to different audiences. The fourth, with military hardware and a model of Tokyo, was a place for men and boys; the third, with fabrics, for women and girls. Down around the pond, meanwhile, a massive Renaissance-style building for foreign exhibitors was flanked on its right by the Mitsubishi and machinery buildings, on its left by the Chinese-style Taiwanese building, with the water chute allowing visitors to splash down in the pond. Earlier exhibitions had sought to superimpose a bureaucratic vision of industrial development on spaces devoted to faith and frivolity. Tokyo in 1907 allowed industry and empire to be subsumed

[30] *Asahi Shimbun*, 1907/3/28, 3, for Western-style painters; 4/2, 6, for foreign machinery; 4/30, 5, 5/1, 5, and 5/2, 3, for patents; 5/12, 6, for the party from Korea; 6/7, 6, for the horse race; and 7/8, 4, and 7/10, 11 and 28, for judging improprieties. *Fūzoku Gahō* 361 (1907/4/25), 1, for complaints; 363 (1907/5/25), 23–24, for Mitsubishi; 360 (1907/3/25), 14–15, 361, 25, and 363, 15–17, for the Taiwanese building; 363, 5–6, for Korean products; 361, 23, for the Ferris wheel, the Hall of Wonders, and the round-the-world tour; 365 (1907/6/25), 30–36, for the aquarium and the watershoot; 361, 9–10, for visiting dignitaries, and 367 (1907/7/25), 3–12, for imperial visits; 360, 17–18, for light bulbs, and 361, 28–29, 363, 34–35, 365, 5–14, and 367, 29–32, for the exhibition at night. Ziomek, *Lost Histories*, 59–65, as well as Kwon, 'An Analysis', 24–27 and 31–34, for the employment of and reactions to the Ainu and Koreans.

and its agents to pursue their ends within a landscape in which, like its predecessor, diversion and distraction had come to the fore. The decision paid dividends. The exhibition was less than half the size of Osaka 1903, ran for nineteen days fewer, and boasted only a third of the exhibits, but attracted nearly seven million paid admissions, an increase of over 25 per cent on its Kansai rival.[31]

The media reported the numbers and chronicled the change in the look and shape of the exhibition. Increasingly, though, their attention was turning to the way in which the space was being used. *Fūzoku Gahō* devoted five special issues to the exhibition, dutifully recording every last detail, but its illustrations tended to linger on the visitors as much as the exhibits. In the daily press, too, it was social markers that caught the eye: the strong accents and out-of-date accessories worn by visitors from the country; the Western-style clothing of government officials and young men-about-town; the more traditional pleated skirts of exhibition judges and students. Clothes were not the only focus, however. In the end, customs (*fūzoku*) had as much to do with sex. Many of those doing the seeing in the early weeks seem to have been young male students, who were drawn by the 305 women, between the ages of 15 and 30, who had been employed to supervise the exhibits. But the gaze was reciprocated. Already by early April, female students had discovered the joys of watching boys work out in the gym, the sole occupant of the exhibition's third site, to the west of the museum.[32] The exhibition's authorities had stopped insisting that its visitors look at certain kinds of exhibits in a

[31] Hashizume, *Nihon no Hakurankai*, 44–49, for an overview and illustrations. *Fūzoku Gahō*, 360 (1907/3/25), 6–17, for a map, description, and photographs: 361, 18, and 363, 11, for gender differentiation; 361, between 24 and 25, for the Taiwanese and foreign products buildings, 365, between 12 and 13, for the watershoot and other attractions, and between 20 and 21, for the scene around the pond. Kuni, *Hakurankai no Jidai*, 267, for the numbers.

[32] For social observation, *Fūzoku Gahō*, 360, cover, 20–21, 22–23, and 32–33; 361, between 10 and 11, 14 and 15, and 18 and 19; 363, between 10 and 11, 12, 13, 20 and 21, and 28 and 29; 365, between 32 and 33; and 367, frontispiece and between 25 and 26. For fashion, *Asahi Shimbun*, 1907/4/1, 6. For the male gaze, *Fūzoku Gahō*, 361, 2; Hashizume, *Nihon no Hakurankai*, 51; and Inoue Shōichi, 'Konpanion ga Jokanshu to Yobareta Koro: Hakurankaijō ni okeru Josei Setsugūin no Seiritsu to Tenkai', in Sano Mayuko (ed.), *Bankoku Hakurankai to Ningen no Rekishi* (Shibunkaku Shuppan, 2015), 385–405. For female reciprocation, *Asahi Shimbun*, 1907/4/8, 6. For a picture of the boys working out, *Fūzoku Gahō*, 365, between 38 and 39. For the earlier history of *fūzoku*, Suzanne O'Brien, *Customizing Daily Life: Representing and Reforming Customs in Nineteenth-Century Japan* (Harvard University Asia Center, 2014).

FIGURE 3.3 In Tokyo in 1907, the city took matters into its own hands, creating an electric wonderland at night around Shinobazu Pond.
Source: *Tōkyō Kangyō Hakurankai Shinobazu Chihan Yakei no Jikkyō* (1907), NOMURA Co., Ltd.

particular way. Its chroniclers had begun to notice how it was a space within which social change itself might be observed.

The most perceptive realized that the exhibition was also a place to reflect. Natsume Sōseki had been born in 1867 and sent to England in 1900 on a government scholarship. He had been depressed by London and overwhelmed by the Paris Exposition Universelle, but on his return to Japan he began to write and publish, achieving success with his first novel in 1905. Two years later, he used the Tokyo exhibition to cast a jaundiced eye on his preoccupation, the modern self, in a serialized novel in the *Asahi Shimbun* newspaper, which was printed alongside daily reports from the exhibition. Contemporary civilization was about stimulation, suggested Sōseki, and stimulation was about being in the now (*tōsei*). There was nowhere to be more in the now than the exhibition, and the most stimulating thing there was the exhibition itself at night (Figure 3.3). Standing at the south end of Shinobazu Pond, looking over the Kangetsu Bridge, to the buildings on the far side of the water, his characters gazed at the Taiwan Building, glinting like a dragon palace, and the Foreign

Building, the most beautiful thing they had ever seen. The exhibition was a place to escape the everyday and suspend disbelief. Sōseki himself was characteristically suspicious, but hardly immune. Writing to a friend, he noted that 'I'm thinking of going to the exhibition and riding the water chute but haven't done it yet'.[33]

FIN DE SIÈCLE

By the noughties, then, exhibitions had come a long way from the simple tool for industrial promotion, observed and imagined half a century before. Overseas, they had become a place for imperial projection. In St Louis in 1904 and London in 1910, the government spent big and did its best. Trade remained a concern, given ongoing deficits with both countries, but diplomacy was paramount. In the US, the exhibition was an opportunity to buttress its image, even while confronting the expansionist Russian bear; in the UK, to 'guarantee the future of the alliance', given that some were beginning to question Japan's own imperial ambitions. In both cases, the Diet loosened its purse strings, and the commissioners produced an impressive show, following the template established in the previous decade: invoking Japanese tradition, through horticulture and history, art and craft; demonstrating modern achievement, through industry and empire. In both places, however, their efforts were challenged by protests from their Asian neighbours, and undermined by showmen, brokers, and organizers, who made their money by catering to the popular demand for the exotic. They provided an older story about Japan, as a place set apart, but thereby also an opportunity for both participants and viewers to do and see different things.[34]

[33] Yoshida, *Bankoku Hakurankai*, 106–07, for Sōseki in Paris in 1900; Natsume Sōseki, *Natsume Sōseki Zenshū*, vol. 4 (Chikuma Shobō, 1988), and Wada Atsuhiko, 'Hakurankai to Dokusho: Miseru Basho, Mienai Basho, Gubijinsō', *Sōseki Kenkyū* 16 (2003), 48–51, for Tokyo 1907. For illumination, see also Kentaro Tomio, 'Visions of Modern Space: Expositions and Museums in Meiji Japan', in Helen Hardacre and Adam L. Kern (eds.), *New Directions in the Study of Meiji Japan* (Brill, 1997), 731–32; Hashizume, 'Toshi no Denka'; and Elise Mizuta, 'Luminous Environment: Light, Architecture and Decoration in Modern Japan', *Japan Forum* 18.3 (2006), 340–41.

[34] For overviews of St Louis 1904, Rydell, *All the World's a Fair*, 155–83, and James Gilbert, *Whose Fair?: Experience, Memory, and the History of the Great St. Louis Exposition* (University of Chicago Press, 2009). For Japanese participation, Itō, *Meiji Nihon*, 127–58. For overviews of London 1910, ibid., 159–208, and Ayako Hotta-Lister and Ian Nish (eds.), *Commerce and Culture at the 1910 Japan-British Exhibition: Centenary Perspectives* (Global Oriental, 2013). For trade, *Teikoku Gikai Shugiin giji sokkiroku*, vol. 27 (Tokyo Daigaku Shuppankai, 1979), 336–37, and *Nichiei*

In 1904 in St Louis, there was a slew of art, to make the case that Japan was the 'sole guardian' of Asia's artistic heritage, and an Imperial Japanese Garden, with replicas of two buildings in Kyoto: a reception hall from the imperial palace, and Kinkakuji, serving as a tea pavilion. There was American admiration for the efficiency and modernity of the industrial exhibits, which included models of the Miike colliery and the Ashio copper mine, among the 'largest in the world'. There was Taiwanese architecture, camphor, and tea, together with a handbook to distinguish between Chinese despotism and 'just and righteous' Japanese rule, though a Chinese vice-commissioner contested the claim, insisting that China could 'march in company with the other powers'. But there were also other Japans on display. Kushibiki Yumito had moved to the US in 1885, created a Japanese garden in Atlantic City in 1898, and brought the Kawakami dance troupe to America prior to their appearance in Paris. Now he built 'Fair Japan' in the entertainment zone, the Pike, with a 100-foot-high reproduction of the Yōmeimon gate at Nikkō, 80 actors, 15 geisha, pottery workers, and 'natives in harness' to haul the 'jinrikishaws'. Sure enough, the live human bodies 'fascinate[d] ... the visitor more than the buildings, the streets, and the inanimate objects'. Close by, they were joined by the Ainu, requested by the authorities and recruited by Frederick Starr, from the University of Chicago. They not only proved a hit but also demonstrated to some the 'germs of ... those material arts', which underpinned Japan's ascent.[35]

Hakurankai Jimukyoku Jimu Hōkoku (Nōshōmushō, 1912), vol. 1, 4. For diplomacy, Ayako Hotta-Lister, *The Japan-British Exhibition of 1910: Gateway to the Island Empire of the East* (Japan Library, 1999) and Kawamura Kazuo, 'Meiji Yonjūsannen no Nichi-Ei Hakurankai Kakei Shiryō ni tsuite', *Gaikō Shiryōkan Hō* 6 (1993), 71.

[35] Imperial Japanese Commission to the Louisiana Purchase Exposition, *The Exhibition of the Empire of Japan, Official Catalogue* (International Exposition, St. Louis, MO, 1904), 5–6, for the garden, 27–28, for trade, 46–63, for art, 202–4, for Taiwan, and 247–63 for mining. Hoshi Hajime, *Handbook of Japan and Japanese Exhibits at World's Fair, St. Louis, 1904* (1904), 5–8 and 38–52, for trade, 114–15 for the garden, 116–18 for art and craft, 118–19 for mining, and 120–21 for Taiwan. For art, see also Carol Ann Christ, '"The Sole Guardians of the Art Inheritance of Asia": Japan at the 1904 St. Louis World's Fair', *Positions: East Asia Cultures Critique* 8.3 (2000), 675–709, and Miya Elise Mizuta, '"Fair Japan": On Art and War at The Saint Louis World's Fair, 1904', *Discourse* 28.1 (2006), 28–52. For the ambivalent American reception, exacerbated by the competing exhibits, Harris, 'All the World', 49–54, and Marietta Holley, *Samantha at the St. Louis Exposition* (GW Dillingham Co., 1904), 91–92, 197–200, and 238–42. For Chinese efforts, Susan Fernsebner, 'When the Local Is the Global: Case Studies in Early Twentieth-Century Chinese Exposition Projects', in David Raizman and Ethan Robey (eds.), *Expanding Nationalisms at World's Fairs* (Routledge, 2017), 174–77. For the vice-commissioner, Christ, '"Sole Guardians"', 693. For Kushibiki, Hashizume, *Jinsei wa*

London in 1910 amplified the dilemma. The government went all out. Two gardens, twelve 1:20 scale models, and an impressive haul of 'priceless treasures', showed Japanese traditions of horticulture, architecture, and art to be alive and well. Elaborate historical tableaux laid out the development of the nation and its military, now equipped with dreadnoughts built 'from her own workshops'. The Palace of the Orient revealed how Japan was transforming Korea and Taiwan, with dioramas contrasting the life of mountain aborigines with that of docile workers picking oolong tea in orderly rows. Manchuria was there too, despite the reservations of the Chinese consul in London, given that it was still a Chinese possession. Again, the local press acknowledged both the achievements of industry and the sophistication of art. Again, though, a different Japan was visible elsewhere. Kiralfy insisted to the protesting Japanese officials that he knew 'what the British Public wants', and drafted Kushibiki to make sure they got it. Well over 200 performers populated another 'Fair Japan', while 'rural workers' busied themselves in the fields of 'Uji Village', and sumo wrestlers wrestled twice a day. Not far away was a 'Formosan Sha', with 'warrior inhabitants', but it was the 'Ainu Home', with 'classic Ainu types', who became the sensation of the summer, a screen on which their visitors and the press projected what they wanted. Some saw little difference between them and Formosan 'cannibals', others used them to lampoon English types. One Japanese reporter came away convinced that 'sustaining the Ainu race is a duty that the Japanese people bear for the people of the world'. As in St Louis, the diversity on display complicated the official efforts to make the empire mean one thing.[36]

Hakurankai, 36–59. For the Ainu, Frederick Starr, *The Ainu Group at the Saint Louis Exposition* (Open Court, 1904), for his part, and Ziomek, *Lost Histories*, 69–90, for their experience and reception. *The Universal Exposition of 1904: Exhibits, Architecture, Ceremonies, Amusements* (Louisiana Purchase Exposition Co., 1904), 9, for 'natives in harness' and 42, for 'germs of ... material arts'.

[36] *Official Report of the Japan British Exhibition*, 242–49 for gardens, 143–45, for art, 169–71, for architecture, 199–202, for history, 283–92, for the Palace of the Orient, and 293–302, for military. *Official Guide*, 21–24, for history, 36–40 and 77–80 for gardens, 45–49, for empire, 51–56 for military, 59–61, for art and architecture, and 81–88 for attractions. Hotta-Lister, *Japan-British Exhibition*, 180, for the London consul. (For a similar dispute in China at the Nanyang Exhibition, also in 1910, Michael R. Godley, 'China's World's Fair of 1910: Lessons from a Forgotten Event', *Modern Asian Studies* 12.3 (1978), 520, and Susan Fernsebner, 'Objects, Spectacle, and a Nation on Display at the Nanyang Exposition of 1910', *Late Imperial China* 27.2 (2006), 104.) Mutsu Hirokichi, *The British Press and the Japan-British Exhibition* of 1910 (Routledge, 2001), for the press. For Kiralfy's knowledge of the British public, 'Letter to Mutsu

Back home, too, by the end of the decade, exhibitions were evolving. The Great Japan Exhibition was already on the rocks. In 1908, it had been pushed back to 1917, nominally to commemorate the fiftieth year of the Meiji era, practically to reallocate money and attention to London. Desultory preparations continued, but sure enough it would be axed in 1911, with a minister noting to the Diet the difficulty of organizing exhibitions at the national level. Instead, organizations or individuals should take the lead, with the government in a supporting role. He was describing a fait accompli. Local authorities had long assumed the mantle of industrial promotion. The tenth and final edition of the 'Kansai' Combined Product Promotion Exhibition (*kyōshinkai*), in Tsuruma Park in Nagoya (not in Kansai) in the spring of 1910, pulled in well over 2 million visitors and some 200,000 exhibits from 31 local authorities, which for the first time included Tokyo. Trade and industrial associations and other bodies now joined local government in adapting the form to sectoral ends. The previous year had seen the first in a series of exhibitions staged by what would soon become the Imperial Invention Association, 1910 similar initiatives by the Japan Industrial Association and the Tokyo Industrial Promotion Association, as well as the seventh in a series of 'National Manufactures' (*zenkoku seisanhin*) exhibitions staged by the indefatigable Kyoto Exhibition Company, albeit with more exhibits than visitors.[37]

In the wake of the war with Russia, too, private companies had turned to exhibitions as a way of encouraging consumer demand rather than industrial supply. Newspaper companies used them to cultivate a national, rather than regional, market. In 1906, the *Osaka Jiji Shinbun* sent a steam train up north, partnering with the Takashimaya department store to advertise their respective wares to a rural audience, while the *Hōchi Shinbun* sent the Rosetta Maru, which had been used as a hospital ship during the war, on a nationwide cruise, from Tokyo to Kagoshima.

Hirokichi, 1909/8/16', *Eikyō Rondon ni okeru Nichi-Ei Hakurankai Kaisetsu no Ikken*, vol. 2, Gaikō Shiryōkan Shozō. Ziomek, *Lost Histories*, 112–43, for the Taiwanese group, 66–111, for the Ainu, and 143–50, for Japanese reactions. *Daily Express* 1910/5/16, for a representative tabloid; Punch 1910/4/27, 293, for lampooning; Uchigasaki Sakusaburo, *Eikoku yori Sokoku e* (Hokubunkan, 1911), 247, for duty.

37 Furukawa, *Kōki, Banpaku, Orinpikku*, 32–51, Kuni, *Hakurankai no Jidai*, 263–67, and Itō, *Meiji Nihon*, 162–72, for the Great Japan Exhibition; and Yoshida, *Bankoku Hakurankai*, 141–49, for national difficulty. Nagoya-shi Hakubutsukan, *Nagoya no Hakurankai*, 48–9, for Nagoya 1910; and Nicholas, 'The Origins', 285–88, for its impact.

Soon enough, exhibitions were beginning to target particular demographics. The year 1907 saw not only the exhibition in Ueno but also a concurrent Women's Exhibition in Shiba Park, which sought to cater to their interests, with arts and crafts and famous doll collections, and thereby to encourage their consumption. The Shirakiya department store exhibited material for obi, while childcare products could be found on the second floor. Meanwhile, children's exhibitions were spreading like weeds from Tokyo, Osaka, and Kyoto in 1906, to Hikone in 1907, Fukuoka in 1908, the Mitsukoshi department store in Tokyo in 1909, and Shimane and Okayama in 1910. Last but not least, private railway companies were beginning to get in on the act. Down in Osaka, the Minō branch line opened in 1910, ending in a loop around a new public hall. The following year, a Forest Children's Exhibition took over the hall, the adjacent park, and the new zoo a little future north. There was plenty for the kids to do: a miniature train and airships, camel and donkey rides, a monkey music troupe, and elephant acrobatics, together with more didactic exhibits about technology and nature, and children around the world. Meanwhile, their parents saw the possibilities of moving out of town, along the line, where their children might be closer to restorative nature. The corporate enthusiasm for exhibitions would only increase in the following decade.[38]

By December 1910, therefore, exhibitions in Japan were strangely poised, between familiar Edwardian bombast about industry and empire, and newer, proto-modernist attention to the possibilities and textures of everyday life. There were no exhibitions that month. Winter was a time for taking stock. There were things to look forward to. The newspapers were still providing weekly updates on the Great Japan Exhibition, even though it was now seven years in the future. They were documenting the preparations for the Espozione Internationale dell'Industria e del Lavoro, which would open in Turin at the end of April. But they were also

[38] Hashizume, *Nihon no Hakurankai*, 42–43, for the travelling exhibitions, 50–51, for the 1907 women's exhibition. For children's exhibitions, Koresawa Yūko, 'Meijiki ni okeru Jidō Hakurankai ni tsuite', *Tōkyō Kasei Daigaku Kenkyū Kiyō* 35.1 (1995), 159–65, and 37.1 (1997), 127–37, Nakamura Kiyoko, 'Kindai Nihon ni okeru "Kodomo" Imēji to Kodomo Hakurankai', *Bijutsu Kyōiku Gaku* 18 (1997), 215–24, and 19 (1998), 223–34, and Kawaguchi Hitoshi, 'Meiji Sue no Chihō ni okeru Kodomo Hakurankai ni tsuite', *Kyūshū Zōkei Tanki Daigaku Kiyō* 22 (2000), 33–46. For department stores, Hatsuda Tōru, *Hyakkaten no Tanjō* (Sanseidō, 1993), 129–45. Minō-shi, 'Rakettokei Senro Jidai no Minō to Sono Ato' (2019), www.city.minoh.lg.jp/soumu/shishi/column/topic-1.html, for an introduction to the Minō line.

reviewing the year. The consensus was that the Japan-British Exhibition had been a mistake. There was little evidence that the official efforts had done that much for trade, though there had been an 'enormous increase in the export of kimonos'. Another consolation was the return of the Imperial Japanese Military Band, which disembarked at Kobe on 1 December, having 'garnered considerable applause' in London. Bands had been banned during the first week of the exhibition back in May, in deference to the death of Edward VII one week earlier. When they were allowed to play, the Japanese had struck exactly the right note, with the 'Dead March' from Handel's *Saul*, long-established as a staple at state funerals, including that of Nelson, a century before.[39]

In London, though, the band had been simply one in a panoply of attractions, not least the entrepreneurial Ainu, who had exploited the opportunities for networking, earning, and learning offered by the show. They stopped in Tokyo two weeks after the band, on their way back home, having watched as locals dived for the cash thrown by passengers as their ship docked in British colonial ports, and with cash to spend in the Mitsukoshi department store. They regaled journalists with the tales of the bear ceremony they had performed twice a day (except when it was raining), the well-insulated houses, the huge horses, the parks and gardens, and the other sights they had seen. They also had a new three-month-old baby with them. Hidehiro's birth, during the exhibition, at the end of August, had been a cause for celebration in west London. His name incorporated the Chinese characters for both 'England' and 'exhibition', though he would die when he was only twenty-three years old. Another colonial subject, however, had been left behind. Eighteen-year-old Ruji Suruchan, from the southern tip of Taiwan, had been recruited to appear in the Formosan village, but died four days before the end of the exhibition and was buried in a pauper's grave in Hammersmith, a couple of miles to the south.[40]

39 *Yomiuri Shimbun* 1910/12/6, 2, 1910/12/14, 2, 1910/12/20, 2, and 1910/12/27, 2, for the Great Japan Exhibition; 1910/12/6, 5 and 1910/12/7, 3, for Turin. Itō, *Meiji Nihon*, 192–97, and Hotta-Lister, *Japan-British Exhibition*, 131–38, for Japanese reaction to the Japan-British exhibition. Ibid., 154–63, for trade and kimonos. *Official Report of the Japan British Exhibition*, 74–75, and *Yomiuri Shimbun* 1910/12/4, 3, for the band.

40 *Yomiuri Shimbun* 1910/12/18, 3, for the Ainu regaling the journalists. Ziomek, *Lost Histories*, 105–8, for sightseeing in London, 103–4, for Hidehiro, and 112–14, for Ruji Suruchan.

4

Modern Times

At both ends of Eurasia, the second decade of the new century has long provided the punctuation with which we conclude our stories about Victorian certainty and begin a new chapter about the fickle modern world. In May 1910, the death of Edward VII not only cast a pall over the opening of the Japan-British Exhibition in London. In combination with other signs and wonders later in the year, it would famously prompt Virginia Woolf to muse (in retrospect) that on or about December 1910, human character changed. In July 1912 in Tokyo, the death of the Meiji emperor proved equally portentous. The shock has been well described. At the time, it drove the commentariat to reflect on how things had changed over the half century of his reign. In retrospect, historians have marked a shift from the public-spirited nation-building of Meiji (1868–1912) to the factionalized politics and privatized culture of the new Taishō era (1912–26). Finally, obviously, the outbreak of World War I in 1914 brought an end to the long European peace of the nineteenth century, plunging that continent into a thirty-year-long conflict and the world into a short twentieth century of extremes.[1]

[1] Peter Stansky, *On or about December 1910: Early Bloomsbury and Its Intimate World* (Harvard University Press, 1997) for the UK. Harry Harootunian, 'A Sense of an Ending and the Problem of Taishō', in Bernard S. Silberman and Harry D. Harootunian (eds.), *Japan in Crisis: Essays on Taisho Democracy* (Princeton University Press, 1974), 3–28 and Carol Gluck, *Japan's Modern Myths: Ideology in the Late Meiji Period* (Princeton University Press, 1985), 213–46, for the end of Meiji. Harry Harootunian, *Overcome by Modernity: History, Culture, and Community in Interwar Japan* (Princeton University Press, 2000), Sharon Minichiello, *Japan's Competing Modernities: Issues in Culture and Democracy, 1900–1930* (University of Hawai'i Press, 2001), and Narita Ryūichi, *Taishō*

Death is deceiving, however, as exhibitions can suggest. On the one hand, the war did consume the energies of Europe and its colonies. In its wake, there was little enthusiasm at the west end of the Eurasian continent for more great exhibitions, though it took a while to question the convictions by which they had been underwritten, about industrial progress, imperial hierarchies, and civilizing missions.[2] Similar beliefs had long migrated across the pond, of course, morphing into claims of American exceptionalism and manifest destiny, which had been rehearsed at turn-of-the-century fairs in the Midwest. By then, too, the Pacific was the preferred arena for the projection of American power. The West Coast had been quick to embrace exhibitions, which cemented its own place, and revealed Japan's shifting one, in the new networks of trade and empire. In San Francisco in 1894, the Midwinter Exposition provided an opportunity to condescend to the 'little brown men', who were immigrating in numbers following the Chinese Exclusion Act in the previous decade. In Seattle in 1909, the Alaska-Yukon-Pacific Exposition saw both reassurance from Japanese officials that immigration restrictions were a good thing, and the archipelago's incorporation on the official seal to symbolize the potential of the region's markets. Back in San Francisco in 1915, the Japanese government built another replica of Kinkakuji, to advance its claims to international parity, but their efforts were again undermined by Kushibiki Yumito, who put a huge papier mâché Buddha on top of the entrance to 'Japan Beautiful' in the amusement zone. Even in North America, though, the war would take its toll. It took time for exhibitions to re-emerge, and when they did they had changed.[3]

Demokurashī (Iwanami Shoten, 2007), for the novelty of Taishō. Harry Harootunian, *History's Disquiet: Modernity, Cultural Practice, and the Question of Everyday Life* (Columbia University Press, 2000), for Japan as a variation on a global theme. E. J. Hobsbawm, *The Age of Extremes: A History of the World, 1914–1991* (Pantheon, 1994), for the short twentieth century.

2 Hew Strachan, *The First World War* (Free Press, 2006), xv–xvii, for how long it took to realize that the war had been a watershed.

3 *Official Guide to the California Midwinter Exposition in Golden Gate Park, San Francisco* (G. Spaulding & Co., 1894), 63 and 115, for the Japanese exhibit, and Rydell, *All the World's a Fair*, 51, for 'little brown men'. Ibid., 185 for the Seattle 1909 seal and 202–5 for ambivalence, hysteria, and official reassurance. For San Francisco, Itō Mamiko, '1915-Nen Panama Taiheiyō Banpaku to San Furanshisuko Hainichi Undō', *Mediashi Kenkyū* 23 (2007), 25–41, Abigail Markwyn, 'Economic Partner and Exotic Other: China and Japan at San Francisco's Panama-Pacific International Exposition', *Western Historical Quarterly* 39.4 (2008), 439–65, and Tseng, *Modern Kyoto*, 144–50. For Kushibiki, Hashizume, *Jinsei wa Hakurankai*, 54–57. The year 1915 also saw a smaller fair in San Diego, exploited by the Japanese immigrant community to prove they

On the other, neither the death of an emperor nor a world at war could derail the enthusiasm for exhibitions in Japan itself. On 1 October 1912, the Hokkaido Exhibit Association opened a Colonial Exhibition in Ueno, dedicating it to the deceased monarch and his eponymous reign, but focusing on the way a new, multi-ethnic Japan could become a 'global nation' (*sekai-teki kokumin*), not least to its colonial subjects. On the ground, this meant extensive exhibits of primary resources from the various colonies, to demonstrate their potential for industry; a film of native and Japanese residents eating together in Karafuto, to appeal to settlers; and a popular human display of colonial subjects, to stimulate tourism. Not everyone was convinced. A Karafuto Ainu questioned Japanese territorial claims, while the press noted that even Taiwan had 'nothing to see as far as industry is concerned'. The crowds kept coming, though, and colonial craft – a Taiwanese spindle, Manchurian ceramics, and Korean pots – proved captivating to two young potters. Soon enough, Tomimoto Kenkichi and Bernard Leach would use their epiphany to help define Japanese and English folkcraft.[4]

The Tokyo exhibition was not the only one that autumn, however, and it presaged a stream of events in the next couple of years. Osaka alone saw three exhibitions in the first six months of 1913: travel in February and March; industry in April and May; and the colonies from late April to June, again dedicated to Meiji's 'holy reign', with the imperial navy cruising in Osaka Bay, and again featuring human displays. Tokyo hit back two months later with a more focused commemoration of the late emperor, with eighteen dioramas to recapitulate the key moments of his

belonged. See Richard Amero, 'The Making of the Panama-California Exposition, 1909–1915', *The Journal of San Diego History* 36.1 (1990), 1–47, and Matthew Bokovoy, *The San Diego World's Fairs and Southwestern Memory, 1880–1940* (University of New Mexico Press, 2005), 17–139. Rydell, *World of Fairs*, for the way North American fairs changed after the war.

[4] Hashizume, *Nihon no Hakurankai*, 60, for an overview and images. *Takushoku Hakurankai Jimu Hōkoku* (Takushoku Hakurankai Zanmu Toriatsukaijo, 1913), 1–3, for the objectives. For the human displays, Tsuboi Shōgorō, 'Takushoku Hakurankai Shutsujō no Sho Shuzoku', *Tōyō Gakugei Zasshi* 29.370 (1912), 590–92; Yamaji, *Kindai Nihon no Shokuminchi Hakurankai*, 51–73, and Ziomek, *Lost Histories*, 151–76, including the Ainu critique. Kate McDonald, *Placing Empire: Travel and the Social Imagination in Imperial Japan* (University of California Press, 2017), 42, for Taiwanese visitors. *Asahi Shimbun*, 1912/10/4, 5, and 1912/10/8, 4, for the media reaction. Tomimoto Kenkichi, *Tomimoto Kenkichi Chosakushū* (Gogatsu Shobo, 1981), 472, for colonial craft. Yuko Kikuchi, *Japanese Modernisation and Mingei Theory: Cultural Nationalism and Oriental Orientalism* (Routledge Curzon, 2004) for the subsequent formation of folkcraft theory.

reign.[5] The change of sovereign, in other words, did little to interrupt the increasing stream of exhibitions. Japan had seen roughly an exhibition a year from the early 1890s to the Russo-Japanese War, on average attracting perhaps 100,000 visitors. The next two decades produced a quadrupling in the number of exhibitions and a doubling of average attendance, with number and scale jumping again in the late 1920s.[6]

The raw numbers do not tell the whole story, of course, but they do point to two of its key features. First, unlike the initial embrace of exhibitions in the 1870s, which was coordinated by the state, the growing enthusiasm for them in the early twentieth century was driven by the strengthening pulse of urbanization, rippling out from Tokyo. This had begun to beat in the 1890s, was supercharged by the wave of industrialization that followed the Russo-Japanese War and World War I, and would be sustained through the 1930s by imperial adventure, colonial development, and informal empire. City, industry, and empire combined to transform the local landscape and everyday life, increasing the demand for exhibitions. Second, averages conceal, though their roughness hints at, the diversity and development of exhibitions. It was the protean ability of an exhibition to accommodate difference, in the demands of organizers, the interests of exhibitors, the kind of exhibits, and the appetites of visitors, which ensured their utility as imperial Japan became more modern. Modernization was not an easy or even thing. But the unevenness of development itself exacerbated the demand for spectacle: to stimulate a local economy or to promote a private business; to provide distraction from an underwhelming everyday life; or to try to convince colonial subjects that they could be Japanese. Come the early twentieth century, exhibitions were primed to come into their own.

COMMERCIAL NOVELTY

The spawning of exhibitions across the empire in the early twentieth century can easily prompt a search for what was new, to underline the distinction between the novelty of modern times and the conservative

[5] For the colonial exhibition, *Meiji Kinen Takushoku Hakurankai Hōkoku* (Meiji Kinen Takushoku Hakurankai, 1913), 1 for holy reign; *Meiji Kinen Takushoku Hakurankai Annaiki* (Yūshinsha, 1913), 37 for the navy; and Ziomek, *Lost Histories*, 176–88, for the human displays. Hashizume, *Nihon no Hakurankai*, 61, the 1913 commemoration of Meiji in Tokyo.

[6] For a partial list, Hashizume, *Nihon no Hakurankai*, 232–33. For the numbers, Nomura, *Hakurankai Shiryō Collection*.

certainties of Meiji. There is no shortage of candidates. The first decade of the new century had already brought several new players into the game, each seeking to turn exhibitions to their own novel account. The most obvious were private companies, which saw in exhibitions an opportunity to advertise their wares, captivate potential customers, and thereby capture the emerging market.[7]

Department stores were among the earliest. Mitsukoshi and the rest had appeared at exhibitions, both at home and overseas, from the 1870s, but by the end of the Meiji period they were turning from restrained Victorian clothing stores to American-inspired consumer wonderlands. The resonance with exhibitions was evident. Two years after its famous 1904 declaration, which announced its transformation from a dry goods to a department store, Mitsukoshi produced a guidebook to the city. It asked, rhetorically, whether anyone who came to Tokyo was not going to see that year's exhibition. It suggested, pleadingly, that anyone who went to the exhibition should also come to Mitsukoshi. By the end of the decade, its competitors were bringing exhibitions in-house, sometimes in a dedicated exhibition space, sometimes taking over the whole store. Claiming to contribute to the greater good – a city (Osaka in 1925), an industry (electricity in 1933), a demographic (women and/or children, recurrently) – was also a way to secure footfall, customers, and sales.[8]

Private rail companies also quickly discovered how exhibitions could draw middle-class demand into the suburbs. The pioneer here was Kobayashi Ichizō, who had bought his first railway in 1906 and gradually

[7] Yoshimi, *Hakurankai no Seijigaku*, 158–76, provides the starting point for what follows.

[8] For department stores in general, Hatsuda, *Hyakkaten no Tanjō*; Brian Moeran, 'The Birth of the Japanese Department Store', in Kerrie L. Macpherson (ed.), *Asian Department Stores* (Curzon, 1997), 141–76; Louise Young, 'Marketing the Modern: Department Stores, Consumer Culture, and the New Middle Class in Interwar Japan', *International Labor and Working-Class History* 55 (1999), 52–70; and Elise K. Tipton, 'The Department Store: Producing Modernity in Interwar Japan', in Roy Starrs (ed.), *Rethinking Japanese Modernism* (Brill, 2012), 428–51. For exhibitions, Fujioka Rika, 'Takashimaya Sōsōki ni Okeru Hakurankai no Yakuwari', *Keiei Kenkyū* 50.1–2 (1999), 249–66; Nishizawa Tamotsu, 'Hyakkaten Keiei ni Okeru Dentō to Kakushin', and Tsuganesawa Toshihiro, 'Hyakkaten no Ibento to Toshi Bunka', in Yamamoto Taketoshi and Nishizawa Tamotsu (eds.), *Hyakkaten no Bunkashi: Nihon no Shōhi Kakumei* (Sekai Shisōsha, 1999), 64–65 and 130–54; and Saitō Hironobu, 'Hyakkaten to Hakurankai: 1930 Nendai o Chūshin ni', in Hashizume Shinya (ed.), *Nihon no Hakurankai: Terashita Tsuyoshi Korekushon* (Heibonsha, 2005), 136–39. Hashizume Setsuya, *Dai-Ōsaka Imēji: Zōshokusuru Manmmosu, Modan Toshi no Genzō* (Sōgensha, 2007), 25, for Osaka 1925, Hashizume, *Nihon no Hakurankai*, 132, for electricity in 1933.

built up what would become the Hankyū empire in northern Osaka, resting on three pillars: a department store at the in-town terminus (from 1925), housing developments along the way (beginning in 1910), and an attraction at the end of the line. The most famous would be the Takarazuka all-female revue, which complemented the older lure of hot springs with the newish charms of transgendered spectacle. Kobayashi's first choice, however, was exhibitions: he staged one for children at the end of another branch of his network in 1911; he launched the revue during a Women's Exhibition in 1913; and he would persist, with exhibitions focusing on weddings in 1914, and the home in 1915.[9] The association of rail, with both suburban middle-class housing and urban transformation more generally – and the utility of exhibitions, in preaching the need for both – would endure.[10] In 1922, the Japan Architectural Association held a small home remodelling (*jūtaku kaizō*) exhibition in 1922, just to the south of the site of Kobayashi's first exhibition in Minō. In 1923, Osaka commemorated the twentieth anniversary of intra-urban light rail in Tennōji, using the park and the buildings that had been left following its first great exhibition in 1903.

Above all, exhibitions became a preferred medium for newspaper companies. Like department stores, from the 1870s, newspapers had made much of exhibitions, using the paper to report on exhibitions and exhibitions to advertise themselves. By the turn of the century, they were doing everything they could to create readers and capture revenues: putting a press on display at an exhibition, setting up branch offices near exhibition sites, publishing guidebooks and supplements and even separate exhibition newspapers. Now, as they turned from high-brow, partisan political advocacy to 'profit-oriented, news-based, [people]-centred journalism', 'gimmicks, stunts, and promotion schemes' became par for the course. Exhibitions and sport proved to be tailor-made media events.

[9] Jordan Sand, *House and Home in Modern Japan: Architecture, Domestic Space, and Bourgeois Culture, 1880–1930* (Harvard University Asia Center, 2003), 133–37, for Kobayashi's vision, and 150–52, for the review. Matsumoto, Kazuaki, 'Senzenki Tetsudō Kigyōka no Kankō, Goraku Senryaku: Kobabayashi Ichizō o Chūshin ni', *Kōtsūshi Kenkyū* 62 (2007), 39–53, for rail, and Shuntaro Nozawa and Jo Lintonbon, 'Suburban Taste: Hankyu Corporation and Its Housing Development in Japan, 1910–1939', *Home Cultures* 13.3 (2016), 283–311, for housing. See also Jennifer Robertson, *Takarazuka: Sexual Politics and Popular Culture in Modern Japan* (University of California Press, 1998), for the review.

[10] Yamaji Katsuhiko, *Chihō Toshi no Kakusei: Taishō Shōwa Senzenshi, Hakurankai hen* (Kansei Gakuin Daigaku Shuppankai, 2017), 121–37, for the intimate link between exhibitions, rail, and regional transformation.

Following a couple of early experiments post the Russo-Japanese War, mid-ranking newspapers, including some regional outlets, started to stage small exhibitions during the 1910s. Come the end of the decade, the big boys were beginning to show up.[11]

A pioneer was the *Yomiuri Shimbun*, which took over Ueno in 1917 to commemorate the fiftieth anniversary of the transfer of the capital from Kyoto to Tokyo, attracting some with a panorama of the Tōkaido, which included the palanquin used by the Meiji emperor, and others with the first long-distance relay race (*ekiden*) between the old and new capitals. The next year, it used the same venue during the height of summer to mount a women and children's exhibition. Through the main gate, which was encased in an artificial mountain, the visitors found themselves in a bucolic rural landscape, with pumpkins and cucumbers, a stream and a water wheel, as well as more didactic offerings, for example, about childcare around the world. A few years later, the paper began to stage a regular summer exhibition at the newly built sumo stadium at Ryōgoku, combining the older seasonal tradition of whiling away the evening on the riverbank with the newer technologies of refrigeration and distraction. Following its takeover by Shōriki Matsutarō, a veteran police bureaucrat and anti-communist, summer exhibitions and sport, especially baseball, proved a useful populist foil to its hardening political stance. The Yomiuri was only one among many, however, who turned to exhibitions during the decade. Competition was intensifying and newspapers were commercializing, leading to newly independent project departments (*jigyōbu*), which managed the outreach efforts and coordinated the exhibitions.[12]

[11] Huffman, *Creating a Public*, 232 for gimmicks and 311 for profit orientation; Anthony S. Rausch, *Japan's Local Newspapers: Chihoshi and Revitalization Journalism* (Routledge, 2017), 30, for creating readers. Tsuganesawa Toshihiro, *Kindai Nihon no Shinbun Kōkoku to Keiei: Asahi Shimbun, o Chūshin ni* (Asahi Shinbunsha, 1979), 110, 153, and 202–4, for exhibitions as advertising. Yamamoto Taketoshi, 'Sōsōki no Media Ibento', in Toshihiro Tsuganesawa (ed.), *Kindai Nihon no Media Ibento* (Dōbunkan, 1996), 44–48, for exhibitions as media events. For media events in general, Yoshimi Shun'ya, 'Media Ibento Gainen no Shosō', in Toshihiro Tsuganesawa (ed.), *Kindai Nihon no Media Ibento* (Dōbunkan, 1996), 3–30.

[12] Nakagawa Dōji, *Rankaiya Ichidai: Waga Hakurankai 100 Nenshi* (Kōdansha, 1969), 127–30 and 133–35, for the early Yomiuri exhibitions; Hashizume Shinya, 'Nenchū Gyōji to Shinbun Kigyō: Ryōgoku Kokugikan no Nōryō Ibento o Megutte', in Toshihiro Tsuganesawa (ed.), *Kindai Nihon no Media Ibento* (Dōbunkan, 1996), 169–91, for the events at Ryōgoku; Yoshimi, 'Media Ibento', 7–9, and Simon Partner, *Assembled in Japan: Electrical Goods and the Making of the Japanese Consumer* (University of California Press, 1999), 74–76, for Shōriki. For project departments, Hayashi Toshitaka, 'Kindai Chihō Shinbun no Jigyō Katsudo', and Tsuganesawa

Department stores, railways, and newspapers were not always after quite the same thing. Putting a kimono on display on the top floor of a store may have led, in theory, to purchasing its sibling a couple of floors below. Unveiling an imperial palanquin, by contrast, demonstrated a loyalty to the nation and so cultivated one to the newspaper, as a reliable sounding board for its target market. All three businesses, however, were targeting the newly prominent middle-class consumer. Women in particular, they suggested, needed an ideal home, within which they could raise their children as exemplary imperial subjects. This was implicit in the kiddylands (*kodomo no kuni*; literally, countries for children), which were quickly becoming a staple at almost all exhibitions, but clear above all at Household Exhibitions (*katei hakurankai*), which began with two in 1915 and would continue until the mid-1930s. In Takarazuka, predictably, the focus was on getting the visitors to spend their savings on suburban leisure, not least a 'family-centred house'. In Ueno, however, the *Kokumin Shinbun* tried to link the question of how one might consume – 'What sort of house to live in? What sort of food to have? And what sort of clothes to wear?' – to grander themes and national imperatives. Modern homes, Tokutomi Sōho, its imperialist proprietor, argued, were rooted in ancient traditions. The exhibition itself, however, foregrounded more contemporary concerns. Model interiors provided a concrete illustration of the 'ideal household'; architects presented plans for the home in which this might be realized; and Itō Chūta, the doyen of the Japanese architectural establishment, underlined the nuclear nature of the family for which it was intended.[13]

The emerging focus on household consumption did not displace the earlier attention to industrial development, however, nor did private companies always have to act alone. By the early twentieth century, trade and industry associations had begun to form and were eager to use exhibitions to advance their cause. The Imperial Invention Association,

Toshihiro, 'Ōsaka Mainichi Shinbunsha no "Jigyō Katsudo"', in Toshihiro Tsuganesawa (ed.), *Kindai Nihon no Media Ibento* (Dōbunkan, 1996), 141–68 and 217–48.

13 Yamaji, *Chihō Toshi no Kakusei*, 100–2 for kiddylands. Sand, *House and Home*, 152–53, for Takarazuka, and 163–66, for the Kokumin Shinbun exhibition. See also Yoshimi, *Hakurankai No Seijigaku*, 153–54, and the official publication, Kokumin Shinbunsha, *Risō no Katei* (Nihon Tosho Senta, 2010). John D. Pierson, *Tokutomi Soho, 1863–1957: A Journalist for Modern Japan* (Princeton University Press, 1980), 199–304 for the proprietor. Toshio Watanabe, 'Japanese Imperial Architecture: From Thomas Roger Smith to Itō Chūta', in Ellen Conant (ed.), *Challenging Past and Present: The Metamorphosis of Nineteenth-Century Japanese Art* (University of Hawai'i Press, 2006), 240–53, for Itō.

the earliest adopter, proved to have a populist touch, attracting audiences in the six figures to a series of exhibitions in Tokyo and Osaka between 1909 and 1932. It was only the first of many, however, with the associations for shipping and shipbuilding (1916), heavy and chemical industry (1917, 1926, and 1931), architects (1922), and horses (1928) soon entering the lists. The star, though, was electricity, which had been around since the end of the previous century and had already made a splash at exhibitions. With the appearance of the tungsten filament in 1910, its fortunes were transformed, colonizing three-quarters of Tokyo households in the next fifteen years, enabled not least by its starring role at exhibitions. In 1918 in Ueno, the Electricity Association combined infrastructure and application. Up on the hill, an artificial mountain provided the opportunity for a demonstration of hydroelectric power, with a model plant and a Pelton waterwheel. Elsewhere, mannequins and dioramas showed how electricity could transform the home. The lessons were reiterated over the next decade and a half, as the Association sought to embed electricity in everyday life. In 1926, 3 million visitors flocked to the Great Electricity Exhibition in Osaka, which also provided a chance for the Ajigawa Property Company to unveil its new western suburbs, comprising twenty hectares (or just over forty acres) of landfill. In 1933, the Osaka Electricity Bureau took over an eight-storey department store to commemorate the tenth anniversary of municipal control of electricity supply. Local authorities, the power industry, property developers, and the retail sector all shared an interest in disenchanting the domestic night.[14]

There were new interests behind exhibitions and new things to see, therefore. Still, older concerns and connections tied Taishō exhibitions to their Meiji predecessors. Not least was the didactic tone that continued to pervade many exhibitions, as they sought both to create and to represent a very moral middle class. An economic transaction was about something

[14] Hashizume, *Nihon no Hakurankai*, 70, 90, 118, and 132, for snapshots of the individual exhibitions, and Nakagawa, *Rankaiya Ichidai*, 131–32 for 1918. Hashizume Shinya, *Modan Toshi no Tanjō: Ōsaka no Machi Tōkyō no Machi* (Yoshikawa Kōbunkan, 2003), 111–32, and 'Toshi no Denka', for electricity and exhibitions; and Murase Keiko, '"Katei Denka" no Disupurē', in Fukuma Yoshiaki, Nanba Kōji, and Tanimoto Naho (eds.), *Hakuran no Seiki: Shōhi Nashonariti Mejia* (Azusa Shuppansha, 2009), 37–49, for the household exhibits. The enthusiasm for electricity – and electricity's need for exhibitions – persisted as both spread to the regions in the 1930s. Yamaji, *Chihō Toshi no Kakusei*, 234–44 for Toyama 1936. The same was true in Europe and the US. See Nye, *Electrifying America*, 33–47.

more than the market, they suggested. For an architect, for example, designing a 'family-centred house' was not simply a way of building a sustainable practice. For the newly suburban, nuclear family, moving to one was not simply a way of escaping the city and ensuring one's comfort. Design and occupation were rather contributions to the greater good, through rationality, efficiency, and hygiene. It would be a mistake to make too much of this, of course. It is clear that architects, developers, and the rest were animated by the possibilities of an emerging, resourced, segmented market. Still, the transaction needed dressing up. Buyers needed to believe both that they were not only going to distinguish themselves from the mass of their fellow consumers but also that they represented the best of their place and time. There were few places more effective than exhibitions for disguising the distinctions of class as an invocation to community.[15]

URBAN ASPIRATION

To focus on what was new, however, is to overlook what stayed the same. The mushrooming of exhibitions – alongside the apotheosis of retail, the privatization of rail, and the commodification of news – was one outgrowth of the urbanization that swept over the archipelago from the turn of the century. As cities expanded, so they embraced exhibitions as celebration and spur of development: to mark an anniversary, a grand project, or a new dawn; to encourage the swelling populace to imagine themselves a community; and to display their contemporary credentials to the rest of the country and the world. The pattern had already emerged in the last years of Meiji, in Kyoto in 1895, Osaka in 1903, and Tokyo in 1907, which would continue to mount the vast majority of exhibitions in the next two decades. Together with Nagoya, Kobe, and Yokohama, they were the prime beneficiaries of the first wave of large-scale urbanization in the first two decades of the new century, which nearly doubled the proportion of Japanese living in communities larger than 10,000 people and more than doubled the number of those in the big six.[16]

At the same time, the increasing number and evolving form of exhibitions in the 1910s and early 1920s prepared the ground for their rapid

[15] Sheldon Garon, *Molding Japanese Minds: The State in Everyday Life* (Princeton University Press, 1998), for the moral middle class. Young, 'Marketing the Modern', for the animation and disguising of class at department stores.

[16] André Sorensen, *The Making of Urban Japan: Cities and Planning from Edo to the Twenty-First Century* (Routledge, 2002), 92.

adoption in the regions thereafter. Exhibitions were a place where private interest and public concern could coincide and blur, often facilitated by a Chamber of Commerce, which provided a conduit between municipal authorities, local business, and exhibition producers. Things changed, of course, over time and space. The waning of enthusiasm in some places was counterbalanced by its waxing elsewhere. An initial wave of enthusiasm for new technologies, commodities, and lifestyles was succeeded by the need to promote recovery, exports, or thrift. During the first half of the century, though, exhibitions spread everywhere in Japan, with cities the most important actor-impresarios in their production.

Already in 1907, Tokyo had shown how its fortunes were no longer subordinate to state initiatives. In 1914 and 1922, it reiterated the message. Again, Ueno Hill and Shinobazu Pond formed the stage, nominally to celebrate, first, the accession of the new emperor in 1912 and, second, the conclusion of World War I in 1918. In fact, as the lag between event and commemoration hinted, the Tokyo Taishō Exhibition and the Peace Memorial Exhibition were opportunities to demonstrate the city's increasing pre-eminence, not only within Japan but also on the global stage. At the opening ceremony for the former, Nakano Buei, head of both city assembly and chamber of commerce, drew a line between the state and its partisan representatives, on the one hand, and city and nation, on the other. It was the 'free spirits' of the business world who strove to serve the country, he claimed, transcending racial barriers and national borders. And it was Tokyo where such spirits were in greatest evidence. Up on the hill, a Tokyo Pavilion underpinned the claim. Panoramas, scale models, photographs, and statistics demonstrated how the city could accommodate its ever-growing population and support an ever-expanding global network, with municipal provision ranging from sewage plants to sports facilities. Its transformation into an imperial metropolis was subsequently supercharged by the war, which saw Japan's industrial production multiplying fivefold, its exports trebling, and the economy as a whole expanding by some 50 per cent in only four years. The change was evident in the crowds at the exhibitions: over 7 million, double the city's population, in 1914, but 11 million in 1922, prompting grumbling from dignitaries about the crush on the opening day and general discontent about insufficient toilets.[17]

[17] Hashizume, *Nihon no Hakurankai*, 62–67 and 74–79, for overviews and illustrations. Yamaji, *Chihō Toshi no Kakusei*, 19–32, for their foregrounding of the city and, in 1922, their imbrication with urban planning. For accounts of 1914, Takemura Tamio, *Taishō*

In both shows, as with exhibitions in general, it is tempting in retrospect to focus on what was new. It is hard to ignore the increasing prominence of the military. In 1914, over in Aoyama, the army showed off its latest aircraft and the navy a huge mock-up of Mikasa, the resurrected flagship of the Japanese fleet during the Russo-Japanese War, supplementing this with their own aircraft, which they flew by the bay in Shibaura. It is easy to notice the shifting fashions of the architectural establishment. In 1914, the top of the hill was taken over by Secessionism, the Austrian variation of Art Nouveau, which annoyed conservatives, who complained that traditional architecture would have better represented 'our national character' (*waga kokuminsei*). It proved enduring, however, reappearing in 1922, alongside a 'culture village', produced by the Society of Japanese Architects. Fourteen model houses suggested that modern life could be lived in about 700 square feet and for under 200 yen, with 'familiar Japanese taste' on the façade and Western convenience inside. It is also hard not to notice an increasingly overt male gaze. In 1914, the marvel of electricity revealed 'an island of beautiful women', where visitors could gawp at a mysterious well, where the women swam with carp, and watch fourteen scenes, including a double suicide, for only twenty *sen*. Commentators, of course, harumphed. Elsewhere, 600 female guards in 1914 were succeeded by 2,000 in 1922, who suffered repeated harassment, were rumoured to have caused a number of marital breakups among the great and good, but also caused frustration, for example, when officials in the Korean Building prevented them talking to male visitors.[18]

Novelty, however, was by now a predictable feature of an ever-evolving show. Exhibitions had long been a gauge of development and a place for distraction. But they were also, increasingly, somewhere to

Bunka: Teikoku no Yūtopia (Sangensha, 2004), 40–58, including Nakano's speech; *Fūzoku Gahō*, no. 457 (1914/3/5), 18–19, for the Tokyo Pavilion. See also Tseng, *Modern Kyoto*, 140–44, and Harootunian, *Overcome by Modernity*, 6–7. For crowds and toilets in 1922, Nakagawa, *Rankaiya Ichidai*, 150–51.

18 For 1914, *Fūzoku Gahō*, no. 457 (1914/3/5), 2–3, for conservative harumphing about architecture and attractions, 60–61 for the island of beautiful women, and 70 for the military. *Asahi Shimbun*, 1914/3/18, 5, and Nakagawa, *Rankaiya Ichidai*, 103–5, for the island of beautiful women. *Asahi Shimbun*, 1914/3/20, 6–7, for Secessionist buildings. For 1922, Yoshida, *Bankoku Hakurankai*, 156–163, for architecture. For the culture village, Uchida Seizō, 'Kenchiku Gakkai no Katsudō Kara Mita Taishō 11-nen Kaisai no Heiwa Kinen Tōkyō Hakurankai Bunkamura ni Kansuru Ichikōsatsu', *Nihon Kenchiku Gakkai Keikakukei Ronbunshū* 529 (2000), 263–70, Sand, *House and Home*, 228–33, and Yamaji, *Chihō Toshi no Kakusei*, 37–49. Ibid., 52–61 for attractions in both years. Nakagawa, *Rankaiya Ichidai*, 139–44, for the 1922 expo as a 'country of beautiful women'. Inoue, 'Konpanion', for female guards.

observe the economy and society from which they emerged. In 1914, a cable car carried visitors across the pond and Japan's first escalator up the hill, allowing them to gauge the progress of industry: exports were booming, mining was making great strides, and Manchurian steel was promising, but the chemical industry remained in its infancy, hops were in short supply, and Japanese products remained inferior to foreign ones. The visitors may have been more drawn by the 'cannibals' in an unofficial Burma-themed South Seas Pavilion, by whom, they were likely reassured to know, they were considered unappetizing, given the vegetarian Japanese diet. But for some, the petrified remains of a 2,000-year-old Indian priest were inappropriate in an exhibition dedicated to the new imperial era. And a visitor from the countryside was annoyed by all the traffic, shouted at for having the wrong ticket for the Ferris wheel, and confused by paintings of naked women next to a statue of General Nogi in the Art Building.[19]

Come 1922, there was widespread export potential, 'remarkable' improvements in machinery, and welcome progress in chemicals. Admittedly, industry remained myopic: the textile sector was overly focused on domestic demand, and the judges doled out more awards for sake than for diesel engines. Still, flying boats at day and floodlights at night provided a decorative accent to the scene around the pond, an official South Seas Pavilion revealed the full colonial potential of Southeast Asia, and a Street of Nations (*bankoku machi*) lured passers-by with more beautiful women and much piano-playing. The crowds, too, excited comment. The expo itself exacerbated social divisions: the well-heeled only had to pay sixty *sen* on weekdays, when they could afford to visit, but workers were asked for eighty on Sundays and holidays, the only time they had off. It also provided opportunity: country girls, who arrived in Tokyo wearing worker's clothing (*monpe*) in March, could be found in the height of fashion by July; delinquents were pestering visitors for money and targeting out-of-towners, with a hundred arrests by the end of the first month (Figure 4.1). The exhibition not only celebrated peace and promised progress but also showed that development remained an uneven, unequal thing.[20]

[19] Yoshida, *Bankoku Hakurankai*, 150–54, for an overview. *Asahi Shimbun*, 1914/3/17, 5, for the South Seas Pavilion, 3/25, 6, for rural befuddlement, 3/27, 4, for the chemical industry, 3/29, 4, for hops, and 4/17, 4, for mining. *Fūzoku Gahō*, no. 457 (1914/3/5), 2 for the mummy, 53–54 for foreign products, and 30–37 and 60–66 for attractions, with the escalator on 61 and cannibals on 62.

[20] Yoshida, *Bankoku Hakurankai*, 154–56, for an overview. *Asahi Shimbun*, 1922/2/26, 2, for exports, 3/20, 4, for the chemical industry, and 4/8, 4, for machinery; 3/14, 4, for myopic textiles, and 7/11, 4, for awards; 3/10, 5, for solicitation, and 3/15, 5, 16, 16, 3/

FIGURE 4.1 In Tokyo in 1922, cherry blossom and visitors were as worthy of depiction as buildings and exhibits.
Source: *Heiwa Kinen Tōkyō Hakurankai Ueno Kōen Kakukan no Zu* (1922), NOMURA Co., Ltd.

The year 1922, however, proved to be the last such spectacle in Tokyo for a while. This was not due to the huge deficit the exhibition left behind. The governor insisted that the influx from the countryside and the impetus to industry more than justified the red ink.[21] But on 1 September 1923 the Great Kantō Earthquake brought Tokyo to its knees, prompting a decade of reconstruction, and leaving the stage clear for Osaka to shine. Eighteen months later, on 1 April 1925, Japan's second city realized the long-held dream of its mayor, Seki Hajime, to incorporate its sprawling suburbs, increasing its total area by a factor of more than three, its population by some 800,000, and becoming overnight the largest city in the Japanese empire and the sixth largest in the world.[22]

27, 2, for response; 3/13, 3, 3/21, 2, 3/28, 2, 4/24, 3 and 4/30, 5, for delinquency; and 4/3, 2, for complaints about pricing. Nakagawa, *Rankaiya Ichidai*, 140, for country girls. Yoshimi, *Hakurankai No Seijigaku*, 150, for the Street of Nations. Hashizume, *Nihon no Hakurankai*, 74–79, for planes and floodlights.

21 *Asahi Shimbun*, 1922/8/1, 2.

22 Jeffrey Hanes, *The City as Subject: Seki Hajime and the Reinvention of Modern Osaka* (University of California Press, 2002), 243, and Hashizume, *Dai-Ōsaka Imēji*, 17.

What better time for the *Osaka Mainichi Shimbun* to make a splash, commemorating not only the expansion but also its own 15,000th issue, with a lecture series, various baseball-related initiatives, and a Greater Osaka Exhibition? A massive advertising poster covered the south side of the paper's headquarters in Dojima, with biplanes overhead and the sprawling main site in Tennōji Park at the bottom, but dominated by the looming bulk of the not-yet-reconstructed Osaka Castle. The castle served as a second site, linking the city to its past, celebrating the recent re-emergence of Toyotomi Hideyoshi, the late sixteenth-century warlord, as the first great Osakan. In Tennōji, however, it was the city's present that took centre stage: rendered as a panorama in a garden at the centre of the park; disaggregated in the main building, with twenty-seven different renditions of the city, ranging from 'Osaka on the water' and 'Osaka in the sky' to 'Osaka in cinema' and 'electrified Osaka'; and summarized in city plans and historical materials in the museum, which was repurposed as a Reference Building (Figure 4.2). The exhibition was an attempt both to announce Osaka's arrival to the world and to assuage the concerns of its residents about the consequences of growth. The *Mainichi*'s contribution to the exhibition had included a healthy injection of cash, part of which was used to set up the Osaka City Association, charged with

FIGURE 4.2 The Greater Osaka Exhibition put the enlarged city front and centre, including a panorama of the city itself.
Source: postcard, NOMURA Co., Ltd.

investigating the social problems that were acknowledged as the price of development.[23]

The mid-1920s marked not only the passing of a torch, which Tokyo would soon seize back, but also the beginning of a larger shift, as exhibitions began to spread beyond the big three into other emerging metropolises and provincial centres. The nearly 2 million who flocked to Osaka in 1925 were soon supplemented by 1 million in Kobe for a silk industry exhibition and, a year later, by nearly 500,000 in Himeji for a national industrial exhibition. These were only the most visited of a raft of initiatives, however, prompted both by ongoing urbanization and increasing economic uncertainty. By 1940, the number of cities in Japan would double, their population almost triple, and their area increase over six-fold. The gulf between the big six and the rest would also diminish, with over forty cities with a population in the six figures.[24] And as they grew, so an exhibition became the preferred way for a city to announce its arrival on the metropolitan stage: a way to connect to nation and empire, to differentiate itself in an increasingly crowded market, and to navigate the chop of the 1920s, not least by promoting exports. 'National product promotion' (*kokusan shinkō*) made its first appearance as a raison d'être in 1926, at an exhibition in Sapporo organized by the Hokkaido Times, and continued to resonate for the next decade. Industry remained a constant at exhibitions as the Shōwa era got under way, with local authorities seeking to secure their own economic development.[25]

Nowhere was this clearer than in 1928. The enthronement of the new Shōwa emperor was not the first time a royal occasion had provided an opportunity for commercial or municipal branding. In 1915, Kyoto had seized the opportunity to celebrate the delayed enthronement of the Taishō emperor; in 1924, the Japan Industrial Association the marriage of the crown prince; and in 1926, the *Tokyo Nichi Nichi Shimbun* and its

[23] Hashizume, *Nihon no Hakurankai*, 82–83, for an overview and illustrations, and *Dai-Ō saka Imēji*, 24–31, for the exhibition, and 56–77, for the rebuilding of the castle.

[24] David Kornhauser, *Urban Japan: Its Foundations and Growth* (Longman, 1976), 2; '2-11, Jinkō Kaikyū betsu Shichōson Kazu oyobi Jinkō (Meiji 21 nen – Heisei 12 nen', 'Dai 2 Shō, Jinkō, Setai', *Shinpan Nihon Chōki Tōkei Sōran*, JapanKnowledge, https://japanknowledge.com.

[25] Hashizume, *Nihon no Hakurankai*, 85, for silk in Kobe, 89, for Himeji, 92, for Sapporo, and 102–6 for industrial promotion in early Shōwa. Yamaji, *Chihō Toshi no Kakusei*, 63–97, for national product promotion. Ishida Ayū, 'Shōwa Zenki to Kokusan Yōfuku Hakurankai', in Fukuma Yoshiaki, Nanba Kōji, and Tanimoto Naho (eds.), *Hakuran no Seiki: Shōhi Nashonariti Mejia* (Azusa Shuppansha, 2009), 72–85, for the evolution of the movement in the 1930s.

Osaka sibling the birth of the first imperial grandchild. These paled in comparison to the rash of exhibitions in 1928, however, with some eighteen enthronement-themed exhibitions in twelve cities, of all shapes and sizes. The small town of Anjo in Aichi was the first out of the blocks, at the beginning of March; while the year would culminate with 1 million flocking to Tennōji Park in Osaka for a combination of coronation, transport, and electricity; 2 million to Tsuruma Park for Nagoya's first big exhibition; and 3 million to Okazaki Park in Kyoto, which foregrounded its millennium of royal heritage, not least with a new, historically informed dance performance of 'Twelve Scenes of the Capital'.[26]

Elsewhere in the archipelago, however, the monarchic focus grew more diffuse: national celebration provided a useful gloss for more local concerns and interests, nowhere more visibly than Tokyo, emerging from its prostration in the earthquake. The *Household Culture Association*, a local evening newspaper, and the Imperial Equine Association would all jump on the bandwagon over the course of the year, but it was the Tokyo Chamber of Commerce that set the pace in the spring, attracting well over 2 million to Ueno with imperial commemoration yoked to national product promotion, and a raft of attractions: an eight-cornered commemorative building, with models of the enthronement ceremonies; the first national defence (*kokubō*) building, with gas masks; a distinctively Japanese model house, with folkcraft produced in Japan; as well as New Heaven and Earth (*Shin Tenchi*), where women turned into skeletons and back again. The star, Ōtake Reiko, outraged the moralists by showering the city with pamphlets and tickets, angelically, from a plane. Outside the grounds, the police were busy trying to eradicate an infestation of scoundrels (*bakko-suru akkan*), who were yet again trying to take advantage of the unsuspecting provincials who had come to Tokyo looking for work.[27]

In these enthronement-themed cases, cities and others were using exhibitions to find their place by proclaiming municipal loyalty to the imperial line. But they were only half of the thirty-six exhibitions that pockmarked the calendar that year. Department stores promoted movies

[26] Tseng, *Modern Kyoto*, 150–60, for Kyoto 1915 and 1928. Hashizume, *Nihon no Hakurankai*, 68, for Kyoto 1915, and 81 for Kyoto 1924. Nakagawa, *Rankaiya Ichidai*, 157–58, for Tokyo 1926.

[27] Hashizume, *Nihon no Hakurankai*, 113–19, for Nagoya, Kyoto, and Osaka, and 107–11, for Tokyo. Nakagawa, *Rankaiya Ichidai*, 173–79, for an overview, including Ōtake. Yamaji, *Chihō Toshi no Kakusei*, 83–102, for a measured assessment. *Tokyo Asahi Shimbun*, 1928/3/7, evening, 2, for scoundrels.

in Nagoya, kimono in Osaka, and continued to target women and children nationwide. Sendai celebrated its industry, Beppu its hot springs, and Hakata its connections to the navy.[28] There was no one way to be a modern city or to fill the exhibition that branded the distinction of the place. Exhibitions promised a solution to the generic modern problem, how to avoid being reduced to a particular instance of a more general phenomenon.[29] By the same token, though, they always did more or less the same kind of thing. They talked about development. They provided distraction. They also provided employment for a burgeoning industry of exhibition professionals, adept at translating the attraction from place to place. By the 1910s and 1920s, their crews were also at work in the colonies, whose authorities were eager to retrace the path already taken by the home islands.

COLONIAL PROJECTION

Modern Japan's imperial aspirations had been clear at exhibitions from the start. Hokkaido was already on display in Kyoto in 1875, Tokyo in 1877, and Berlin in 1879, showing how colonial 'cultivation' (*kaitaku*) was making the most of the island's raw potential. By the early 1900s, it had been joined by Okinawa, Taiwan, Korea, and Manchuria, as evidence of Japan's arrival and aspirations as an industrial, imperial power; cultivation had given way to 'colonization' (*takushoku*), which meant not just planting, but settlement; and colonial exhibits in Japan and overseas were beginning to be supplemented by exhibitions in the colonies themselves. Following the early Meiji template, Japanese colonial overlords seized on exhibitions to promote development, demonstrating the desirability of progress to the audience at home (i.e., in the colony), exhibiting the results abroad (the metropolis), and so encouraging trade between the two (as well as investment and immigration, back into the colony). Soon enough, as back home, colonial governments were joined by other interested parties, including industrial associations and Japanese-language newspapers. The bureaucratic focus on agriculture and industry was joined by appeals to the consumer, not least tourists from Japan.

Hokkaido remained the model, perhaps. Agricultural exhibitions had started up on the island in 1878. Furniture from the island won prizes in

[28] Yamaji, *Chihō Toshi no Kakusei*, 147–56 for Sendai, 224–29 for Beppu.

[29] Naoki Sakai, 'Modernity and Its Critique: The Problem of Universalism and Particularism', *South Atlantic Quarterly* 87 (1988), 475–504.

Tokyo in 1881. Thirty years later, a local newspaper organized a travelling exhibition around the island by train. Come 1918, therefore, the Hokkaido Agricultural Association was ready to organize a massive exhibition, commemorating the colony's fiftieth anniversary. A poster showed exhibition and island watched over by the goddess of colonial development, with the key to unlock Japan's northern gate in her right hand. In Sapporo, there was a whaling diorama in the fisheries building and fake bears in the forestry building; in Otaru, a model whale to walk inside and tigers in forests in the Manchuria-Mongolia building. There was participation from Tokyo, Kyoto, and Nagoya, and from Taiwan, Korea, and Manchuria. There were nearly 1.5 million visitors. In the short term, the exhibition helped inoculate the island against the rice riots then sweeping the main islands, in protest at low wages, wartime inflation, rampant profiteering, and government inaction. In the longer term, more exhibitions were needed, including both smaller events, focused on industrial promotion and social hygiene, and occasional larger initiatives. In 1926, the Hokkaido Times organized a national products promotion exhibition in Sapporo, to celebrate both its own anniversary and the government's second colony-wide development plan. It was not as big as the 1918 event, but it still attracted over 500,000 visitors.[30]

From the start, however, on the island and at exhibitions, in Tokyo and the West, the story of its development had been troubled by the presence of the Ainu, the descendants of the original inhabitants of the northern parts of what would become Japan, who raised questions about the extent to which the new nation-state could ever become one. As the nation became an empire, so the problem was replicated. Each new acquisition brought different peoples within the imperial pale. Exhibitions not only allowed the state to demonstrate how it was incorporating territory and people, with maps, statistics, and reports, but also afforded a spectacle of enduring environmental and ethnic difference, evident not least in human displays. At exhibitions overseas and on the Japanese mainland, the problem was manageable. Before long, the exotic had been domesticated, the audience had moved on, and the colonies had been reduced to tropes: Taiwan was tropical, Karafuto cold, Manchuria would provide abundant resources,

[30] 'Bakumatsu-Meiji Zenki Kōbunsho Sashie Korekushon' (2021), www3.library.pref.hokkaido.jp/digitallibrary/content02/86-2020-12-24-06-02-46.html, nos. 37, 42, and 53, for early agricultural exhibitions, Berlin 1879, and Tokyo 1881. Sapporo-shi Kyōiku Iinkai, *Shin Sapporo Shishi* (Hokkaido Shinbunsha, 1986), vol. 3, 661–71 for 1918, 622 for its inoculating effects, and vol. 4, 735–36, for 1926, and 796–97 for hygiene exhibitions.

Korea, for the most part, little of note.[31] In the colonies themselves, however, the dilemma persisted. Imperial incorporation required both colonial subjection, that is, the overwriting of what had gone before, and ongoing subordination. Both were evident at exhibitions, together with the police surveillance required to make sure that the show came off. But they also revealed each colony to be its own place, with a distinct history and diverging present. However hard they tried, colonial officials could never quite disguise colonial difference.

There were ways to exploit it. The acquisition of Taiwan in 1895 had allowed its new overlords to parade their imperial ambitions in Osaka in 1903 and economic vision in St Louis in 1904. Development on the island itself had been encouraged, too, from the turn of the century, by a number of small, agriculture-focused product fairs (*hinpyōkai*). By 1908, there were signs of progress, with a travelling exhibition on the newly built rail line linking Taipei to Tainan, and another product fair, expanded to include ceramics and textiles. During these first two decades, however, the authorities were preoccupied by the ongoing resistance to colonial rule, and by implementing colonial control. The temporary end of armed resistance in 1915 prompted the government to celebrate, with an 'exhibition' (*kyōshinkai*) the following year. They had 'mopped up' operations against the 'wicked bandits' and 'satisfactorily' put the mountain aborigines in their place. An 'aboriginal tribe building' (*banzokukan*), with representatives from the Atayal and Tsou peoples, revealed their now-pacified everyday lives. A tourist building and a blizzard of publicity materials suggested that the island was the place for Japanese urbanites to recreate. The focus of the exhibition, however, was the ongoing need to foster development. Mounds of goods from the island and the rest of the empire filled the four floors of the newly built Government-General building. The 1,500 exhibits in the China and South Seas Building suggested the potential of connections to points beyond. Despite the celebration, it would be a while before peace was finally secured on the island. In the meantime, occasional exhibitions argued the need for police and hygiene (1925) or for product promotion (1926). It was not until 1935 that an honest-to-goodness exhibition (*hakurankai*) would be able to claim that the colony had, in fact, been transformed.[32]

[31] See Chapter 3 for the Ainu and gazumping. Yamaji, *Kindai Nihon no Shokuminchi Hakurankai*, 84, for tropes, 108–10 for their subsequent rehearsal for Korea, and 147–58 for Manchuria. *Asahi Shimbun*, 1922/4/2, 4, repeats the stereotypes.

[32] Yamaji, *Kindai Nihon no Shokuminchi Hakurankai*, 198–224, for the early exhibitions in Taiwan. Aso, *Public Properties*, 97–100, for 1908 and exhibitions in Japan. Harry Lamley, 'Taiwan under Japanese Rule, 1895–1945: The Vicissitudes of Colonialism', in

The authorities would eventually pretend that Taiwan was a model Asian colony. They could never quite conceal the fractures in Korea, but by the same token, quickly turned to exhibitions to try to cover them up. The first attempts, in 1907, during the Protectorate, and 1915, commemorating five years since annexation, were product fairs (*kyōshinkai*), but the latter clearly had greater ends in sight. The story was simple enough. The peninsula had been a source of regional instability, with suffering people, devastated agriculture, and impoverished industry. True, there were ongoing 'hostilities', but they were the result of a 'basic misunderstanding'. The exhibition would reveal an 'exemplary colony', showing 'Japanese capitalists ... what is really going on' and the Korean people the 'benevolent' nature of colonial administration. To do so, it took over the Gyeongbokgung Palace, the main residence of the Joseon dynasty, both dating from the late fourteenth century, but now under Japanese control. The Gwanghwamun Gate, the main entrance to the palace, now led to various Renaissance- and Secession-styled buildings, stuffed with 40,444 exhibits, from 13,717 Korean exhibitors, 3,917 settlers, and various colonial bureaucrats. The focus was clearly on economic development, including both a survey of progress achieved, a 'potential-evaluating building' (*shinseikan*), and the industrial machinery, from Japan and beyond, with which to realize it. The pill was sweetened by exhibits of imperial benevolence, which had improved education, mitigated unemployment, and alleviated famine; and by the inevitable attractions, including a train ride through a panorama of the peninsula and illumination, courtesy of a couple of searchlights and 6,000 bulbs. The official Japanese- and Korean-language newspapers pitched in, with a concurrent Household Exhibition, preaching the virtues of modern middle-class life and underlining 'the large number of points on which the Korean household needs to be improved'.[33]

Murray A. Rubinstein (ed.), *Taiwan: A New History* (M. E. Sharpe, 1998), 203–15, for the first two decades of colonial administration.

[33] Chōsen Sōtokufu, *Chosen Sōtokufu Shisei Nenpō, 1915* (Kuresu Shuppan, 1991), 18–19, for the official version, and 22–23, for some details. Li, Tae-Moon, '1915-nen "Chōsen Bussan Kyōshinkai" no Kōsei to Naiyō', *Keiō Gijuku Daigaku Hiyoshi Kiyō, Gengo, Bunka, Komyunikēshon* 30 (2003), 26–55, and Todd Henry, *Assimilating Seoul: Japanese Rule and the Politics of Public Space in Colonial Korea, 1910–1945* (University of California Press, 2014), 92–113, for comprehensive accounts. See also Kim, 'Industrial Exhibitions', 185–94; Hong Kal, *Aesthetic Constructions of Korean Nationalism: Spectacle, Politics, and History* (Routledge, 2011), 13–31; and, for the expo's place in a broader attempt to stimulate tourism from the archipelago, Kate McDonald, 'The Boundaries of the Interesting: Itineraries, Guidebooks, and Travel in Imperial Japan' (PhD thesis, UC San Diego, 2011), 103–14. Yamaji, *Kindai Nihon no*

The Japanese take managed to persuade some Koreans. The exhibition was presented as a collaboration, financed for the most part by the city of Seoul, with the Japanese Government-General and the captive Korean royal family providing token contributions. The Korean elite could also be found on support committees, not least Lee Wan-Yong, who had been advocating westernization for over two decades, had served as puppet Prime Minister during the Protectorate, and would be posthumously exhumed and dismembered following Independence in 1945. He was not alone, however. Jang Ji-yeon had famously written a 'loud lament for losing my country' in 1905, when such things were possible, but now produced a series of twenty-two articles, arguing that the exhibition would develop industry and improve welfare. The crowds came too, with 1 million total visitors, pumped up by 80,000 groups from the countryside and well over 100,000 free admissions. The ground had been prepared, of course. The police had arrested 'vandals' and vagabonds in advance and monitored the reaction of the 'ignorant' classes during the show. The confection was enough to convince most Japanese outlets that the exhibition had worked, increasing trade and ending the recession. But some remained sceptical. Newspaper proprietor Tokutomi Sōho insisted that Japan's purposes were not 'aggressive', but that Koreans still had to be satisfied 'materially and psychologically'. He was not wrong. Strikes were on the increase and some Korean intellectuals were convinced that development was destroying the economy, making independence urgent.[34]

It would be too much to ask an exhibition to assuage their concerns, but it did not stop the colonial government and its collaborators from trying. The loyal *Maeil Shinbō* newspaper organized one in Gyeongbuk, south of Seoul, in the autumn of 1918 and was planning one for South Hamgyong, on the northeast coast, when the colony exploded in protest on 1 March 1919. Provoked by repressive colonial administration and inspired by US President Woodrow Wilson's advocacy for the right to national self-determination at the Paris Peace Conference, more than a thousand demonstrations throughout the peninsula called for independence, to be met by violent suppression by the Japanese authorities. In its

Shokuminchi Hakurankai, 114–22, for the household exhibition. In-Hwan Oh and George Won, 'Journalism in Korea: A Short History of the Korean Press', *Transactions of the Royal Asiatic Society, Korea Branch* 51 (1976), 3–9, for the newspaper industry.

34 Li, '1915-nen "Chōsen Bussan Kyōshinkai"', 25, for financial contributions and 55–60 for visitors. Kim, 'Industrial Exhibitions', 196–209 for collaboration, visitors, supportive and critical Korean intellectuals, Tokutomi, police, vandals, and strikes. For Japanese reservations, *Chōsen Oyobi Manshū* 100 (1915/11), 16.

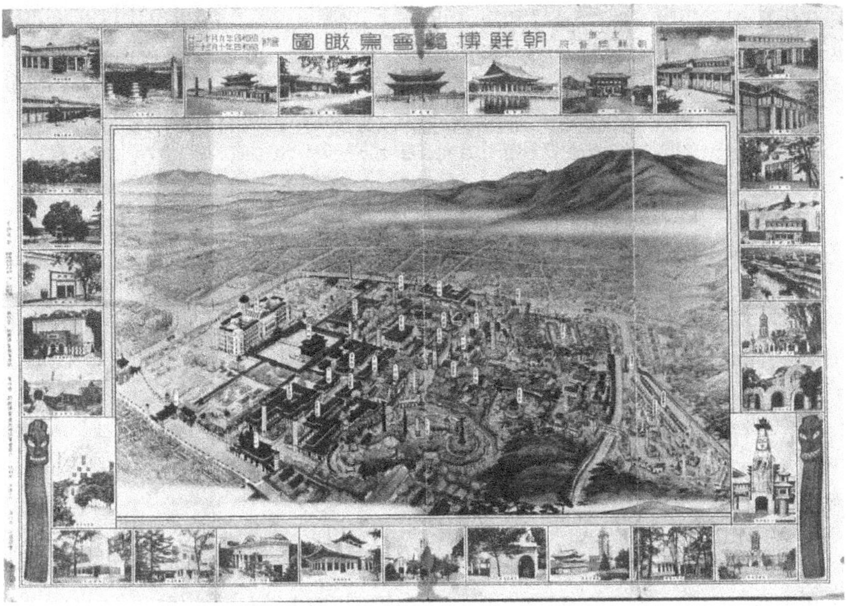

FIGURE 4.3 Bird's eye view and exhibition showed how the Korean royal palace was now hidden behind the new Government-General Building, with colonial capital and peninsula sprawling beyond.
Source: *Chōsen Hakurankai Chōkanzu* (1929), NOMURA Co., Ltd.

wake, famously, the Government-General decided to rebrand its modus operandi as 'cultural' rather than 'military' rule. Three years later, the Peace Memorial Exhibition in Tokyo provided an opportunity to rehabilitate metropolitan perceptions of the economy, not least the terrible reputation of Korean rice, and to convince some middlemen that the colony was still worth their business. The 1920s also saw a steady stream of exhibitions on the peninsula, organized by the usual mix of local authorities, chambers of commerce, industry associations, and the occasional newspaper.[35]

Come 1929, therefore, the authorities were ready to commemorate twenty years of rule, back at the Gyeongbokgung Palace, now overshadowed by the looming Government-General Building, which had replaced the Gwanghwamun, relocated to the east (Figure 4.3). The Korea Exhibition tried again to reveal the progress of the colony, fusing the 'particular culture' of Korea to the general logic of development.

[35] Yamaji, *Kindai Nihon no Shokuminchi Hakurankai*, 190, for 1918, 110–14 for 1922, and 122–23 for the 1920s on the peninsula.

A corridor of Korean-style buildings led to a large, modernist Homeland Building. A Rice Building, not far from Army and Navy Buildings, limned the foundations of empire. The oppressive hierarchy of pre-colonial judicial practice was contrasted to the disciplined authority of the modern colonial state. And again, the didactic sticks were balanced by distracting carrots of congress, consumption, illumination, and attraction: cultural housing, courtesy of the Korean Architectural Association; performers from India, France, and the United States; a toy train for children, which this time took them around the world. Korean intellectuals again proclaimed the benefits of enlightenment. One million again came through the gates. The exhibition could hardly disguise, however, the increasing crime and homelessness marking cities throughout the empire.[36]

It would only get worse. Come the 1930s, exhibitions would have to work even harder, in the wake of global depression and Japanese aggression, to pretend that the empire's aims were pacific. In the 1920s, it was still just possible to believe in the promise of progress, buoyed by not only acquisition of actual colonies but also the fact of informal empire in China. The Japanese commitment to the continent had long been clear, with Japan–China trade buildings, for example, at the Tokyo Taishō Exhibition in 1914, and by a whole exhibition devoted to the same topic in Shanghai in 1919.[37] By 1927, China would attract some 85 per cent of Japan's foreign direct investment, of which 80 per cent was flowing into Manchuria. On the surface, therefore, the industrial promotion exhibition in Dairen in 1925 could celebrate the progress achieved since the city's incorporation in 1915 and its expansion the previous year. It illustrated the industry enabled by Dairen's linking of archipelago and continent; provided the usual run of inducements and attractions, from illuminations to treasure hunts to female divers (*ama*); and capitalized on the Manchurian tourist boom, tying itself into the circuit by including Russo-Japanese War battlefields in its own guidebooks. It attracted nearly

[36] *Chōsen Hakurankai Kinen Shashinchō* (Chōsen Sōtokufu, 1930) for photos and *Chōsen Oyobi Manshū* 263 (1929), 326–30, for a contemporary account. Yamaji, *Kindai Nihon no Shokuminchi Hakurankai*, 123–37, and Henry, *Assimilating Seoul*, 113–29, for overviews. Kal, *Aesthetic Constructions*, 32–43, and Wybe Kuitert, 'Kyonbokkun kara Chōsen Hakurankaijō e no Kūkan Henbō', in Sano Mayuko (ed.), *Bankoku Hakurankai to Ningen no Rekishi* (Shibunkaku Shuppan, 2015), 255–59, for the use of the space. See also McDonald, 'The Boundaries of the Interesting', 144–49.

[37] Yamaji, *Kindai Nihon no Shokuminchi Hakurankai*, 85–86, for the 1910s, and 168–73, for 1925.

500,000 paying visitors, nearly double the size of the city itself, of whom almost half were Chinese.

Behind the scenes, though, the exhibition was animated by local anxiety about the uncertain economic environment, which suggested the unstable foundations of the empire itself. Dairen's monopoly on trade between Japan and the Chinese northeast, established in 1905, had been undisturbed by the opening up of direct trade between Korea and Manchuria in the 1910s. Recession had struck the city in 1920, however, and Japanese settlers were struggling both to compete with Chinese traders and to implement effective self-governance. From 1923, too, their monopoly was increasingly threatened by the support of the local warlord, Chang Tso-lin, for alternative channels for trade. The organizers of the exhibition had therefore appealed to the Government-General and the South Manchuria Railway Company for funding. They had also solicited participation from the Guomindang, the Chinese Nationalist Party, but been rebuffed. There was increasing anti-imperial and anti-Japanese sentiment sweeping through the party and its urban strongholds to the south. Nor, finally, did the exhibition succeed in providing the economic stimulus the organizers hoped.[38] In the next decade, the ongoing uncertainty, increasing Chinese resistance, and Japanese military aggression would produce a new kind of colonial landscape and exhibition.

STAGING EXHIBITIONS

By the end of the 1920s, then, empire and exhibitions seemed to be proceeding hand in glove. The latter provided cover, under which local authorities, ambitious entrepreneurs, and questing elites could thicken the sinews connecting their colony, city, or company to the imperial skeleton, while also bulking up the muscle that marked their contribution to the leviathan. Exhibitions could never quite close the gap – between city and country, metropolis and locality, empire and colony. Their attractions were too various. Their audiences were always multiple. The development they sought to display and to catalyse was always on the wing. They were

[38] Emer O'Dwyer, *Significant Soil: Settler Colonialism and Japan's Urban Empire in Manchuria* (Harvard University Asia Center, 2015), 171–80, for Dairen's problems in the 1920s, and 201, for the exhibition's failure. For the regional context, Gavan McCormack, *Chang Tso-Lin in Northeast China, 1911–1928: China, Japan, and the Manchurian Idea* (Stanford University Press, 1977), 6–8 and 92–95, and Yoshihisa Tak Matsusaka, *The Making of Japanese Manchuria, 1904–1932* (Harvard University Asia Center, 2001), 130–32, 162–63, 201–2, and 273–81.

also too generic. There were only so many ways to exhibit industry, to celebrate empire, or, more immediately, to stimulate consumers and attract the crowds. Not that this stopped anyone from trying. The pace of change, the gyrations of the economy, the ructions of society meant that there was a constant demand for people who knew how to put an original face on things. By the early years of the new imperial era, therefore, there was a growing industry of professionals who promised to provide exactly this. Their work showed, however, why they could never quite deliver.

All exhibitions needed advertising and, by the 1920s, Yoshida Hatsusaburō was the go-to artist for a place in search of a brand. Yoshida had started out in 1914 or so by designing railway maps, but quickly worked out how to do something more. A bird's eye view, through a fisheye lens, allowed him to centre city, locality, or attraction in the local topography, threaded through with rail and thereby connected to the world beyond, often with Mount Fuji on the horizon. The recipe was catnip to local authorities and private companies, who were eager to put themselves on imperial and tourist maps. By the second half of the decade he was creating a series of maps centred on exhibitions. An August 1928 bird's eye view of the 'Kyoto Grand Exposition in commemoration of the imperial coronation' became, three months later, a guide to the city's famous places commissioned by and centred on the Daimaru department store. The following year he was in Korea, revealing how the Korea Exhibition dominated the city. But this was only one of over thirty works he produced during his stay, with the magnum opus being a picture of the whole peninsula. His template afforded a pace of production that allowed him to satisfy his multiple clients and to produce a comprehensive record, albeit one where everywhere began to look more or less alike. This may have been the point. Cities and others came to Yoshida for the same reason they held exhibitions: both to stand out and to fit in, as part of a conventionalized, familiar landscape, connected to the greater good, playing their imperial part. The demand would only increase in the following decade.[39]

[39] Yoshida was rediscovered at an exhibition in Sakai in 1999. For overviews, Yuhara Kimihiro, *Taishō, Shōwa no Chōkanzu Eshi: Yoshida Hatsusaburō no Panorama Chizu* (Heibonsha, 2002), and Hotta Yoshihiro, *Yoshida Hatsusaburō no Chōkanzu o Yomu: Egakareta Kindai Nihon no Fūkei* (Kawade Shobō Shinsha, 2009). For an online gallery, 'Yoshida Hatsusaburō-shiki Chōkanzu Dētabēsu', https://iiif.nichibun.ac.jp/YSD/, which includes the Daimaru commission: 'Daimaru o Chūshin to Seru Kyōto Meisho Annai Chōkanzu', https://iiif.nichibun.ac.jp/YSD/detail/003207602.html. For the exhibition

As it would for exhibitions. That it could be met was due in large part to a newly defined, clearly differentiated, gradually professionalizing breed of showman, the *rankaiya*, the exhibition specialist or, perhaps better, expo dude. At the bottom were the descendants of the Edo showmen, who found space within exhibitions to set up a stall or to mount an attraction, convincing the crowds to hand over a few *sen*. Then there were people who built the site, ranging from the bosses in charge of constructing and decking out the buildings, to their employees, who designed the interiors, built the cases, painted the backdrops, and wired the lighting. Finally, there were the impresarios, descendants of Kushibiki and the like, who would appear in project departments in newspaper companies and tourism sections in local governments, just as they were struck by the exhibitionist itch, promising to conjure the wonderland through which the newspaper's circulation could be secured or the city's future assured. Some were eager to distinguish themselves from the riff-raff, insisting that they were decorators rather than *rankaiya* and tracing their genealogy back to Meiji predecessors. Others came close to charlatans, demanding kickbacks from the constructors, enjoying the hospitality of the exhibitors, and molesting the female attendants, before vanishing just as the exhibition was about to open. More reliable were figures such as Nomura Taisuke, who founded his eponymous industrial art (*kōgei*) company in 1892, got going by displaying chrysanthemum dolls in the 1900s, provided the inspiration for the Yomiuri's summer extravaganzas at Ryōgoku in the 1920s, and so the foundation for a successful post-war business. Its 'DNA since foundation', it says, has been 'creating and vitalizing spaces providing emotion to people'.[40]

A closer look at the mechanics of it all is provided by Nakagawa Dōji, a wannabe oil painter, born in 1904, who fell into life as a *rankaiya*

version, Tseng, *Modern Kyoto*, 117–22. For the Korea exhibition, Tomii Masanori, 'Keijō "Modan" no Chizu o Yomu', *Himoji Shiryō Kenkyū Sentā Newsletter* 37 (2017), 28–30, and for the trip as a whole, Nakanishi Ryōtarō, 'Kindai no Chōkanzu ni Egakareta Chōsen Hantō: Yoshida Hatsusaburō, "Chōsen Daizue" no Moji Jōhō no Bunseki', *Rekishi Jinrui* 47 (2019/3), 108–78.

40 Nakagawa, *Rankaiya Ichidai*, 10–11, for a definition. Ishikawa Atsuko, 'Shiryō Kara Miru Rankaiya to Sōshokugyō no Rekishi', in Sano Mayuko (ed.), *Bankoku Hakurankai to Ningen no Rekishi* (Shibunkaku Shuppan, 2015), 351–55, for the etymology of the term. Hashizume, *Jinsei wa Hakurankai*, for a collection of biographies, underlining the diversity of possible careers, and Hashizume Shinya, *Kindai Nihon no Kūkan Purannātachi* (Haseko Sōgō Kenkyūjo, 1995), 132–44, for Nomura, and 'Nenchū Gyōji', 175–76, for the relationship with the Yomiuri. For Nomura's DNA, 'Solutions', www.nomurakougei.co.jp/english/solutions/index.html.

during the 1920s as a means of making the money he needed to make his way through art school, and whose reminiscences appeared in 1969. Working on exhibitions may have begun as a side hustle, but sure enough he stayed, captured by what he calls their 'magic' (*maryoku*). He graduated from painting signs to making cases to constructing advertising towers (*sōshokutō*). He started off in Tokyo. He had visited the Tokyo Taishō Exhibition in Ueno as a ten-year-old, he worked at the Peace Memorial Exhibition as a teenager, and by his mid-twenties he was back in the same place, building things to commemorate the new emperor's accession. Come 1929, like Yoshida, he was in Korea, running into other *rankaiya* from Tokyo wherever he went, squabbling about who did what work, and getting ill. It was the beginning of a glorious decade, which would culminate with wartime exhibitions in Osaka and Nanjing.

Nakagawa's reminiscences focus, however, not so much on the occasion of the exhibition or the contents of the exhibits, but on the technology of attraction; on what it takes to draw a crowd, whether papier mâché or lighting effect; on the contrivance (*shikake*) that enabled the make-believe. This not only makes for some alarming switchbacks along the way, from imperial palanquins to haunted houses, and water wheels to exploding bombs; it also provides a useful counterpoint to the more official accounts, with their emphasis on municipal (or national) distinction and pride. A *rankaiya* specialized in the transferable skill of mobilizing crowds, which developed considerably over time. Early exhibits had comprised piles of goods and declarative signage, together with tables, maps, and charts, considered sufficient to do the didactic job. By the 1920s, with technological marvels among the exhibits and live, sometimes nude, bodies among the attractions, exhibitions had moved beyond three-dimensional dioramas to immersive environments and dynamic exhibits. Such techniques might be deployed for a time under a particular rubric, for example, product promotion. Soon they would be turned to the ends of national mobilization and total war. But they rested on tried-and-tested party tricks, valuable to the extent that they never quite produced the same effect twice. The *rankaiya* – and, by extension, the exhibitions they made – were, if not uninterested, then indifferent to the world they were asked to put on show.[41]

[41] Nakagawa, *Rankaiya Ichidai*, 105–6, for contrivance, and 163 for magic.

5

War and Peace

The official liturgy of modern Japan insists that we break the story at 1945, with Hiroshima as the sacrificial lamb, the Peace Memorial Park the cathedral, and the annual ceremonies on 6 August the sacrament. The history of the city, however, is more complex and troubling than the standard version allows, as the story of its exhibitions suggests.[1] After four centuries as a fishing village and three more as a castle town, Hiroshima started to transition, at the end of the nineteenth century, to regional hub of industry and empire. The city threw in its lot with the military early, setting up what would become the headquarters of the Fifth 'Carp' Division of the Imperial Army in 1871. The navy soon arrived too. A modern port made it the preferred choice for the Imperial Headquarters during the first Sino-Japanese War. The Kure Naval District, just down the coast, built Japan's first capital ship at the end of the Russo-Japanese one. The city and its suburbs were not all guns. Electric light arrived in 1893, rail in 1894, water supply in 1899, gas in 1910, and light rail in 1912. Three years later, the prefecture opened an Industrial Promotion Hall, a hundred metres or so southeast of the t-shaped Aioi Bridge, where two branches of the Ōta River delta converged. Exhibitions confirmed, however, the extent to which civil and military were imbricated. In 1929, a Shōwa Industrial Exhibition sprawled over three sites: agriculture and industry, folkcraft and colonies, as well as the statutory kiddyland, on the West Drill Field, surrounded by the Fifth Division's buildings, just to the south of the castle; army aircraft

[1] Lisa Yoneyama, *Hiroshima Traces: Time, Space, and the Dialectics of Memory* (University of California Press, 1999).

and the relics of General Nogi, the hero of the Russo-Japanese War, in Hijiyama Park; and the navy in Ujina Port. A couple of years later, the tone shifted. An Emergency (*jikyoku*) Exhibition in 1932, also on the West Drill Field, celebrated the novelty of Manchuria and commemorated the fiftieth anniversary of the Imperial Rescript to the Military. It was succeeded by a National Defence and Industry Exhibition in 1935 and a China Incident Exhibition in 1938, both in Kure, and a Rising Asia Construction Exhibition in 1939, in front of Hiroshima station. One year later, the Kure dockyard would launch the largest battleship ever built. It would also become the Imperial Navy's largest base and arsenal during the Asia-Pacific War.[2]

At the end of that war, the battleship was sunk, the city destroyed, courtesy of an atomic bomb targeted on the Aioi Bridge, and the economy in ruins. It did not take long, however, for the authorities to begin planning its recovery or for exhibitions to start back up. One of the earliest post-war exhibitions, organized by the local Industrial Promotion Association, opened in late spring 1947 in Fukuyama, 100 kilometres up the coast. Two years later, the Diet designated Hiroshima itself a peace memorial city, and the ruins of the Industrial Promotion Hall became the iconic Atomic Bomb Dome. The National Sports Festival followed in 1951, in the grounds of the reconstructed castle, with a mock wooden keep, accompanied by a small exhibition devoted to physical culture (*taiiku bunka*). The travelling Atoms for Peace exhibition, sponsored by the United States Information Agency (USIA), came to town in 1956, occupying the Peace Memorial Museum, which had opened the previous year at the southern end of the park. Finally, in 1958, the city staged a Grand Reconstruction Exhibition, with just under a million visitors flocking to the Peace Memorial Park and orthogonal Boulevard. The exhibition included the usual suspects – local history, modern industry – as well as evidence of the brave new world that was coming into being, with an artificial satellite provided by the Soviet Union. But it also had space for atomic energy, and for the atomic bomb.[3]

[2] Ran Zwigenberg, 'Hiroshima Castle and the Long Shadow of Militarism in Postwar Japan', *Japan Review* 33 (2019), 198–99, provides a summary of Hiroshima's pre-war exhibitions, Yamaji, *Chihō Toshi no Kakusei*, 156–76, a detailed account of 1929.

[3] Morris Low, *Visualizing Nuclear Power in Japan: A Trip to the Reactor* (Palgrave Macmillan, 2020), 82–87, for the planning of the Atoms for Peace exhibition. Ibid., 138–39 and Ran Zwigenberg, '"The Coming of a Second Sun": The 1956 Atoms for Peace Exhibit in Hiroshima and Japan's Embrace of Nuclear Power', *The Asia-Pacific*

The local transformation points to a larger story, which echoes and troubles some assumptions about the relationship between militarist pre-war and pacific post-war Japan. Nowadays, it is clear, at least among historians, that there was as much continuity as change.[4] It remains tempting, of course, to explore what went wrong before the war – the sources and causes of the progression from modernism to militarism. Exhibitions provide a convenient site. It is easy enough to find ones where the military was a dominant presence in the early 1930s, as well as ones later in the decade that sought to celebrate mobilization, sacrifice, and 'holy war'. They were balanced, however, for longer than one might think, by aspirations to peace, not least overseas. By 1940, admittedly, the contrast between pacific protestation, to the world, and missionary mobilization, at home, argues for a Japan that was at least two-faced. Through it all, however, as in Hiroshima, exhibitions continued to insist the seeming contradiction could be resolved, by being subsumed under the rubric of development. Economic recovery remained the basso ostinato of exhibitions throughout the 1930s and 1940s, from earthquake, economic upheaval, and the devastation that rained from the skies. Before the war, exhibitions allowed local authorities and the state to insist that development would contribute to the good of the whole, country and world. After the war, exhibitions allowed people to believe – as it emerged from the ashes of defeat, under the Americans – that Japan had broken with the past and development might resume.

DEFENDING THE NATION

Given that the modern nation-state was designed for war, it made sense for military hardware and accomplishment to feature heavily at exhibitions. The link between military capacity and great power status had been clear in Paris in 1867, where the star attraction had been a massive cannon, exhibited by Krupp, and Japanese observers had puzzled out the mechanics of mines. By the end of the century, modern Japan had come of

Journal: Japan Focus 10.6.1 (2012), for its stop in Hiroshima. Zwigenberg, 'Hiroshima Castle', 208–13, for the Grand Reconstruction Exhibition.

4 For two very different early arguments for continuity, Chalmers Johnson, *MITI and the Japanese Miracle: The Growth of Industrial Policy, 1925–1975* (Stanford University Press, 1982), and Yasushi Yamanouchi, J. Victor Koschmann, and Ryuichi Narita (eds.), *Total War and 'Modernization'* (Cornell University, 1998). For a recent synthesis, Gordon, 'Japan's Transwar Political Economy', in Laura Hein (ed.), *The New Cambridge History of Japan*, vol. 3 (Cambridge University Press, 2023), 401–31.

age, courtesy of the Sino- and Russo-Japanese Wars, making the military an established presence at subsequent exhibitions. In 1903, the Fifth Domestic Industrial Exhibition in Osaka provided the occasion for an imperial fleet review, with some seventy warships unveiling the military muscle that had secured the acquisition of the Taiwanese bounty displayed on site. In 1906 and 1907, booty from the war against Russia, some provided by the army, appeared in two travelling exhibitions, sponsored by newspaper companies. By the beginning of the Taishō period, the military was an established and welcome presence at exhibitions. In 1913, the navy once again cruised around Osaka Bay, complementing a colonial exhibition commemorating the late emperor. In 1914, the two services faced off in Tokyo, albeit at a distance from the main exhibition site.[5]

It was in the late 1920s, however, that the army and navy began to use exhibitions more systematically as a medium through which to advance their cause. Some of the reasons for this had been percolating for a while: the military's sense of terminal uniqueness, given its constitutional loyalty to the emperor, rather than the state; its increasing alienation from modern society, which seemed incapable of providing the raw male material needed for soldiering; and the precarious international environment following World War I, which had exacerbated a Japanese sense of both grievance and vulnerability. At Versailles in 1919, Japan had been awarded control of only some of Germany's Pacific colonies. At the Washington Naval Conference in 1922, its navy had been restricted to three-fifths those of the UK and US. Chronic insecurity was exacerbated by a crisis of confidence in mid-decade, as the military came under attack from a fiscally minded government and a diplomatically inclined foreign policy establishment. The arrival of a militarist premier in 1927 provided an opportunity to reboot the military brand, paving the way for a series of public relations initiatives, which included exhibitions. But the initial campaign hardly solved the problem. The perception of crisis was ongoing. The Great Depression, beginning in 1929, and the London Naval Conference the following year confirmed the perception of an inhospitable global environment, prompting adventurism in Manchuria

[5] Miyake Yūji, 'Kuruppu-sha no 19 Seiki: Futari no Kuruppu, Bankoku Hakurankai, Nihon', in Yoshida Mitsukuni (ed.), *Bankoku Hakurankai no Kenkyū* (Shibunkaku Shuppan, 1986), 87–110, for Krupp, and Sumi, *Tokugawa Akitake*, 55–56, for mines in 1867. See the previous chapter for 1903, 1913, and 1914, and Fujitani, *Splendid Monarchy*, 121–49, for Meiji military reviews.

and a series of campaigns to mobilize the populace for national duty. From 1931, the military inaugurated a campaign to convince people, with pamphlets and petitions, lectures and exhibitions, of the need for national defence (*kokubō*).[6]

The ground had been prepared at the end of the previous decade. The first National Defence Buildings appeared at two exhibitions in 1928 and quickly became a staple. In Nagoya, the army and navy shared the same building, modelled after a battleship, but in Tokyo, more typically, given the inter-service rivalry, each built its own. The army exhibited its latest kit in the National Defence Building, alongside gas masks from around the world, while a tank used a pontoon bridge to trundle over the pond. The navy showed off its new aircraft in the Aviation Building, some of which took to the skies. In Seoul in 1929, the National Defence Building was even larger than the Rice Building. The strategic anxiety produced by Chinese instability was more pressing than the agriculture that was the colony's economic raison d'être. Finally, the twenty-fifth anniversary of the Russo-Japanese War in 1930 saw the services going all out. In Tokyo, there were biplanes at the main entrance to a Sea and Sky Exhibition in Ueno and a demonstration of torpedoes on the pond. Over in Yokosuka, the Mikasa sat alongside a tank constructed out of the glass used on warships, in which female abalone divers (*ama*) plied their craft. In Osaka, the army's Fourth Division sponsored an exhibition at the Takashimaya department store, to demonstrate how it was 'the people (*kokumin*) … who must defend the country', given that 'perfect peace' was never guaranteed. But it was the navy in Kobe that attracted over a million, piggy-backing on plans to commemorate the sixtieth anniversary of the port and promote trade. A fleet review of 200 warships, under the imperial gaze, together with a human cannonball and a strongwoman, proved irresistible.[7]

Just over a year later, the invasion of Manchuria upped the ante still further, catalysing enthusiasm for national defence, encouraging commercial interests to invest in the idea, and providing new opportunities for

[6] Edward Drea, *Japan's Imperial Army Its Rise and Fall, 1853–1945* (University Press of Kansas, 2009), 146–50, for the crisis of confidence following World War I. Louise Young, *Japan's Total Empire: Manchuria and the Culture of Wartime Imperialism* (University of California Press, 1998), 130–40, for the campaign.

[7] Hashizume, *Nihon no Hakurankai*, 107–13, for 1928, 121–25 for 1930. Nakagawa, *Rankaiya Ichidai*, 177–79 for Tokyo 1928, 185 for Seoul 1929, and 191–96 for Tokyo 1930. Yamaji, *Chihō Toshi no Kakusei*, 97–100, for Tokyo 1928. See also Shimazu, *Japanese Society at War*, 241–48.

them to do so. The year 1932 saw a minor rash of exhibitions devoted to Manchuria, with seven out of the year's crop of twenty explicitly focused on the new acquisition. National Defence Buildings became increasingly prominent at all exhibitions, often featuring a 'defence of the north' panorama, with Kwantung Army soldiers bundled up in warm clothing, looking out at the Soviet border town of Blagoveshchensk. Admittedly, the attraction diminished over time. There were four national defence exhibitions in 1934, two in 1935, and one in 1936. But other shows, too, argued that economy and society should be subordinated to military demand. A Great Industrial Mobilization Exhibition in Osaka in April 1935 turned its main site into an urban battlefield, complete with tanks, while the National Defence Industries Building took the form of a battleship.[8] The military was not the only thing on display in the 1930s, but it was never far away.

Not that it had to do everything itself. The increasing prominence of the military was predictable, given that the nation was at war, but also driven by a deepening relationship between the services themselves, particularly local authorities, and various professional and commercial concerns, all eager to exploit the possibilities of Manchuria. It made sense for Nagoya in 1932, Sendai in 1933, and Osaka in 1934 to stage national defence exhibitions. They were the respective bases of the Third, Second, and Fourth Infantry Divisions. The local economy was heavily dependent on the army.[9] Cities and services could also rely on the increasing number of specialist exhibition companies, who competed to provide the biggest bang and so secure the military buck. The rivalry on site between the navy and the army was spiced behind the scenes by that between Hirose Shōkai, which had started business in 1914 and worked mainly for the former, and Kenseisha, founded in 1933, who serviced the latter.[10] Kenseisha also employed Nakagawa Dōji, who started building attractions for the military in the 1930s, and who provides a glimpse into the networks through which they came into being.

Nakagawa worked on a national defence building in Sapporo in 1931 but had to scramble to research and source his exhibits. Most of the hardware was out of date. He only realized that Japan had

[8] Nakagawa, *Rankaiya Ichidai*, 216, for the panoramas. Hashizume, *Nihon no Hakurankai*, 143, for Osaka 1935.

[9] Stewart Lone, *Provincial Life and the Military in Imperial Japan: The Phantom Samurai* (Routledge, 2011), for the relationship between local society and the military.

[10] Hashizume, *Jinsei wa Hakurankai*, 132–35, for Kenseisha's work for the military.

paratroopers from a German aviation magazine. But then he met a Major Sakuma, who transformed the displays from dreary exhibits of military hardware on dirt floors to interactive exhibits, which even children could understand. Nakagawa's sources also improved. He visited Manchuria itself in 1932, acquiring some army materials about the border with Soviet Russia. Then, at the end of 1935, Sakuma introduced Kenseisha to a Major Tsutsui, from the headquarters of the Tenth Division in Himeji, and a Mr Fukushima, from the local chamber of commerce. With the backing of the Central Army District in Osaka, they transformed a run-of-the-mill industrial exhibition into a huge national defence expo in 1936, taking over a baseball stadium and more besides. The Imperial Military Building at the Great Shining Japan Exhibition of 1936 featured: a battleship made out of the food consumed on board battleships; an electrically operated demonstration of a submarine attack on an enemy ship; and a re-creation of the Manchurian plain, with trenches, through which visitors could walk, and tanks, which crushed the enemy fort. Immersive environments promised to recruit audiences to the military cause more effectively than wall panels ever could. An invitation from the Seventh Division in Hokkaido soon followed. By this point, Nakagawa was also in cahoots with a decidedly right-wing, university-based research group, eager to prove that trade required the navy, allowing Nakagawa and his colleagues to further up their game.[11] Military spectacle was in demand. But it was not the only thing on show.

MODERNIZING THE EMPIRE

Even when it was running amok, the military was only one among many attractions. The Manchurian campaign provided useful grist for the exhibitionary mill for a couple of years. Soon, though, the perception of crisis had subsided.[12] And even while it was ongoing, different places were using exhibitions in different ways, to promote their own recovery

[11] Nakagawa, *Rankaiya Ichidai*, 213–18 for the early 1930s, 231 and 236 for his collaboration with the research group, and 241–52 for his activity in the mid-1930s. The group was most likely the Kokubō Kenkyūkai at Keiō University, one of many under the umbrella of Minoda Muneki's Genri Nihonsha, founded in 1925. Andrew Barshay, *State and Intellectual in Imperial Japan: The Public Man in Crisis* (University of California Press, 1988), 36–48, for Minoda.

[12] Sandra Wilson, *The Manchurian Crisis and Japanese Society, 1931–33* (Routledge, 2002), 218–24.

and development. While Nagoya, Sendai, and Osaka were celebrating their military connections and extolling national defence in 1932, 1933, and 1934, Kanazawa, Nara, and Nagasaki were all promoting tourism, the last capitalizing on the recent official designation of Unzen National Park. In 1935, Kure trumpeted its importance as a naval base and arsenal, with a Great National Defence and Industry Exhibition, but it still found space for three circuses, an underwater revue, and the mummified corpses of five criminals, recently discovered in Peru. It was also only one of nearly thirty exhibitions that year, with a handful preaching the importance of the military, the police, and mobilization, but the majority running the familiar gamut, from industrial endeavour to municipal commemoration, forestry to footwear, tobacco to candy. The biggest was Yokohama, with over 3 million flocking to a '1-million-yen expo', which had been underwritten by the great and good, public and private (*kanmin yūryokusha*), to commemorate its recovery from earthquake and depression. The exhibition included the military, of course, but only at the far end of a narrow site, which provided a comprehensive account of the empire and the world beyond, suggested by the funnels of the liner that adorned posters and postcards, and embodied in the American rodeo.[13]

As much as a militarizing nation, in other words, exhibitions in the 1930s were product and producer of a modernizing empire. Exhibitions in the colonies, too, encouraged the ongoing weave of war and peace, not least in the latest acquisition. Japan had been active in Manchuria, and Manchuria had provided exhibits for Japan, for a quarter of a century, in large part courtesy of two organizations established following the Russo-Japanese War. The South Manchuria Railway Company had started out by taking over the Russian network, but had expanded massively, both by building new lines and diversifying into almost all areas of the economy, from power generation to agricultural research. The Kwantung Army, analogously, had started as a garrison in the railway zone, but after two decades had developed its own hard-right visions of geopolitical truth, strategic necessity, and national destiny. Its invasion of Manchuria in 1931, followed by the establishment of the nominally independent,

[13] Hashizume, *Nihon no Hakurankai*, 129 for Kanazawa, 131 for Nara, 134–35 for Nagasaki, 142 for Kure, and 140–41 for Yokohama. Yamaji, *Chihō Toshi no Kakusei*, 210–19, for Kanazawa. Hara Yasunori, 'Taishō Shōwa Shoki Nagasaki-Ken ni okeru Kankō Kaihatsu to Kōtsū Ami no Keisei: Unzen no Jirei', *Keizaigaku Kenkyū* 70.4 (2004), 73–81, for Unzen. Kure-shi, *Kokubō to Sangyō Dai Hakurankai Shi* (Kure-shi Yakusho, 1936), 571 for Kure. *Yokohama Shishi, 2, Shiryō hen* (Yokohama-shi, 1993), vol. 1.1, 8–10, 264–71, and 1018–24, and Nakagawa, *Rankaiya Ichidai*, 226–30, for Yokohama.

effectively puppet state of Manchukuo in 1932, provided not only new attractions for exhibitions in Japan but also new possibilities for exhibitions. They too might promulgate the fiction by which the new state was underpinned, that a Confucian-inflected 'kingly way' would produce an ordered society and harmony among the 'five races' (Manchu, Mongol, Han Chinese, Korean, and Japanese).[14]

In 1933, Dairen staged a Great Manchuria Exhibition, commemorating not only the invention of the country the previous year but also revealing the anxiety of the city. Mementoes of fallen South Manchuria Railway employees showed the sacrifices that had been made, displays of natural resources the potential of the territory acquired, and military hardware in a huge National Defence Building the commitment to defend it. The local press hymned 'the paradise of the kingly way'. The Colonial Minister, Nagai Ryūtarō, insisted that the new state would help Asia overcome 'white imperialism'. The exhibition also showed it to be a confection, however. Life-sized 'Manchurian' and Japanese dolls, at the entrance to the main commemorative building, tried to give the fiction some substance; but the Local Customs (*dozoku*) Building revealed that the region comprised not only five distinct peoples but also various minority groups. There were also more immediate concerns in play. Dairen had lost its monopoly on business in the region and was now subordinate to the new capital at Shinkyō, though still three times as large. The city was eager to show that it remained both a conduit for 'economic cooperation between China and Japan', and a tourist gateway, an 'international paradise' with an 'exotic aroma' all its own. The local crowds stayed away, though some Japanese were convinced. On Manchukuo Day, 15 August, 100 leading Japanese business figures and 12 local officials, including the Mayor of Osaka, showed up. The Japan-Manchuria Business Council was soon in business.[15] In 1936, in Toyama and in 1937 in Nagoya, Manchuria was the star at exhibitions

[14] Matsusaka, *The Making of Japanese Manchuria*, for developments prior to the creation of the puppet state, Prasenjit Duara, 'The New Imperialism and the Post-Colonial Developmental State: Manchukuo in Comparative Perspective', *The Asia-Pacific Journal: Japan Focus* 4.1 (2006), for its ideology, and Young, *Japan's Total Empire*, for its effects back in Japan.

[15] Yamaji, *Kindai Nihon no Shokuminchi Hakurankai*, 175–93, for the exhibition. *Manshū Nippō* 1933/7/23, 1 and 7, and 1933/7/24, evening edition, 1, for reporting. Young, *Japan's Total Empire*, 192–99 for the Business Council, 251–53 for economic statistics. For Nagai, Peter Duus, 'Nagai Ryūtarō and the "White Peril," 1905–1944', *The Journal of Asian Studies* 31.1 (1971), 41–48. For Dairen, O'Dwyer, *Significant Soil*.

(a)

(b)

FIGURE 5.1 By 1935, Yoshida Hatsusaburō was a master at revealing the connections between exhibition, city, colony, and empire. The Taiwan Exhibition occupies the centre of Taipei, with Tokyo and Fuji on the centre right of the horizon, Manchuria to their left, and southern China on the other side of the Grass Mountain.
Source: Yoshida Hatsusaburō, *Shisei Yonjū Shūnen Kinen Taipei-shi Chōkanzu* (1935), National Museum of Taiwan History.

back in Japan, albeit at the former as part of a paean to regional electrification.[16]

Even the original, model colony, Taiwan, revealed the fissures that ran through the empire. The year 1930 had seen the Wushe uprising, which resulted in the deaths of 130 Japanese settlers and nearly 10 times that number of Seediq, supposedly a 'tamed' tribe. The colony's fortieth anniversary in 1935 provided an opportunity to show that order had been restored. Taiwan participated in five exhibitions in Japan that year, but the main event was the massive Taiwan Exhibition, spread over four sites in Taipei and complemented by ten smaller events outside the capital. The exhibition, promised the press, would do it all: 'introduce the region, promote industry, contribute to culture, and spread national defence thinking'. The usual creative suspects got to work to make it so. Yoshida produced a ravishing poster (Figure 5.1) and Nakagawa a much-lauded Sugar Building, a paean to 'the mother of Taiwanese culture', 'a delight for children, and a solace for the aged', which reappeared in an exhibition in Hyōgo the following year. The authorities drilled the colonial lessons home, demonstrating the progress made from 'savage management in the old days', through 'the daily life of the Takasago

[16] For Nagoya, Yamaji, *Kindai Nihon no Shokuminchi Hakurankai*, 148–50. For Toyama, ibid., 155–58 and Yamaji, *Chihō Toshi no Kakusei*, 234–44.

people', to 'savage management now'. At least one aboriginal visitor was convinced. His first impression of the exhibition, as simply a pile of stuff, was transformed, courtesy of the military hardware in the National Defence Building, to the conviction that Japan could not lose in a war. His initial impression, though, was echoed elsewhere.[17]

The scale of the ambition and the complexity of the colony meant that the exhibition sprawled across four sites. Even so, the first was crammed, as a cartoon in the leading newspaper suggested. The cramped ten-acre lot next to the city hall was devoted to imperial industry, but the welter of stimulus served to fragment any larger theme, with visitors streaming through the architectural jumble, individual buildings thrusting at various angles in an attempt to attract attention, and occasional voices caught in bubbles as they try to find their way. The second occupied the twenty acres of Taipei Park, which, the cartoon suggested, provided an ordered space for colonial administration, but was dominated by the Monopoly Bureau and Japanese cities – Osaka with a neon sign on a forty-seven-foot spire, Nagoya the familiar silhouette of its castle – leading to complaints from local merchants about unfair competition. It was the third site, though, in Dadaocheng, a Chinese neighbourhood, that veered most clearly away from the colonial script. The result of a petition from the Chinese community, it made space not only for imperial aspiration, with maps of rival empires in the Southern Building and military canines in another, but also for pacific partnership. There were special buildings for Thailand and Fujian, and a barrage of events and attractions, including performances by a 150-member Peking Opera troupe, a procession of the Bodhisattva Mazu through the streets, 'raw savage dances', and five treasure hunts. The crowds were captivated: less than 10,000 from Japan, but nearly 3 million from Taiwan itself, catalysing an insular tourist boom, some of which found its way to the fourth site, Grass Mountain, to the north of the city, which had already been established as a resort and would soon be designated as a National Park.[18] The empire remained

[17] *Taiwan o Daihyōsuru Mono* (Taiwan Shinbunsha, 1935), 1006–8, for the press promise. Masuda Keiichiro, *Utsukushiki Kyushu: 'Taisho Hiroshige' Yoshida Hatsusaburo no Sekai* (Kaichōsha, 2009), 92 and 102, for Yoshida. Nakagawa, *Rankaiya Ichidai*, 231–41, for the Sugar Building. *Taiwan Nichinichi Shinpō* 1935/10/17, 9, for the comments on stuff and defence.

[18] Soyama Takeshi, *Shokuminchi Taiwan to Kindai Tsūrizumu* (Seikyūsha, 2003), 223–32 and Yamaji, *Kindai Nihon no Shokuminchi Hakurankai*, 231–70, for overviews, including Dadaocheng, attractions, and tourism. *Taiwan Nichi Nichi Shinpō*, 1935/10/14, 4, for the cartoons. Ha Sae-bong, 'Mohyeongui Jeguk', *Tongyang Sahak Yŏn'gu* 78 (2002),

subject to continuous surveillance and occasional violence, but its development was being driven in large part by the demands of its subjects, which seemed to be biased to peace.

In Japan, too, the middle of the decade saw a series of attempts by cities to keep the pacific dream of empire alive, right up until the country stumbled into war in July 1937. The first half of that year saw the familiar gamut of exhibitions, comprising everything from ideological didacticism to sprawling distraction. The *Osaka Mainichi Shimbun* was responsible for two. In Tokyo, a dutiful 'Politics Exhibition' in the old Diet Building commemorated the completion of the new one and the seventieth anniversary of the Meiji revolution, with architectural plans and a slew of documents from around the world. In its hometown, meanwhile, 'Daimai Fairland' purported to commemorate the Anti-Comintern Pact, recently signed between Japan and Germany, with a central memorial hall, courtesy of Kenseisha, which included ten dioramas of the Hitler Youth. The main draw, however, could be found in the Hankyū railway company's new Nishinomiya Kitaguchi stadium, which serviced an extended line between Osaka and Kobe. Soon it would be occupied by a new baseball team, the Braves, but for now an American circus pulled the crowds. Further away from the big cities, the bid for attention took on distinctive local colour. On the south coast of Shikoku, a new rail line connecting Kōchi to the north coast and, via ferry, the main island prompted an exhibition promoting local industries, such as papermaking, and attracting tourists to 'spiritual Tosa', which was embodied on the poster by a loin-clothed male figure brandishing a thigh-sized tuna. On the east coast of Kyushu, Beppu staged a Great International Hot Springs and Tourism Exhibition, to advance its claim to be Japan's premier resort, with models of spas from around the world, dioramas of thirty representative Japanese *onsen*, a virtual tour of Kyushu's famous places, and a huge model of Beppu itself, as well as an open-air bath, open to visitors.[19] Exhibitions were a way to transform long-standing bathing culture into mass-marketed consumer good.

It was Nagoya, however, that owned the spring of 1937. The city, together with local business and media, had embraced the exhibition

165–99, for Japanese cities and unfair competition. McDonald, 'The Boundaries of the Interesting', 209–11, for the exhibition's treatment of Taiwan's indigenous peoples.

[19] Nakagawa, *Rankaiya Ichidai*, 253–61, for Daimai Fairland. Hashizume, *Nihon no Hakurankai*, 159 for Tosa. Yamaji, *Chihō Toshi no Kakusei*, 229–34, for Beppu. See also Hashizume, *Jinsei wa Hakurankai*, 165–90, for its development as a resort.

game for good in the 1920s, with electricity in 1923, hygiene in 1926, imperial enthronement in 1928, national products in 1929, colonies in 1930, and military affairs in 1932. But 1937 was different. The thirtieth anniversary of the opening of the port was an opportunity to celebrate its coming of age, with a seven-figure population and the infrastructure to match. The Pan-Pacific Peace Exhibition marked the occasion on a new 500,000-square-metre site on the waterfront. Yes, the world was changing, and the future was murky. In the official account, the great and good gave their various ponderous takes. A retired general, Matsui Iwane, insisted that it would be natural for Japan to advance into South China and Southeast Asia, 'build[ing] a new culture … by fusing cultures east and west'. The governor of Aichi, however, sermonized about international friendship and the mayor of Nagoya about its openness to the world. The exhibition would realize 'amity and friendship' among 'interested nations'. The organizers produced a slew of foreign language publications, while radio programmes rehearsed potential conversations with English speakers. The foreign exhibits are 'over that fountain. Shall I go with you?' Manchukuo and a couple of other Chinese puppets; Thailand, Indonesia, Australia, South Africa, and a scattering of Central and South American states; and a potpourri of subnational participants, including the Rangoon, Mexican, and Los Angeles chambers of commerce, the city of San Francisco, and the University of British Columbia, gave some substance to the claims, which were embodied in a forty-five-metre-high Tower of Peace directly behind the main gate. The combination was enough to attract nearly 5 million visitors by the time the exhibition closed, though only 6,625 of these were from outside the empire.[20] However implausible in retrospect, even in the first half of 1937 the possibility of peace was a fervent belief, born of municipal pride and

[20] Hashizume, *Nihon no Hakurankai*, 113, for 1928. Ikawa Mitsuo, 'Manshū Jihen Zengo no Nagoya Shinbun no Ibento', in Tsuganesawa Toshihiro and Ariyama Teruo (eds.), *Senjiki Nihon no Media Ibento* (Sekai Shisōsha, 1998), 113–27, for 1930. For 1937, Nishio Rintarō, 'Kokusai Hakurankai Toshite no Nagoya Han-Taiheiyō Heiwa Hakurankai: Sono Hikari to Kage', *Hibarino* 23 (2001), 49–60; Yamaji, *Chihō Toshi no Kakusei*, 249–307, and Nathan Hopson, '"A Bad Peace?': The 1937 Nagoya Pan-Pacific Peace Exhibition', *Japanese Studies* 38.2 (2018), 137–51. *Shin Aichi Shinbun* 1937/1/1, 10–11, for waffling by the great and good. Torsten Weber, 'The Greater Asia Association and Matsui Iwane, 1933', in Sven Saaler and Christopher Szpilman (eds.), *Pan-Asianism: A Documentary History. Volume 2: 1920–Present* (Rowman & Littlefield, 2011), 137–48, for the retired general. *Nagoya Han Taiheiyō Heiwa Hakurankai Kaishi* (Han Taiheiyo Heiwa Hakurankai, 1938), 16 for amity and friendship, 359 and 381 for publicity. Hashizume, *Nihon no Hakurankai*, 154–58, for foreign visitors and images.

commercial interest, that development might continue and concord could reign.

GASLIGHTING THE WORLD

One week before the exhibition in Nagoya closed, the Japanese government would make the same argument in Paris. It had been trying to do so for a while. Japan had been a regular feature as international exhibitions had picked back up in the 1920s. Traditional architecture framed emerging industry in Rio in 1922 (accompanied by a Japanese naval squadron); in Paris in 1925 (counterpointing the deco elsewhere on site); and in Philadelphia in 1926 (providing both the largest exhibit in the Foreign Building and a gift to the city of 1,600 flowering trees). The creation of the Bureau Internationale des Expositions (BIE) in 1928 would eventually bring some order to the proliferation of events, but it took a while to impose its mark.[21] In Belgium in 1930, Japan produced a scientific exhibit in Liège and a smaller colonial one in Antwerp. It was a no-show in Paris the following year, having had to decline on fiscal grounds.[22] The invasion of Manchuria in September not only made the decision a sensible one but also made for more work at international exhibitions, which were becoming adept at veneering the old violence of industry and empire with the new future perfect promised by science.[23]

[21] For the BIE's own account of its origins and importance, see Bureau International des Expositions, *Commémoration Du 75ème Anniversaire de La Convention de Paris* (BIE, 2004). For a history, Iwata Yasushi, 'Kokusai Hakurankai no Rekishi ni Hakurankai Kokusai Jimukyoku (BIE) ga Hatashita Yakuwari', in Sano Mayuko (ed.), *Banpakugaku: Bankoku Hakurankai toiu, Sekai o Haakusuru Hōhō* (Shibunkaku shuppan, 2020), 131–46.

[22] Lazzaro Rezende Livia, 'The Raw and the Manufactured: Brazilian Modernity and National Identity as Projected in International Exhibitions (1862–1922)' (PhD thesis, Royal College of Art, 2010), 255 and 287–90, for Rio 1922; Thomas H. Keels, *Sesqui! Greed, Graft, and the Forgotten World's Fair of 1926* (Temple University Press, 2017), 105 and 146 for Philadelphia 1926; and Itō Mamiko, '1931-nen Pari Kokusai Shokuminchi Hakurankai', in Fukui Norihiko, Itō Mamiko, and Muramatsu Kōichi (eds.), *Sekai no Shūshū: Ajia o Meguru Hakubutsukan Hakurankai Kaigai Ryokō* (Yamakawa Shuppansha, 2014), 245–48 for Paris 1931.

[23] For interwar exhibitions as imperial veneer, Robert Kargon, Karen Fiss, Morris Low, and Arthur Molella (eds.), *World's Fairs on the Eve of War: Science, Technology, & Modernity, 1937–1942* (University of Pittsburgh Press, 2015). Rydell, *World of Fairs*, for the US, Daniel Stephen, *The Empire of Progress: West Africans, Indians, and Britons at the British Empire Exhibition, 1924–25* (Palgrave Macmillan, 2013), for the UK, and James D. Herbert, *Paris 1937: Worlds on Exhibition* (Cornell University Press, 1998) for France.

Similarly, they were also one of the few places where Japan could try to convince the world that its aggression on the continent was in fact a contribution to peace, however threadbare the case.

The problem was already evident in Chicago in 1933. Tucked away on the western edge of the Century of Progress Exhibition, the main Japanese building reassured its American observers that long-standing tropes about the country remained intact. The throwback building was 'a typical example of her architecture'. The exhibits included 'countless examples of world-famous Japanese handicraft', a demonstration of the process of making silk 'from cocoon to finished article', and the 'charming ceremony of tea drinking ... enhanced by dainty Japanese ladies'. But the reassurance was undermined by a building one-tenth its size, just to the side. The local Chinese residents' association had protested, the Foreign Ministry had raised concerns, and the American organizers had got involved, but iconography, exhibits, and publicity professed the claims of Manchukuo as a sovereign state. In official publications, the building was overshadowed by a replica of a golden lama temple from Jehol Province, north of the Great Wall, less than 100 miles from Beijing. On the ground, however, the province was being overrun by the Kwantung Army.[24] Their operations would taper off towards the middle of the decade, but the problem – how to suture the traditional story about a pacific Japan with the mounting evidence of an aggressive power – would persist.

By 1937, in Paris, a simple solution seemed to be at hand: ignore geopolitics, modernize the architecture, and focus the exhibits on tradition. Sakakura Junzō had learned his lessons from Le Corbusier. His impeccably international, modernist design for the Japanese pavilion eschewed the grand statements elsewhere on site, not least the face-off between Nazi Germany and Soviet Russia, the former topped with an eagle and swastika, the latter with a male worker and a female peasant holding hands. For foreign observers Sakakura's restrained design evinced his 'double artistic culture', which represented what was 'most

[24] Rydell, *World of Fairs*, 82–84 and 93–106 for an overview. *Progress* 4.10 (March 1933), 33, for the arrival of the Japanese exhibit on site, and Century of Progress International Exposition, *Official Guide Book of the Fair* (A Century of Progress, 1933), 95–96, for a description. Itō Mamiko, 'Hakurankai to Shokuminchi: Shokuminchi Teikoku Nihon no Kenjiyoku', in Ugai Masashi and Kawaguchi Akihiro (eds.), *Kinō no Nihon: Kindai Shakai to Bōkyaku Sareta Mirai* (Yūshisha, 2012), 64–74, and Kari Shepherdson-Scott, 'Conflicting Politics and Contesting Borders: Exhibiting (Japanese) Manchuria at the Chicago World's Fair, 1933–34', *The Journal of Asian Studies* 74.3 (2015), 539–64, for the Manchuria exhibit and the Jehol temple.

modern in present-day Japan, while respecting carefully the ancestral influences which, in this country more than any other, inspire and direct national life'. The exhibit did more of the same. After perusing tea, textiles, and handicrafts on the ground floor, the visitors made their way up an exterior ramp to a 'hall of science'. Any threat the latter might suggest was mitigated by the rugs, ceramics, and dolls that served as 'specimens of Japanese industry', backed by a photomural of Mount Fuji and kimono-wearing women.[25]

In Tokyo, however, things were a little trickier. Plans were advancing for Japan's first international exhibition, to be held in 1940, in conjunction with the Olympics, to commemorate the putative 2,600th founding of the Japanese nation. There had been some quarrels along the way. The Home Ministry and its allies were determined that the anniversary should be observed with requisite solemnity, and so turned to domestic, didactic ends. The city, however, wanted to stimulate economic demand, meaning something more international, with a lottery to secure the funding and entertainment to attract the crowds. Eventually, a compromise was reached. The exhibition would comprise two sites, on adjacent artificial islands in Tokyo Bay. The first, in 'sublime' Japanese architectural style, would place 'spirit and culture' on one side, industry and nature on the other of a Memorial Hall, modelled on a shrine. The second, in a 'free, modern' style, would accommodate agriculture, chemical and manufacturing industries, foreign participants, entertainment, and parking. The composite, somehow, would promote 'the fusion of cultures east and west, the development of global industry, and the advance of international peace', showing how an unbroken royal line had spawned Japanese modernization and so imperial international harmony – 'eight corners of the world under one roof'.[26]

[25] *Livre d'Or Officiel de l'Exposition Internationale des Arts et des Techniques Dans la Vie Moderne, Paris, 1937* (Éditions Spec, 1937), 465, and Kargon et al., *World's Fairs on the Eve of War*, 25–27 for the pavilion and exhibits. Inoue, *Senjika Nihon no Kenchikuka*, 208–21 and Fujiki Takao and Toyoda Kentarō, '1937-Nen Pari Bankoku Hakurankai Nihonkan no Sekkei Ikisatsu ni Tsuite', *Nihon Kenchiku Gakkai Keikakukei Ronbunshū* 514 (1998), 217–23, for the architecture.

[26] Angus Lockyer, 'Expo Fascism? Ideology, Representation, Economy', in Alan Tansman (ed.), *The Culture of Japanese Fascism* (Duke University Press, 2009), 276–95, for a detailed account. *Kigen 2600-nen Kinen Nihon Bankoku Hakurankai Gaiyō* (Kigen 2600-nen Kinen Nihon Bankoku Hakurankai Jimukyoku, 1938), 18–19, for the site, 1 for cultural fusion, and 4 for eight corners. *Banpaku* 1 (1936/5), 4–5 for unbroken line, and 13 (1937/5), 6–7 for economic effects. For the planning, see also Furukawa, *Kōki, Banpaku, Orinpikku*, 62–84, and Mashiyama Kazushige, 'Maboroshi no Hakuran Toshi

In July 1937, however, war broke out. It took a while for the government to admit as much and for plans to change. In Paris, the official in charge of the Japanese exhibit continued to insist that Japan's 'greatest ambition is to work in close collaboration with the Occident'.[27] In Tokyo, the planning for 1940 continued. In August, the Diet agreed to provide some financing. In March 1938, the association announced that a high school teacher in Chiba had won the competition to write the lyrics of the official expo song.[28] But on 15 July 1938 the cabinet announced that both expo and Olympics were postponed indefinitely. At home, material and cultural resources were to be devoted to the war in China, even though the government continued to insist it was nothing more than an 'incident'. The disavowal of the proper name was spineless, but is useful, pointing to the contradictions being exposed in the body politic and its international standing. By the same token, though, it was imperative to insist to that world that nothing had changed. International exhibitions remained the last best hope of pretending that war could be passed off as peace.

The New York and San Francisco fairs, which opened in early 1939 and finally closed in autumn 1940, provide an uncanny reflection of a Pacific world falling apart.[29] The Japanese authorities did not hide industrial achievement or imperial ambition. In San Francisco, they placed a statue of three girls in front of a map of the Pacific, 'representing Japan, Korea, and Manchukuo' and symbolizing 'cooperation and peace between the three Asiatic races'. In New York, they persuaded the Commissioner to the Far East on a trip to Manchuria that it was 'the greatest story of progress in the world', though he failed to convince his colleagues, who insisted on a 'purely railway exhibit'. Still, the Hall of Nations included not only a 27-foot-high, 108-foot-wide photomural of Mount Fuji but also smaller ones charting Japan's progress in everything from aeronautic engineering to education and fishery, and a massive

Keikaku: Tōkyō Tsukishima, Nihon Bankoku Hakurankai', in Sano Mayuko (ed.), *Bankoku Hakurankai to Ningen no Rekishi* (Shibunkaku Shuppan, 2015), 267–95, and 'Kigen 2600-nen Kinen Nihon Banpaku no Keikaku to Sono Shūhen: 1930 Nendai no Kokusai Hakurankai Nihon Tenji o Meguru Renzokusei', in Sano Mayuko (ed.), *Banpakugaku: Bankoku Hakurankai toiu, Sekai o Haakusuru Hōhō* (Shibunkaku shuppan, 2020), 160–90. For an account focused on technology, Kargon et al., *World's Fairs on the Eve of War*, 83–107. Sandra Collins, *The 1940 Tokyo Games: The Missing Olympics: Japan, the Asian Olympics and the Olympic Movement* (Routledge, 2007), for the Olympics.

27 Kargon et al., *World's Fairs on the Eve of War*, 26.

28 Furukawa, *Kōki, Banpaku, Orinpikku*, 122, and *Banpaku* 22 (1938/3), 3.

29 Rydell, *World of Fairs*, 84–91 and 106–35 for overviews.

bronze statue of a Japanese workman pouring molten steel. Given the circumstances, though, the main task was to counter the growing American opposition to Japanese aggression in Asia.

On both coasts, therefore, the story was one of innocuous modernity and unchanging tradition, exemplified by the work of women and the production of silk. In San Francisco, this meant a mash-up of a 'feudal castle', a samurai residence, and a Buddhist temple, set in a landscape garden, with lake. Inside, two of the 200,000 young women employed in the industry showed how silk was produced, while a diorama of a boudoir in a modern Japanese home suggested how to use it for clothing, furnishing, and decoration. In New York, the Japanese pavilion was modelled on the Ise Shrine, 'the purest expression of Japanese architecture', surrounded by a Heian-period garden, and included a Grand Hall with silk-covered walls, 'authentic temple lanterns', and the largest lacquer screen ever made; a Diplomatic Room with the telegraph key presented by Perry in 1854, a photomural of the 'happy' history of Japanese–American relations, and a silver-and-pearl replica of the Liberty Bell; and a Silk Room with daily demonstrations in a 'miniature factory in which eternities of silken strands are spun from cocoons'. The Japanese commissioners suggested that American newspapers might want to feature the pavilion on their 'woman's page', given the 'particular appeal' of its 'pretty tea ceremonies and artistic flower arrangements' to female visitors.[30]

[30] For San Francisco, *Golden Gate International Exposition* (Japanese Committee for the Golden Gate International Exposition, 1939), 22–24, for the pavilion, and 27, for the statue. For the New York exhibits, *The Japanese Pavilion, New York World's Fair, 1939* (Japanese Commission to the New York World's Fair, 1939), 5–7, 10–11, 29, and 41–52. For the commissioner, Herbert S. Houston, 'Seeking a Way to Peace in Asia: An Address Delivered before the University of Oregon, at Eugene, September 23, 1937', Box 17, and 'Suggested Plan for the Participation of the South Manchuria Railway Company in the New York World's Fair of 1939', Box 534; for his colleagues, W. H. Standley, 'Letter to Kaname Wagasuka,' 1938/6/18, Box 534; and for the suggestion, '"Japanese Pavilion Oasis in Midst of World's Fair", Release #209', Box 101, New York World's Fair 1939 and 1940 Incorporated Records, Manuscripts and Archives Division, New York Public Library (hereafter NYWF). For the photomurals, Helena Čapková, 'Transnational Networkers: Iwao and Michiko Yamawaki and the Formation of Japanese Modernist Design', *Journal of Design History* 27.4 (2014), 370–85; and Shirayama Mari, 'Senji Senden to Shashin Hekiga: Yamahata Shōgyoku to 1939-nen Nyū Yōku Bankoku Hakurankai no "Yakushin Nihon" o Chūshin ni', in Sano Mayuko (ed.), *Banpakugaku: Bankoku Hakurankai toiu, Sekai o Haakusuru Hōhō* (Shibunkaku shuppan, 2020), 191–207. For American antipathy, Warren Cohen, *America's Response to China: A History of Sino-American Relations* (Columbia University Press, 2020), 132–35.

The bad faith was made incarnate in 1939 by one Tsukimoto Akiko, 'Miss Japan', who carried across the Pacific a 'Flame of Friendship', which was lit from the '1500-year-old sacred fires of Izumo Shrine', consecrated in Tokyo, and reconsecrated in San Francisco and Los Angeles, before finally arriving in New York. 'Japan Day' comprised her presentation of the flame to the American president of the fair; a procession to and the enshrinement of the flame in the pavilion; and a command performance of the 'Grand Cherry Show' by the forty 'performers (all girls)' of the 'Takarazuka Theatrical Chain Corporation'. One year later, three years into the Sino-Japanese War, another 'Japan Day' was enough to convince one New York congressman that there could be no objections to a Japanese 'Monroe Doctrine for the Orient'. The US alone possessed colonial prerogatives in the Americas. So Japan might exclude Europe from Asia. Three months later, however, in late September 1940, in the last week of the San Francisco fair and a month before the closing of the New York one, a five-day skirmish broke out between French and Japanese troops in northern French Indochina. The day after the fighting died down, Japan entered a tripartite pact with Germany and Italy, forming the Axis alliance that would persist for the next five years. Events in the real world revealed the insubstantial pageant at the exhibition for the flimsy thing it was.[31]

MOBILIZING FOR WAR

Back in Japan, there had been little doubt that the empire was at war, whatever it was called. The invasion of China in July 1937 had produced a clean break with the variety of exhibitions that had been evident through the first half of the year. There were a couple of small exhibitions devoted to the war later in the year, but 1938 saw a surge of interest, with ten exhibitions referring to the China 'Incident' by name, a further ten devoted to the war or the military, and only six choosing to focus elsewhere. The tone was set early, in March, by the Great National

[31] For the flame and Japan Day, *The Japanese Pavilion*, 8; 'Japan Day Schedule, June 2, 1939', Box 230, 'Contract between Takarazuka Theatrical Chain Corporation and New York World's Fair, 1939, Inc., May 1939', and Leo Casey, 'News Release No. 1475', both Box 534, NYWF. For Indochina, Ikuhiko Hata, 'The Army's Move into Northern Indochina', in James Morley (ed.), *The Fateful Choice: Japan's Advance into Southeast Asia, 1939–1941* (Columbia University Press, 1980), 155–63, and John Dreifort, 'Japan's Advance into Indochina, 1940: The French Response', *Journal of Southeast Asian Studies* 13.2 (1982), 279–95.

Spiritual General Mobilization and National Defence Exhibition, which opened at Ueno one day after the passing of the National Mobilization Law. Hoshino Shaku, its eighty-five-year-old President, explained that its ambition was to 'perfect the nation's defence in such a way that military and people were as one' (*gunmin itchi no kokubō kansei*). Exhibitions were both a propaganda medium and a commercial opportunity, not least for newspaper companies. They therefore continued unabated for a while, with twenty-three in 1939 and twenty in 1940, rehearsing the 'incident', celebrating the 'holy war' it had precipitated, urging the 'spiritual' mobilization by which this was underpinned, or previewing the 'new Asia' which it would enable. There was a drop-off thereafter, but only in 1945 and 1946 were there no exhibitions in Japan.[32]

Wartime exhibitions did a number of different things.[33] First, they brought the war home, with exhibition companies crafting immersive environments so that visitors might commune with the greater good. The 'Incident Building' at Hoshino's exhibition was one example, nearly three times the size of the next largest structure and stuffed to the gills with some thirty panoramas and dioramas, reproducing the events of the war. It paled, however, next to the China Incident and Holy War Exhibition, which opened in Nishinomiya a week later. The Osaka Asahi paid for and Kenseisha effected the wholesale transformation of the baseball stadium into a 20,000-square-metre panorama (Figure 5.2). A life-size replica of Beijing's Zhengyangmen gate was placed at the entrance, the bleachers were turned into the North China mountains, and the pitch provided a stage for the recreation of the Battle of Shanghai, the first major engagement following the invasion of China proper. Outside the stadium there was a full-scale field camp, with life-size pillboxes and underground bunkers, as well as bumper-cars-turned-tanks, parachute drops, and miniature fighters for the children. The exhibition was a triumph, and the same crew showed up at the same site at the same time the following year, 1939, with the Building Greater East Asia Exhibition. This time, the stadium staged the Battle of Wuhan, which

32 Fukuma Yoshiaki, 'Senji Hakurankai to "Seisen' no Hokorobi', in Fukuma Yoshiaki, Nanba Kōji, and Tanimoto Naho (eds.), *Hakuran no Seiki: Shōhi Nashonariti Mejia* (Azusashuppansha, 2009), 278–79 for Hoshino and 294–300 for the media bonanza. Yamaji, *Chihō Toshi no Kakusei*, 309–11, for 1937 as turning point. See also Max Ward, 'Displaying the Worldview of Japanese Fascism: The Tokyo Thought War Exhibition of 1938', *Critical Asian Studies* 47.3 (2015), 414–39, for the ideology of early wartime exhibitions.

33 Following Fukuma, 'Senji Hakurankai'.

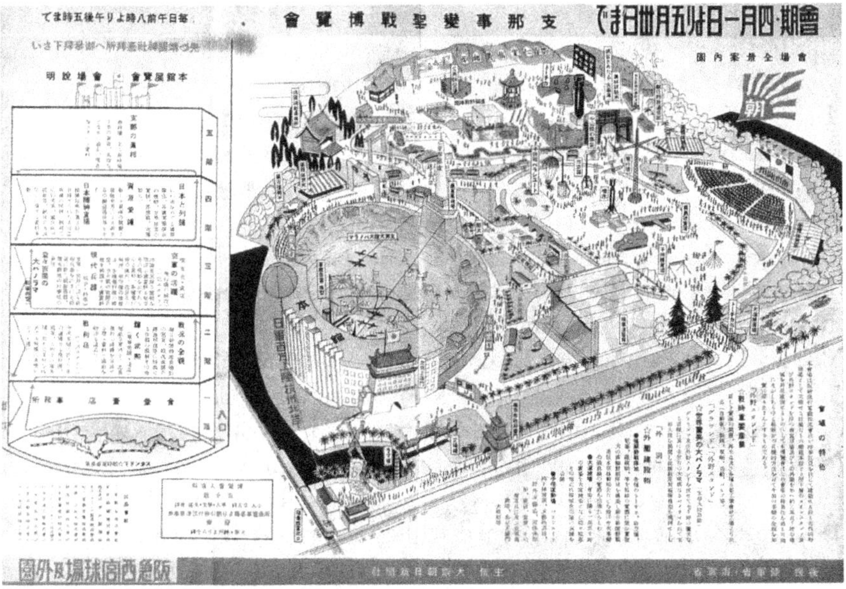

FIGURE 5.2 The year 1938 saw the transformation of a baseball stadium into a stage for recreating, teaching, and encouraging war.
Source: Hankyū Railway leaflet, NOMURA Co., Ltd.

had sprawled over much of the interior of southern China the previous summer. Tanks trundled over barbed wire, the infantry followed on behind, 142 planes were suspended over bombed-out villages, and the scoreboard displayed the lyrics to 'Father, you were strong', the winner in an Asahi competition for a song of gratitude to the imperial forces. A 'New East Asia' tour included some twenty sites, from Hsinking, the new Manchurian capital, 'proud of its modern beauty', through Shanhai Pass and Badaling on the Great Wall, to Beijing, 'the old capital with a new destiny'. Reminiscing, the head of Kenseisha noted that this had been the high point of his career. The war provided an opportunity to perfect the expo dude's craft.[34]

It was not enough, however, to immerse, to entertain and, perhaps, to compel. Second, exhibitions provided space for commemoration. This too was familiar territory. Mementoes of the fallen – from photographs,

[34] Yamaji, *Chihō Toshi no Kakusei*, 311–20, for immersive environments. Fukuma, 'Senji Hakurankai', 281–83, for Ueno and Osaka 1938; Nakagawa, *Rankaiya Ichidai*, 262–69, Hashizume, *Jinsei wa Hakurankai*, 136–46, and *Nihon no Hakurankai*, 162–66 for Osaka 1938 and 1939. Sharalyn Orbaugh, *Japanese Fiction of the Allied Occupation: Vision, Embodiment, Identity* (Brill, 2007), 236, for the lyrics to the song.

swords, uniforms, and medals, to letters, wills, and last words – had been on display for well over fifty years. As the war in China bogged down, however, honouring the fallen became routine. The 1938 exhibition in Ueno included 188 items from twenty-eight bereaved families. The one in Nishinomiya put the effects of well over 100 deceased officers and men on display in a room of 'radiant military deeds' (*kagayaku bukun shitsu*). The panoramas provided a way to submerge the self in the collective and the cause, but the mementoes allowed the viewer to connect to the fact of individual sacrifice. A visitor from Toyama came to Nishinomiya seeking to mourn his deceased younger brother, found him in a photograph on display at the exhibitions, and burst into tears. A widow and her children in Ueno wept when they saw the Ki-93 fighter, riddled with bullet holes, in which their husband and father had died. Private suffering was the obverse of public service, but both could be mobilized for the greater good and turned to profit. The widow had been consoled by Prime Minister Konoe, who happened to be passing by, allowing the Tokyo Asahi to report a 'pitiful play without words' (*kanashiki mugengeki*).[35]

From the beginning, though, the authorities knew that they also had to teach the public how war was to be waged. Third, therefore, exhibitions had to instruct. In 1938, Hoshino was explicit about the need for the nation to grasp the revolutionary logic of total war, requiring not only economic regulation but also civilian defence against air raids, fire, and gas. As the war expanded, the lessons became ever more explicit. The Wartime Industrial Mobilization Exhibition in 1940 attracted 1 million to Ueno with a relentless focus on national defence. The Great National Defence Science Exhibition in Nishinomiya in 1941 included not only battlefield recreations and military hardware, as before, but also detailed instructions for the home front. There was the ongoing need to manage facilities, materials, and processes effectively. No longer was it enough, however, to conserve resources. Now was the time to find substitutes, which would soon become necessities: sardine oil as industrial lubricant, for example. Three years into the quagmire in China, it was also time to acknowledge the human cost. A building for the wounded included exhibits of artificial limbs, training for the newly blind, and more. This was not welfare. 'Having received an honourable wound, a loyal, brave, imperial soldier' should expect to serve again as 'an industrial warrior'.[36]

[35] Fukuma, 'Senji Hakurankai', 283–88 and 300–2.

[36] Fukuma, 'Senji Hakurankai', 279, for Hoshino, and 288–94, for Nishinomiya 1941. See also Hashizume, *Nihon no Hakurankai*, 172–73.

Beneath the paeans to unity, bravery, and sacrifice, therefore, wartime exhibitions were pushed by their organizers and often pulled their visitors in different ways. The tension was evident in 1940, with exhibitions throughout the empire revealing both the willingness to invest in the imperial anniversary and the impossibility of realizing its promises. Consumerism was still alive and well. In Tokyo, six retail companies joined forces to mount simultaneous exhibitions in seven department stores, pulling in nearly 5 million with a kaleidoscope of ancient history, cutting-edge technology, and imperial achievement. In the colonies, while tourism was at an all-time high, the need for imperial loyalty was acute and the contradictions of wartime empire on full display. At the Great Keijō Exhibition, on the outskirts of Seoul, the Keijō Nippō and its collaborators tried to convince the Korean public that they belonged in the story. A half-size version of a tower recently erected in Miyazaki purported to show how the eight corners of the world were being gathered under one roof. Around it, the Imperial History Hall used dioramas to rewrite the Korean past as Japanese mythistory; the Colonial Rule Commemoration Hall ignored everything before 1905 and showed the underwater tunnel that would link Pusan and Shimonoseki in 1970; and the Heroic Deeds Hall displayed the belongings of the fallen. Elsewhere, visitors, many of them school children dragooned on class trips, could view military hardware and taste military rations. What they could not do, however, was see themselves in the exhibits. A pilot conscription project had started in 1938, but 'only twenty-two ... of [the] 12,600 peninsulars who had [so far] lost their lives in the war were of Korean descent'.[37]

More than most, the exhibition was a thing of make-believe, though the future it limned was not quite a thing of fantasy. By the end of the war, the number of fallen Korean soldiers would skyrocket, by a factor of a thousand. An underwater tunnel was under construction, albeit between Honshu and Kyushu, and would be celebrated at an exhibition dedicated to the 'sacred task of energizing Asia' the following year. As the war went on, however, the dream of building – tunnels, empires, worlds – would be revealed to be the flimsy thing it was. The late summer of 1942 saw huge

[37] Kenneth J. Ruoff, *Imperial Japan at Its Zenith: The Wartime Celebration of the Empire's 2,600th Anniversary* (Cornell University Press, 2010), 74–79 for department store exhibitions, 84 for tourism, and 37–41 and 86–97 for Miyazaki. Todd Henry 'Celebrating Empire, Fighting War: The 1940 Exposition in Late Colonial Korea', *Asea Yŏn'gu* 51.4 (2008), 72–112, and *Assimilating Seoul*, 171–75 for the Great Keijō Exhibition.

crowds at two exhibitions in Manchuria, in Tianjin, the old treaty port, and Shinkyō, the new capital. By September, Nakagawa was on his way to China, trying to avoid American submarines, to set up yet another Greater East Asia exhibition in Nanjing, nominally on behalf of the administration of Wang Jingwei, but paid for by his Japanese puppet masters in the Central China Expeditionary Army. The city seemed normal, with candy in the shops and music in the streets; but there was gunfire in the hills, Japanese guards at the city gates, and barbed wire around the exhibition site. Nor did the military's money stretch far. The exhibition comprised a grand total of two pavilions, one devoted to the war and one to the co-prosperity in which it would result. A two-dimensional reproduction of the battleship Izumo blew over in the wind. Wang attended the opening ceremony, accompanied by bodyguards with automatic weapons. On the streets, however, the Chinese soldiers had fakes. There was no way of knowing if they might turn them on their nominal head of state.[38]

The year 1943 saw a last flurry of wartime exhibitions – as well as Nakagawa's last hurrah. The Decisive Battle Air Defence Exhibition back at Nishinomiya brought together a familiar public-private crew – the Central District Army, Hyōgo Prefecture, and the *Osaka Mainichi Shimbun* as sponsors, Kenseisha as producers. Nakagawa was able to produce some coloured diagrams but could not find any booze in what had once been a sake-drinking town. Where visitors had earlier marched through replicas of Chinese gates, now they found a five-metre-long air raid shelter at the entrance. From mobilizing for war, exhibitions now had to prepare people for surviving its effects. Nakagawa worked on a Greater East Asia exhibition in Harbin in July, and one in Daegu in October, to commemorate the beginning of military conscription on the peninsula. It was his last exhibition. The crossing back to Japan was fraught. There were no bento on the train from Hakata to Tokyo. Nakagawa spent the rest of the war with his wife's family in Niigata. By 1945, he was blind from drinking bad sake.[39]

[38] Yamaji, *Kindai Nihon no Shokuminchi Hakurankai*, 193–94 for Hsingking. Nakagawa, *Rankaiya Ichidai*, 279–82, and Yamada Atsushi, 'Ou Seiei Nanjing Seifu Shita no Dai Tōa Sensō Hakurankai', in Shibata Tetsuo and Yamada Atsushi (eds.), *Chūgoku to Hakurankai: Chūgoku Nisenjūnen Shanhai Bankoku Hakurankai ni Itaru Michi* (Seibundo, 2014), 95–112, for Nanjing.

[39] Nakagawa, *Rankaiya Ichidai*, 283–87. Brandon Palmer, 'Imperial Japan's Preparations to Conscript Koreans as Soldiers, 1942–1945', *Korean Studies* 31.1 (2007), 63–78, for conscription in Korea.

RECOVERING IN PEACE

Two long years later, however, exhibitions were starting back up, and Nakagawa's old cronies and competitors piling back into the game. Japan's surrender on 15 August 1945 had been followed within two weeks by the arrival of its occupiers, for the most part American. The Supreme Commander for the Allied Powers (SCAP), nominally Douglas MacArthur, soon became shorthand for the occupying authorities, who quickly got to work rooting out militarism and promoting democracy. For the Japanese people, recovery would be slower and harder, as cities tried to rebuild, communities to accommodate a flood of refugees from the continent, and families to find enough to eat, often in the absence of the traditional male bread-winner. By 1947, there were a few signs of a new normal, with a constitution, labour standards, and adult education programmes. There were also some small shoots of indigenous recovery, ranging from the forty-year-old Honda Sōichirō motorizing bicycles to the country's first Western-style beauty pageant. Behind the scenes, though, SCAP policy was shifting to a focus on political stability, economic recovery, and military build-up, in service of the developing Cold War, thereby creating an environment within which exhibitions could again thrive.[40]

The first post-war exhibition saw Nomura taking over the basement of the Takashimaya department store in Osaka in April 1947. 'New Japanese Tourism', it suggested, was one key for economic recovery, with a tourist 'album' comprising Nara in the autumn, hot springs on the Japan Sea coast in the winter, apprentice geisha in Kyoto in the spring, and a three-dimensional map of the recently designated Ise-Shima National Park, to represent summer. One year later, seeking to exploit its position as the gateway to the new park, the town of Uji Yamada staged its own Tourism and Peace Exhibition. Back in 1930, the local authorities had piggybacked on the periodic rebuilding of Ise Shrine with an exhibition foregrounding their place as the epicentre of state Shinto and the Japanese empire. An Imperial Treasure Building had showcased the objects used in the enthronement ceremonies. A History Building had

[40] John Dower, *Embracing Defeat: Japan in the Wake of World War II* (Norton, 1999), 48–53 for repatriation, 89–96 for the food crisis, 151 for Miss Ginza, and 533–34 for Honda. Andrew Barshay, *The Gods Left First: Imperial Collapse and the Repatriation of Japanese from Northeast Asia 1945–56* (University of California Press, 2013), for repatriation.

recreated the age of the gods in diorama. Now, in 1948, the disestablishment of Shinto meant that they had to pivot, rebranding the region from a linchpin of imperial ideology to an icon of pacific nature. The exhibits included not only the usual suspects – trade and industry, machinery and transportation – but also new pretenders. Rather than imperial treasures, there was the US, with the shrine's history museum remodelled into an American Cultural Centre. Rather than colonial fantasy, there was a competition to design ideal reconstruction housing. And rather than local geisha performing traditional dances, to bring the myths to life, there was a mass matchmaking event, to restore some demographic vigour. The combination was a winning one, pulling in well over 500,000 visitors and well over 3 million yen in profit. The town continued to bolster its tourist appeal once the Americans left, with two more exhibitions in 1954 and 1958, which resituated the shrine as the focal point of the local political economy.[41]

Tourism was not the sole focus of early post-war exhibitions, but the story of Ise suggests the general pattern. As the Occupation dragged on, there was an increasing flurry of activity by local authorities, with four exhibitions in 1947 and seven in 1948, before they headed into double figures. At first glance, their branding suggests that everything must have changed. The pre-war and wartime staples – imperial history and colonial empire; national defence and holy war – were nowhere to be seen. Peace itself headlined many of the shows. In 1949, Nagano complemented the ceremonial, septennial unveiling of the usually concealed replica of the permanently hidden main icon of Zenkōji with a thirteen-metre-tall Peace Tower. Three years later, Yamanashi would combine the coming into force of the San Francisco peace treaty with the 700th anniversary of Nichiren's advocacy of the Lotus Sutra as the path to peace and prosperity. To bring back the crowds, though, an exhibition needed novelty acts. None were bigger than Japan's new overlords. The year 1949 also saw two exhibitions, in Yokohama and Takamatsu, which foregrounded American examples of 'life and technology in the new

[41] For an overview, Hamada Yō, 'Shūkyō to Hakuran no Kindaishi', in Fukuma Yoshiaki, Nanba Kōji, and Tanimoto Naho (eds.), *Hakuran no Seiki: Shōhi Nashonariti Mejia* (Azusa Shuppansha, 2009), 135–38. For 1930, Nakagawa, *Rankaiya Ichidai*, 186–91. For 1948, Hashizume, *Jinsei wa Hakurankai*, 249–78. For the municipal context, Ise-shi, *Ise Shishi* (Yamato Gakugei Tosho, 1982), 630–35. For the shrine's post-war rebranding, Jonathan Reynolds, 'Ise Shrine and a Modernist Construction of Japanese Tradition', *The Art Bulletin* 83.2 (2001), 316–41, and Tze M. Loo, 'Escaping Its Past: Recasting the Grand Shrine of Ise', *Inter-Asia Cultural Studies* 11.3 (2010), 375–92.

age', as well as Japan's first 'state fair', in Nara. The acme of Americophilia would follow a year later in Kobe.

The America Fair of 1950, supported by the Occupation authorities and four Japanese ministries, made it plain that American history, economy, and society was the model for the imminent Japanese future. At the opening ceremony, the head of SCAP's Civil Information and Education Section (CIE) held forth on freedom as a spiritual endowment. At the centre of the site, the Politics and History Pavilion, a plaster replica of the White House, showed how freedom had spread west, from the Mayflower to the surrender ceremonies on the USS Missouri on 2 September 1945, which had formally ended the war. Elsewhere, America was unveiled as power, possibility, and attraction. The power was evident in the latest fighter aircraft, including an F51 Mustang; the possibility in the Blondie-esque lifestyle on display in six model houses, with porches, processed food, and rationalized kitchens; and the attraction in a four-acre panorama, 'American Journey', which took the visitor from west to east coasts and back again, some of it in a life-size mock-up of a Stratocruiser 'Flying Hotel' (Figure 5.3). The moral was driven home by a lottery, with prizes including two round-the-world trips on Pan-Am; three small American-style houses; and the official exhibition song, which eulogized the arrival from America, a century after Commodore Perry, not of a black ship bristling with guns, but a white one, 'full of things to rent'. The second verse underlined the point with a 'silver transport plane . . . flying through the sky'. In the wake of defeat, the standard craving for the new at exhibitions was morphing into a cargo cult. The vision of Japan at the centre of its own Asian empire, inaugurated by and mimicking the endeavours of the Euro-American powers, had been replaced by a mirror provided by the US. [42]

A closer look, however, suggests that much had stayed the same. The organizer remained the *Asahi* newspaper, eager to burnish its newly democratic bona fides. The producers were the same expo professionals, with Kenseisha providing the replica White House. And the site was the Nishinomiya baseball stadium and surrounding park, which had echoed with their recreations of the war in China only a few years before.

[42] For peace and religion, Hamada, 'Shūkyō to Hakuran', 139–43. For the American Fair, Tsuganesawa Toshihiro, 'Asahi Shinbunsha no "Amerika Hakurankai"', in Tsuganesawa Toshihiro (ed.), *Sengo Nihon no Media Ibento, 1945–1960-nen* (Sekai Shisōsha, 2002), 161–86; Hashizume, *Nihon no Hakurankai*, 189–91; and Low, *Visualizing Nuclear Power*, 23–27. Yoshimi, *Shinbei to Hanbei: Sengo Nihon no Seijiteki Muishiki* (Iwanami Shoten, 2007), for Japanese ambivalence towards America.

FIGURE 5.3 A massive panorama of famous places in the US, from the Golden Gate Bridge, through the Grand Canyon, to Manhattan, featured heavily in advertising for the America Fair.
Source: *Amerika Haku*, poster (1950), NOMURA Co., Ltd.

The larger exhibition landscape, too, demonstrated at least as much continuity as change. The other big fair in 1950, a few stops down the line, was the Kobe Fair or, more formally, the Japan Trade and Industry Exhibition, picking up the main concern of expos two decades earlier.[43]

The overriding concern during the early post-war period, as it had been in the early 1930s, was economic recovery. Before the war, the word itself had often been glossed in terms of forward motion – 'promoting' (*shinkō*) local industry, revealing a 'burgeoning' (*shinkō*) region or a 'leaping' (*yakushin*) Japan – but 'recovery' (*fukkō* and its cognates) had also appeared, for example, in Yokohama in 1935, and became a general concern after it, headlining the first large-scale event, the Great Recovery Exhibition in Osaka in 1948. The exigencies of time and place were evident. Model houses were displayed at the exhibition and became actual housing after the fact, to address the continuing shortage of places to live. In other places, like peace, recovery was paired with something else, through which economic activity might be revived: tourism in Matsue in 1948, Takamatsu in 1949, and Yonago in 1950; culture in Odawara in 1948, Okayama in 1949, and Nishinomiya in 1950. In the aggregate, though, economy and particularly industry were the focus of almost all early post-war exhibitions, with trade the most fervent hope. The first large-scale post-war exhibition in eastern Japan, in Yokohama in 1949, made it explicit. The exhibition nodded to the changed circumstances, with a Pan Am exhibit and a model room displaying American family life. In other respects, though, it was simply picking up the threads, connecting city to citizens and Japan to the world, as exhibitions had been doing just over a decade before. Foreign buildings, trade halls, and industry-specific pavilions demonstrated that Japan was again open for business. A scientific invention building, a lie detector demonstration, and an erotic underwater revue provided something for male children and adults alike.

Regional exhibitions thus remained tied to economic development and popular demand, as they always had been. But they took place in a country that remained under America's thumb, even when the occupiers had left. The USIA started up in August 1953, taking up globally where the CIE in Japan had left off, with the rebranding of atomic energy as a key concern. As a resource-poor archipelago, with huge energy needs,

[43] Tsuganesawa, 'Asahi Shinbunsha', for the Asahi, and Hashizume, *Jinsei wa Hakurankai*, 153–59 for Kaji. Hashizume, *Nihon no Hakurankai*, 187–88, and Low, *Visualizing Nuclear Power*, 27–31, for the Kobe Fair.

Japan had good reason to embrace the nuclear option, but the understandable public reluctance was reinforced in March 1954, when a hydrogen bomb test on Bikini Atoll showered a Japanese fishing boat, the Lucky Dragon, with radiation. Luckily for the Americans, they had Shōriki Matsutarō, the proprietor of the *Yomiuri* newspaper, on their side. Released in 1947 after serving twenty-one months as an accused war criminal, he used the paper, together with Nippon TV, to embrace atomic energy and rehabilitate his political fortunes. January 1954 saw a series of articles celebrating the 'capture of the sun', March a scoop of the Lucky Dragon incident, August an exhibit on 'nuclear power that anyone can understand', and the following May a series of public lectures. By then he was also in the Diet, soon to be Home Minister and chair of the newly formed Atomic Energy Commission. Six months later, the USIA's Atoms for Peace exhibition opened in Tokyo, the first stop on a triumphant twenty-two-month nationwide tour. Fifteen sections laid out the story of nuclear power, from pioneers (including Japan's first Nobel Prize winner), through basic physics, through generation, to application (from industry, through medicine, to food preservation), to education and research. The *Yomiuri* banged the drum, collaborating with local newspapers and even, nudged by the USIA, its arch-rival, the *Asahi*. More than 2.5 million visitors flocked to see the brave new nuclear world, with the millionth, conveniently, showing up at the Peace Memorial Museum in Hiroshima. The authorities were not only delighted but also unusually interested in what the visitors thought. Before and after surveys at six of the eleven stops on the tour concluded that 'the technological possibility of the peaceful use of nuclear power was sufficiently understood'.[44]

The Atoms for Peace exhibition pointed both forwards and back, although not quite in the way it suggested at the time. It continued the

[44] For 'Atoms for Peace' globally, Mara Drogan, 'The Nuclear Imperative: Atoms for Peace and the Development of US Policy on Exporting Nuclear Power, 1953–1955', *Diplomatic History* 40.5 (2016), 948–74. For Shōriki, Shun'ya Yoshimi, 'Radioactive Rain and the American Umbrella', *The Journal of Asian Studies* 71.2 (2012), 322–26. For Japan, Low, *Visualizing Nuclear Power*, 67–171. For the exhibition as media event, Yoshimi Shunya, 'Mō Hitotsu no Media Toshite no Hakurankai: Genshiryoku Heiwa Riyō Haku no Juyō', in Tsuchiya Yuka and Yoshimi Shunya (eds.), *Senryō suru Me, Senryō suru Koe: CIE/USIS Eiga to VOA Rajio* (Tokyo Daigaku Shuppankai, 2007), 291–316, and Ikawa Mitsuo, 'Genshiryoku Heiwa Riyō Hakurankai to Shinbunsha', in Tsuganesawa Toshihiro (ed.), *Sengo Nihon no Media Ibento, 1945–1960-nen* (Sekai Shisōsha, 2002), 247–65. For public opinion, Ikawa Mitsuo, 'Genshiryoku to Yōron Kenkyū, Saikō', *Masu Komyunikēshon Kenkyū* 84 (2014), 53–64.

long-standing use of exhibitions to foreground new technology – and of new technology to draw the crowds – conjuring a future which might dissolve the fears of the present. It continued to demonstrate the fluid interplay between commercial interest and political authority, domestic concerns and international ties, which had long produced a steady stream of exhibitions, in the major cities and beyond. It suggested, however, a turn away from the raucous clamour of most exhibitions towards something slightly more designed and didactic. It pointed therefore to the potential for exhibitions to become again, after more than half a century of neglect, an affair of state – not just overseas, as a way of projecting soft Japanese power, but also at home, as a way of dragooning the Japanese people into the cause of development.

6

Big Event

Osaka is a city that has been and continues to be shaped by expos, from the Fifth Domestic Industrial Exhibition in 1903, which turned an old temple district into an urban playground, to Expo 2025, currently being planned for a reclaimed island in the bay. The greatest of them all, though, was Expo 70. Today, its site is occupied by museums and greenery, with a few remaining expo fragments, including the iconic Tower of the Sun. Back in 1970, however, the place looked very different. The top of the tower poked through a massive roof, suspended in mid-air, which stretched from the main entrance over a huge 'Festival Plaza'. Together they comprised the 'theme zone', the heart of the expo, which pumped visitors through the arteries that would disperse them through the site. To the east sprawled evidence of the varying fortunes of first, second, and third worlds: the low-slung air dome wonder of the American pavilion, with a lump of rock recently returned from the moon; a smear of Japanese corporate pavilions to its north; a soaring Russian pavilion, kitty corner from the Americans, housing Sputnik and the like; and the smaller fry – unaligned and underdeveloped – in subsidized, shared housing on lower ground towards the centre of the site. The Japanese government pavilion was on the other side of the site, safely recessed behind more corporate buildings, preserving the self-regard of a place set apart. A sprawling Japanese garden, on the northern edge of the site, anchored the claim in immemorial communion with nature. Finally, on the other side of the expressway, an amusement park, bus parking, and administration buildings provided the necessary coordination, connection, and distraction. The expressway itself had opened just before the expo, part of the infrastructure sanctioned by

FIGURE 6.1 National infrastructure, Cold War ally, corporate Japan, and new dormitory town, superimposed on the bulldozed Senri Hills.
Source: Bettmann, via Getty Images.

the spectacle, although at the time it was only a few miles long (Figure 6.1).[1]

The first international exhibition in Japan, long a dream, marked a watershed in Japan's post-war history, in its use of exhibitions, and in the history of exhibitions generally, and has produced a slew of publications, which shows no signs of slowing down.[2] It was the culmination, first, of a concerted attempt by the Japanese state, following the war, to rebuild the economy and to re-engage the world, albeit on terms set in Washington DC. Domestically, Expo 70 not only enabled and marked a tipping point from the productionism of the 1950s and 1960s to the consumerism that would crest in the 1980s. It also coincided with a growing awareness of the burgeoning environmental and social costs of growth.[3] It therefore

[1] *Nihon Bankoku Hakurankai Kōshiki Gaido* (Nihon Bankoku Hakurankai Kyōkai, 1970). Sandra Wilson, 'Exhibiting a New Japan: The Tokyo Olympics of 1964 and Expo '70 in Osaka', *Historical Research* 85.227 (2012), 159–78, for the expo, with the 1964 Olympics, as a stage for Japanese self-assertion.

[2] Sano Mayuko, *Banpakugaku: Bankoku Hakurankai toiu, Sekai o Haakusuru Hoho* (Shibunkaku Shuppan, 2020), 244–368, for one recent compilation.

[3] Gordon, 'Japan's Transwar Political Economy', for an overview; Laura Hein, 'Growth versus Success: Japan's Economic Policy in Historical Perspective', in Andrew Gordon

became a lightning rod at the time for criticism from the left, of Japanese collaboration in the ongoing Cold War, and of an omnipresent state, an 'administered society', even 'information fascism'. In retrospect, too, the expo has been suborned as witness to the general triumph of capitalism and/or the advancing logic of spectacle.[4]

Second, 1970 marks the adoption by the Japanese state of mega-events (*biggu ibento*) as a way of calibrating its plans for Japan's development. In the late nineteenth century, the state had sponsored exhibitions at home as a way of instructing Japanese producers in the ways of industry. In the early twentieth, it had restricted its focus to exhibitions overseas, to prove its international bona fides. Now, following the proof of concept at the Tokyo Olympics in 1964, international exhibitions at home became a way not only to announce Japan's arrival at the first world table but also to canalize investment, implement plans, and mitigate the distortions of growth. The experiment was a success, generating a slew of public works projects, the highest visitor numbers in history, until 2010, and a substantial profit, which continues to fund 'international cultural exchange and international goodwill' today.[5] It thereby provided a template for subsequent exhibitions in Japan. Third, as the first major international exhibition to be held outside the West, Expo 70 also provided a model for Japan's successors in the development Olympics, soon adopted in East Asia and, most recently, the Middle East.

On these grounds alone, it is worth lingering on the expo, to understand how expos serve not only to pump out the bromides of capitalism, nationalism, and the rest but also to effect specific programmes of

(ed.), *Postwar Japan as History* (University of California Press, 1992), 99–122, for the costs of growth.

[4] For the cold war context, Andrew James Wulf, *U.S. International Exhibitions during the Cold War: Winning Hearts and Minds through Cultural Diplomacy* (Rowman & Littlefield, 2015), 180–203. For contrasting takes on managed society, Tada Michitarō, *Kanri Shakai no Kage: Fukusū no Shisō* (Yomiuri Shinbunsha, 1971), José M. Tortosa, 'Socioeconomic Paradigms: Present Situation and Prospects', *Futures* 22.10 (1990), 1002–12, Tetsuo Kogawa and Douglas Lummis, 'Dialogue: The Managed Society', *AMPO Japan-Asia Quarterly Review* 23.4 (1992), 43–46, and J. Victor Koschmann, 'Intellectuals and Politics', in Andrew Gordon (ed.), *Postwar Japan as History* (University of California Press, 1992), 414–18. Nakase Toshikazu, *Bankokuhaku to Jōhō Fashizumu* (Azekura Shobō, 1970), for information fascism. Debord, *Society of the Spectacle* and *Comments*, for the logic of spectacle.

[5] *Nihon Bankoku Hakurankai Kōshiki Kiroku* (Nihon Bankoku Hakurankai Kinen Kyōkai, 1972), vol. 1, 12, for the first of many mentions of the visitor numbers in the official report. 'What Is the Japan World Exposition 1970 Commemorative Fund?' (2022), www.osaka21.or.jp/jecfund/english/, for the use of the profit.

political economy. From a distance, the event and the plans can seem monolithic, subsuming difference in an implacable machine. A closer look reveals a familiar picture, however, of an expo that came into being through the interaction of multiple actors and agendas, networks and systems, who were sometimes coordinated, but often in conflict, and frequently indifferent to each other and the whole. Event and organizers could never finally incorporate, but had to accommodate, the inevitable loose ends. In its wake, too, the uneven, unfinished business of development ensured both a continuing demand for spectacle and its failure, finally, forever, to convince.

FROM PROMOTING EXPORTS TO PLANNING DEVELOPMENT

Exhibitions were not in short supply in Japan after the war, courtesy of local authorities, as detailed in Chapter 5. It took a while, though, for the economy to revive, international exhibitions to start back up, and the Japanese state to get into the game. The Korean War, beginning in 1950, primed the pump, but by 1953 the resulting boom had produced increasing inflation, a trade deficit, and slowing exports, and its end in 1954 a full-blown recession. The Ministry of International Trade and Industry (MITI), spawned in 1949, quickly stepped to the fore, taking over the Japan External Trade Organization (JETRO) and tasking it to promote exports, with exhibitions a prominent part of the prescription. From 1956, a series of ships mounted a floating trade fair, preaching the virtues of Japanese manufacture around the world. In 1962, the purpose-built Sakura Maru took over the ministry. Its first mission saw '430 booths ... 10,000 items of machinery, durable consumer goods and textiles ... [and] 100 persons' sent to twelve ports in Africa, the eastern Mediterranean, and the Near East.[6] International exhibitions provided not only another venue but also a chance to veil the economic imperative in less grasping, more reassuring garb.

By the time they re-emerged in the post-war period, like regional exhibitions in Japan, international exhibitions looked different. Gone for good were the industrial catalogues and imperial braggadocio, replaced by the twin bromides of development, in both capitalist and communist variations, and of corporate futurology, which had begun to leach into expos between the wars. Gone, too, for the most part, were the

[6] Johnson, *MITI*, 172–97 for the Occupation, 227–32 for the Korean War, MITI, and JETRO, and 248 for the Sakura Maru. *New York Times* 1967/7/1, 48, for the quote.

original metropolitan stages, in London and Paris, replaced by the regional satellites of the capitalist West. Haiti in 1949 was the first post-war expo, but international participation was limited.[7] Brussels in 1958 thus provided Japan's first opportunity to show that it had changed, albeit on the far edge of the site, with a plot one-tenth that of the US. While the expo tried to show how 'scientific civilization' would create 'a more humane world', JETRO told a three-act story about how the country was recovering 'from the edge of the abyss', courtesy of a culture that reconciled 'hand and machine': first history, from *haniwa* to Hiroshima; then industry, including textiles, but emphasizing electronics; and finally life in the future, culminating in a Japanese home. The pavilion, similarly, looked back to the past. In Paris in 1937, Sakakura Junzō had subsumed tradition within international modernism. In Brussels, Maekawa Kunio used steel, concrete, and glass to provide a discrete frame for white exterior panels, dark wooden posts and beams, an interior garden, and more. For its European observers, the whole served to show how tradition allowed 'modernism [to be] stamped with true poetry and dignity'.[8]

Japanese success in Brussels set the pattern for international exhibitions to come, even as the Cold War ramped up and the space race got under way. Three North American expos in the 1960s all saw variations on the theme: traditionalizing architecture and the inevitable Japanese garden argued that industrial progress was indigenous in origin and pacific in intent. In Seattle in 1962, white exterior panels provided a counterpoint to the more contemporary efforts of Japanese architects elsewhere. In New York in 1964 and 1965, the stone walls and moat of an early modern castle, again by Maekawa, contained exhibits ranging from 'one of the world's largest tankers' and 'the world's fastest express train' to the 'charming evidences of the quiet rural Japan of a century ago'. Finally, in Montreal in 1967, a modernist update of the Shōsōin, the mid eighth-century treasure house of Tōdaiji in Nara, housed a story of 'harmony', with nature, tradition, and technology. By this point, however, the seams were beginning to show. Canadian observers noted that the actual

7 Hadassah St Hubert, 'Visions of a Modern Nation: Haiti at the World's Fairs' (PhD thesis, University of Miami, 2018), 99–143.

8 Gonzague Pluvinage, *Expo 58: Between Utopia and Reality* (Editions Racine, 2008), for an overview. *Exposition Universelle et Internationale de Bruxelles, 1958* (Établissements Généraux d'Imprimerie, 1958), vol. 1, 116, for location, vol. 3, 117–22, for exhibit and quotations, and vol. 4, 135–37, for the architecture. Jonathan M. Reynolds, *Maekawa Kunio and the Emergence of Japanese Modernist Architecture* (University of California Press, 2001), 208–14, for an analysis.

exhibits – and the cars and motorbikes displayed outside – smacked of a trade fair, producing a 'Montreal shock', which sent bureaucrats and their consiglieres scrambling as they ramped up preparation for Expo 70.[9]

Exports were essential and exhibitions overseas useful, but it was the domestic economy that was the primary focus of bureaucratic attention and political concern for the first few decades after the war. The need to galvanize domestic consumption culminated in a promise by Ikeda Hayato, the Prime Minister, in 1960, to double income over the following decade, in part by disbursing over 16 trillion yen of public investment. The funds also provided an opportunity to address the distortions of growth – the concentration of heavy industry along the Pacific coast, the increasing pre-eminence of Tokyo – by formulating plans for the 'national land' (*kokudo*). The term had emerged as a policy rubric in the 1930s and the government had passed an act in 1950, but it was in the 1960s that legislation gave way to planning and implementation. Tokyo was the guinea pig. Its successful bid for the Olympics had prompted the investment of 1 trillion yen, five times that of the preceding 1960 games in Rome. Only 4 per cent, however, went to the Games themselves, while nearly 80 per cent was spent on transportation infrastructure, providing Tokyo with new subways and expressways, to support a population that had grown from 11 to just under 17 million in the previous decade, as well as the Shinkansen to Osaka. While the construction was under way, moreover, the first Comprehensive National Development Plan of 1962 expanded the vision to the whole archipelago, aiming to even out development among the regions. In 1969, the second would begin to address environmental and social concerns, while promising roughly 500 trillion yen more investment.[10] By that point, though, preparations for the Osaka Expo were well under way.

[9] For summaries, *Kōshiki Kiroku*, vol. 1, 26–28, and Yoshida Mitsukuni, *Bankokuhaku no Nihonkan* (IMAX, 1990), 35–38. For contemporary descriptions, *Official Guide, New York World's Fair 1964* (Time Inc., 1964), 140, and *Expo 67: Guide Official* (Maclean-Hunter, 1967), 108. For criticism of Japan in Montreal, *Yomiuri Shimbun* 1967/7/9, 14, and 1967/7/14, 8; Komatsu Sakyō, 'Nippon 70 Nendai Zengo: Dokyumento – Orinpikku Kara Bankokuhaku e', *Bungei Shunjū* 49.2 (1971), 220–70, 262, and Ichikawa Fumihiko, 'Montoriōru no Konwaku: 1967-nen Haku no Nihonkan Tenji Mondai', in Sano Mayuko (ed.), *Banpakugaku: Bankoku Hakurankai toiu, Sekai o Haakusuru Hōhō* (Shibunkaku shuppan, 2020), 240–43.

[10] Johnson, *MITI*, for the vision, and 15–16, for exports; Bai Gao, *Economic Ideology and Industrial Policy in Japan: Developmentalism from 1931 to 1965* (Cambridge University Press, 1997) for its assumptions, and 233 for public investment; O'Bryan, *The Growth Idea*, 18–47, for the embrace of planning. Yasuhiko Shima, 'The Income Doubling

Planning alone, it turned out, was not enough. The bureaucrats also needed a way to catalyse the construction, conjure the future, and prompt the behaviour envisioned in the plan. Big international events were the ticket. They not only announced Japan's reintegration into the international community but also provided a medium through which to channel investment, providing a deadline and a global audience to which the nation might be held to account.[11] Staging an expo in Osaka would begin to rebalance attention away from Tokyo; provide a stage on which state and capital could adumbrate the development to come; and supply an excuse to extend the transport network beyond the Tokyo-Osaka corridor.

EXPO ECONOMICS

When the possibility of an expo became public, in March 1964, it sounded as if things were under control. MITI had begun to explore the possibility of holding Asia's first 'world exposition', said the press, for reasons that could have been lifted from the Meiji playbook, a century earlier: to heighten awareness, at home and abroad, of Japanese industry; to galvanize industry further by displaying foreign products; and to promote exports. The actual factors behind the decision were a bit more specific. The original discussions, in 1962, had been informed in part by concern about Japan's rapidly worsening trade deficit. But there was also the BIE in Paris, pressing the government to ratify the international exhibitions treaty; and Toyota Masataka, who as a bureaucrat had worked on the cancelled 1940 expo and now as a member of the Diet was advocating for its resurrection in various committees. By May, a high-level delegation was visiting the BIE in Paris. Come June, the cabinet agreed to formulate plans. MITI set up a research office in September, the government decided

Program and Public Investment', *The Kyoto University Economic Review* 31.1 (1961), 14–52, for a contemporary account. Sorensen, *The Making of Urban Japan*, 178–83 and 197–99, for overviews of the 1960s. Sano Hiroyoshi, 'Zenkoku Sōgō Kaihatsu Keikaku ni okeru Kyoten Kaihatsu Kōsō ni kansuru Kenkyū', *Toshi Keikaku Ronbunshū* 47.3 (2012), 403–8, for the 1962 plan.

11 Komatsu, 'Nippon 70 Nendai Zengo', 227–28, for the Olympics numbers. Tada Osamu, *Okinawa Imēji no Tanjō: Aoi Umi no Karuchuraru Sutadīzū* (Tōyō Keizai Shinbunsha, 2004), 20–33, and Andrea Urushima, 'The 1970 Osaka Expo: Local Planners, National Planning Processes and Mega Events', *Planning Perspectives* 26.4 (2011), 635–47, for big events. Hatanaka Akihiro, *Gorin to Banpaku: Kaihatsu no Yume, Honrō no Rekishi* (Shunjūsha, 2020), 103–50, for the expo as a development apparatus.

to go ahead in November, and a week before Christmas 1964, the Diet ratified the treaty.[12]

The expo was hardly a done deal, however. There were international rivals to vanquish, domestic interests to be mollified, and a bureaucracy to create. The first hurdle was the easiest. There were rumours that Osaka would face competition, certainly from Melbourne, possibly from Philadelphia, but in the event both challenges failed to materialize, and the BIE waved Osaka's application through in September 1965. The second problem was knottier. Inevitably, the prospect of hosting an international jamboree, and attracting the requisite investment, had sparked a bidding war among local governments, which was exacerbated by factional politics in the ruling Liberal Democratic Party. Osaka got its start early, in April 1964, with the support of the Prime Minister, his ministerial allies, and factional confrères, but while the prefecture had its eyes on the Senri Hills, in the north of the city, the city wanted to develop the area around the port. Kobe was also in the early running, but the Deputy Prime Minister had his heart set on Lake Biwa, in Shiga, with a plan to reclaim 6 million square metres from the lake. Come the autumn in Osaka, city and prefecture had agreed that the hills made more sense than the port, given the former's prominence in the national transport plan. But a regional compromise only emerged in April 1965, three weeks before the application was due in Paris, with Senri hosting the expo, but Kobe and Shiga benefiting from the public works bonanza.[13]

Even with the application approved, third, it took time to work out how to organize and fund the expo. The Japan Association for the 1970 World Exposition came into being in October 1965. In retrospect, it was attributed with 'monolithic solidarity', 'which knew no sectionalism nor regionalism', but this was hardly evident at the time. Initially blessed with 3 divisions, 8 sections, and 45 directors, it had swollen by 1970 to 42 departments, 6 offices, and well over 800 functionaries, half of

[12] *Asahi Shimbun*,1964/3/8, 5; 1964/6/9, pm, 1; 1964/8/15, 5; 1964/11/5, 7; 1964/12/4, 7, for developments during 1964. Ikeguchi Kotarō, *Nihon no Bankoku Hakurankai: Sono Igi, Keikaku, Kōka* (Tōyō Keizai Shinpōsha, 1968), iii for the trade deficit. *Nihon Bankokuhaku*, 1 (1966/12), 50; and *Kōshiki Kiroku*, 1972, vol. 1, 34, and vol. 3, 584 for the official version of the timeline.

[13] *Asahi Shimbun*, 1964/6/15, pm, 1; 1964/7/27, pm, 1; 1965/3/21, 7, 1965/4/3, pm, 5; 1965/4/5, 2; 1965/5/13, pm, 1; 1965/8/19, 2; and 1965/9/14, 1–2, for international competition, domestic rivalry, and resolution. *Kōshiki Kiroku*, vol. 1, pp. 34–39 for the official version. Factional tensions were omnipresent in 1964. See Komatsu, 'Nippon 70 Nendai Zengo', 243.

whom were seconded from central and local governments and a quarter from the private sector. Nor was anyone local willing to preside over the mess. Eventually, the Minister of Trade and Industry had to put in a call to the Tokyo-based chair of the Japan Business Federation, who had fled to Cairo, to convince him that he needed to step up to the plate. The expo, in short, was the product of a bureaucracy, the normal inefficiencies of which were compounded by the ephemeral nature of the loyalties required. Contemporary observers noted that collaboration was going to be hard to come by.[14]

The Association's role, however, was not to plan and design the expo, which could be subcontracted, but to keep things on track and to make the numbers add up. This was tricky, given the uncertainty about how much the whole thing was going to cost. Overall budget estimates swung wildly, more than doubling from the 42 billion yen set in 1966, so prompting ongoing debates about how many would come and who was going to pay. On the construction side, Osaka was adamant that the government provide more subsidy than it had for the Olympics. On the other, the ballooning operations budget led to an almost 50 per cent rise in the price of an adult ticket between July 1966 and January 1968, and a public outcry. In the end, the concerns were moot. When the final accounts came in, both budgets had ended up in the black. The state had picked up two-thirds of the construction tab, Osaka contributing its desired modicum, and the balance coming from gambling, golf clubs, and a lottery. Operations, meanwhile, had generated a tidy profit of 20 billion.[15] The 86 billion yen of direct expenses, however, were only a drop in the expected bucket. Two years ahead of time, the bureaucrats were calculating that the expo might generate 2 trillion yen for the economy as a whole, double that of the Olympics, from related projects, visitor spending, and induced demand. The prospect – and the threat of

[14] *Asahi Shimbun*, 1965/7/11, 7, 9/12, 3, 1965/9/14, 2, 1965/10/23, pm, 1, 1965/11/3, pm, 1, 11/16, 1, and 1966/3/8, 1, for contemporary observation and comment. *Kōshiki Kiroku*, vol. 1, 14–15, 36–37, and 43–45, and vol. 3, 317–32, for the official version. Komatsu repeatedly underlines the bureaucratic nature of the Association, for example, Komatsu, 'Nippon 70 Nendai Zengo', 244.

[15] *Asahi Shimbun*, 1964/6/9, pm, 1, 1965/4/3, pm, 5, 1965/8/19, 5, and 1966/3/8, 1, and 1970/9/27, 3 for the budget story. *Kōshiki Kiroku*, vol. 1, 15–16, and vol. 3, 116–49, for the official accounts. The public outrage about ticket prices was mitigated somewhat by the knowledge that anyone who still owned a booklet of twelve tickets for the 'postponed' 1940 Tokyo Expo was welcome to exchange them for one adult or two child tickets to the 1970 Osaka event. *Nihon Bankokuhaku* 1 (1966/12), 54, and *Nihon Bankokuhaku Nyūsu* 2, 4.

distortion in markets nationwide, not least in labour – was such that by 1970 banks were commissioning their own reports, to ensure that they had taken its impact into account. When the numbers were finally in, the expo had been responsible for a third of a per cent of the total gross national product (GNP) between 1967 and 1970.[16]

Seen by the state, in other words, the expo was part of the plan: counterbalancing the pull of Tokyo, buttressing a bipolar economy, and opening up circuits further west. The project proceeded at multiple scales. First, locally, the expo prompted the extension of two existing intra-urban rail lines, which allowed Osaka to expand into the northern hills. The above-ground Senri line had been built in the 1920s and taken over by the Hankyū Corporation during the war, which extended it, first in the early 1960s, to service Senri New Town, and now again, via a temporary station at the expo's west gate. There were 30,000 living there in 1967, almost all in *danchi* – the five-storey, walk-up apartment blocks provided by the Japan Housing Corporation; 100,000 come 1970, enjoying the new infrastructure, perhaps, but also exposed to the high vegetable prices and crowds that the expo brought in its wake. Some of them may also have got into the city via an overground extension to the underground Midōsuji line. Originally built in the 1930s, to link the north and south of the city, it too had become part of the Hankyū portfolio in the 1960s, extended first to the new Shinkansen station, then temporarily to the expo's main gate, and subsequently to a central station for the new town.[17]

Second, regionally, the expo consolidated Osaka's position as the hub of the emerging national expressway network. Discussed during the war and legislated in the late 1950s, the first three expressways had opened in the early 1960s, including a short stretch from northwest Osaka to a few

[16] *Asahi Shimbun*, 1965/9/9, 7, for early MITI predictions, and 1970/2/3, 15, for bank reports. *Nihon Bankokuhaku Jiten* (Marunouchi Risāchi Sentā, 1968), 747–55, for the 1968 prediction and Ikeguchi, *Nihon no Bankoku Hakurankai*, 106–56, for analysis. *Kōshiki Kiroku*, vol. 1, 16–19, for a summary, and vol. 3, 306–8, for labour concerns.

[17] *Yomiuri Shimbun* 1967/6/29, 2, and 1967/10/4, 5, for new lines, and 1971/3/29, 4, for new towns. For the big picture, Sorensen, *The Making of Urban Japan*, 185–88. For new towns, Shinzo Kiuchi and Noboru Inouchi, 'New Towns in Japan', *Geoforum* 7.1 (1976), 1–12. For *danchi*, Laura Neitzel, *The Life We Longed for: Danchi Housing and the Middle Class Dream in Postwar Japan* (MerwinAsia, 2016). For Senri, Michelle L. Hauk, 'Postwar Residential New Towns in Japan: Constructing Modernism', PhD dissertation (Washington University, 2015), 93–136; Nakamaki Hirochika, 'Chūkū Kōzō de Toku Senri Nyū Taun to Ōsaka Banpaku', in Sano Mayuko (ed.), *Bankoku Hakurankai to Ningen no Rekishi* (Shibunkaku Shuppan, 2015), 301–8, and 'What Is Senri New Town?' https://senri-nt.com/ourtown/en/. For high prices, Ann Waswo, *Housing in Postwar Japan: A Social History* (Routledge Curzon, 2002), 36–38.

miles east of Kyoto, which happened to pass a few hundred yards from what would become the expo site. A plan for the national network was enacted in 1966. Nagoya and Tokyo were connected in 1969. Then, two weeks before the opening ceremony, two more expressways opened. Neither went very far. The Chūgoku expressway ran for six miles, starting from a new interchange just west of the expo's main gate and going through the middle of the site. It would take another thirteen years before it was complete. The Kinki expressway was four miles longer, starting from the same interchange, but with only two lanes, meaning that cars could only go one way. It would be another two years before the other two lanes were finished.[18] The expressways were complemented, however, by a new prefectural ring road, which had been completed and linked Ikeda, north of the city and west of the expo, to Sakai, in the south.

Finally, nationally, the expo energized the vision of a high-speed train network, which featured in the government exhibit on site, both in its current incarnation as the Shinkansen and in an imagined future of linear rail. In 1970, there was still only a single line linking Tokyo and Osaka. But the expo provided the nudge for tourists, not simply businessmen, to use it. Visitors from Tokyo could enjoy an expanded schedule, comprising nine trains an hour, each with sixteen rather than twelve cars, then change at Shin-Osaka, onto the Midōsuji line extension, before being deposited at the expo's main gate. They could also look forward to going further. The extension of the line from Osaka to Okayama had been authorized in September 1965, one week before the BIE's formal approval of the expo. Construction started a year and a half later, one day after groundbreaking on the expo site. And the Diet passed the National Shinkansen Development Law, two months into the expo itself.[19] For politicians and bureaucrats in Tokyo, the expo was catalyst and hub of an emerging transport web, a testament to their planning ability and a statement of

[18] *Yomiuri Shimbun* 1966/2/8, 4, for the Chūgoku expressway. John Black and Peter Rimmer, 'Japanese Highway Planning: A Western Interpretation', *Transportation* 11.1 (1982), 29–49, and Dharish David, 'The Japanese Experience with Highway Development', *Journal of Infrastructure Development* 6.1 (2014), 17–42, for overviews.

[19] *Nihonkan* (Tsushō Sangyōshō Kigyōkoku Bankokuhaku Junbishitsu, 1970), 15 and 21, for the government exhibit. *Yomiuri Shimbun* 1969/7/13, 14, Christopher Hood, *Shinkansen: From Bullet Train to Symbol of Modern Japan* (Routledge, 2006), 29 and 212–13, and Hitomi Nakanishi and Fumitaka Kurauchi, 'Japan's Linear Megalopolis: Shinkansen High-Speed Rail as the Spine of a 60-Year Mega-Region Evolution', in Michael Neuman and Wil Zonneveld (eds.), *The Routledge Handbook of Regional Design* (Routledge, 2021), 110–11, for its co-evolution with the expo. Jessamyn Abel, *Dream Super-Express: A Cultural History of the World's First Bullet Train* (Stanford University Press, 2022), for its cultural significance.

developmentalist intent. Even before it opened, its ability to crystallize the fluid potential of plan and capital had proved the worth of the medium for the years to come.

EXPO INTELLECTUALS

With approval, budget, and organization in place, and infrastructure on the way, the bureaucrats could take a step back and let others work out what the expo should be about, look like, and contain. Like local authorities, for whom exhibitions had long been a magnet, intellectuals and other creatives proved not only powerless to resist the siren call, whether to display their craft or hone their critique, but also credulous in believing their view of things would prevail.

The most important intellectual contribution to the expo came from a small group facilitated by the *Hōsō Asahi*, a high-brow outlet for comparative 'cultural research'. The Bankoku-haku o Kangaeru Kai (Group for Thinking About the Expo) initially comprised an anthropologist, a sociologist, and a science fiction novelist, but soon added others, including an architectural critic and a section chief from the Economic Planning Agency (EPA). They were determined to have a good time, meeting first in July 1964 in the main drag of the leading geisha district in Kyoto, and subsequently over copious amounts of seafood at hotels in Mie, which prompted a renaming as the Kaikūkai (Shellfish Eating Group). Their discussions were serious, however, focused on how the expo might advance a two-pronged intellectual agenda, which had already been sketched out by Umesao Tadao, the anthropologist. First, in 1957, he had argued that Japan should be understood not in terms of Western theory, as either a pale Weberian imitation of Western modernization or an Asian variation of Marx's stage theory, progressing from feudalism to capitalism; but, like Western Europe, as an ecological civilization with its own autochthonous dynamic, which had enabled it to reach the capitalist present. Then, in 1963, ten years before Daniel Bell's discovery of 'post-industrial society', he postulated that the contemporary world was shifting from industrial to 'information society', which set off a torrent of speculation, not least by the government, about its potential impact. Together, the two arguments allowed the group not only to narrate the past but also to speculate about the future, prefiguring the official story that would be told on site.[20]

[20] For the group's activities and context, Komatsu, 'Nippon 70 Nendai Zengo', 220–25 and 229–30; Umesao Tadao, *Umesao Tadao Chosakushū* (Chūō Kōronsha, 1989–94),

Umesao and his colleagues were not simply bystanders, however. As their thinking about expos proceeded, so their ideas were adopted and labour co-opted by the bureaucrats. Informal meetings at the prefectural level in early 1965 quickly escalated to formal contacts with MITI officials. They were asked to put together a committee to come up with a theme. They held a public meeting, to clarify their position. Umesao insisted that the group 'wasn't engaged to, just dating' the authorities; Komatsu Sakyō, the science fiction novelist, that their love was merely 'platonic'. In fact and in the public eye, however, their co-optation by the state was clear. Initially, for example, the Theme Committee tried not only to capture development in the round, celebrating its 'light', made possible by multiple 'wisdoms', but also acknowledging 'the dark side of progress'. What emerged as the official line, however, was a simplistic aspiration of 'progress and harmony for mankind', with the latter not a work-in-progress, but an achieved state, based on 'the philosophy of the East'. Umesao was reasonably phlegmatic about becoming the expo's 'stagehands', but Komatsu less so. He had proposed the study group and been enthused by the prospect of an expo in 1964, as he contemplated the way Tokyo had been transformed by the Olympics and wondered how Kansai might be similarly transformed. By the end of 1965, however, he was fed up with bureaucratic surveillance. By the spring of 1966 he was angry and tired. He and his colleagues continued to work for the Association, which had to scramble for intellectual and

vol. 13, 34, and vol. 21, 13; and William O. Gardner, 'The 1970 Osaka Expo and/as Science Fiction', *Review of Japanese Culture and Society* 23 (2011), 28–32. For ecological civilization, Umesao Tadao, 'Bunmei no Seitaishikan Kosetsu', *Chūō Kōron* 72.2 (1957), 32–49, and Angus Lockyer, 'National Museums and Other Cultures in Modern Japan', in Daniel J. Sherman (ed.), *Museums and Difference* (Indiana University Press, 2008), 108–9. For modernization theory, Harry Harootunian, 'America's Japan/Japan's Japan', in Masao Miyoshi and Harry Harootunian (eds.), *Japan in the World* (Duke University Press, 1991), 196–221, and Sebastian Conrad, '"The Colonial Ties Are Liquidated': Modernization Theory, Post-War Japan and the Global Cold War', *Past and Present* 216.1 (2012), 181–214. For Marxism, Hayashi Yūjirō, Komatsu Sakyō, and Umesao Sakyō, 'Miraigaku e no michi', in Katō Hidetoshi (ed.), *Nihon Bunka no Tenbō* (Nihon Seisansei Honbu, 1968), 9–15. In a series of conversations in 1997, Umesao reiterated his antipathy: 'Marx is useless when looking at Japanese modernization'. For information society, Umesao Tadao, 'Jōhō Sangyō Ron: Kitaru beki Gaihaiyō Sangyō Jidai no Yoaka', *Hōsō Asahi* 104 (1963), 4–17; Tessa Morris-Suzuki, *Beyond Computopia: Information, Automation and Democracy in Japan* (Kegan Paul, 1988), 6–16; Alistair Duff, *Information Society Studies* (Routledge, 2000), 3–6; and Youichi Ito, 'Mass Communication Research in Japan: History and Present State', *Annals of the International Communication Association* 10.1 (1987), 65–67. Koschmann, 'Intellectuals', for the broader intellectual context.

artistic cover, following the criticism of the Japanese exhibit in Montreal in 1967. But they were also going their own way, turning their early work on the expo into full-blown futurology, ranging from the shape of the archipelago and its cities to the organizations in which people would work, the nature of human relationships, and more.[21] This was hardly a radical initiative, however. It too was catnip to bureaucrats, who needed to know that planning was possible and wanted the expo to say much the same thing.

Ikeguchi Kotarō was only the most vocal. A MITI bureaucrat, he had been assigned to the expo team in mid-decade and leveraged his experience into a stream of publications, followed by a fifty-year career as expo booster and development apologist, under the pen name of Sakaiya Taichi. We will meet him again. His initial pitch was predictable. The expo would stimulate the economy, redress Tokyo-centrism, build 'Japanese confidence', exhibit 'Asian pride', and foster international exchange. But there was more. The expo not only exhibited the technological possibilities of the emerging post-industrial age: lightweight materials freeing buildings from the constraints of gravity; information technology to ensure the consumer's safety and comfort; massive screens to educate and entertain a docile public. It limned a 'strategy for the future' of increasing economic competition, not least in its organization of increasingly scarce labour: the site was appearing on time, using only a third of the anticipated workforce, thanks to modular design and prefabricated units. Behind this was a general principle that ran through the expo – and therefore the society of the future – like a stick of rock. Hierarchies were out. Networks were in. The expo had multiple producers, linked through the Association, but each at the centre of their

[21] Komatsu, 'Nippon 70 Nendai Zengo', 227–32 for initial enthusiasm, 243 for dating, 246–48 for public meeting and platonic love, 254–60 for disenchantment, and 260–63 for futurology and continuing work. Umesao, *Chosakushū*, vol. 21, 14–15 for stagehands and vol. 12, 479–83, for futurology. For the links between the group and the theme committee, Saotome Kenji, '1970 nen Ōsaka Banpaku no Kihon Rinen: "Bankokuhaku o Kangaeru Kai" ni Yoru Sōan Sakusei no Haikei to Keii', in Sano Mayuko (ed.), *Banpakugaku: Bankoku Hakurankai Toiu, Sekai o Haakusuru Hōhō* (Shibunkaku shuppan, 2020), 246–65. The committee's discussions were transcribed as *Nihon Bankoku Hakurankai Kōshiki Kiroku Shiryōshū Bessatsu*, vol. D-1 (Nihon Bankoku Hakurankai Kinen Kyōkai, 1971). *Kōshiki Kiroku*, vol. 1, 58–65 for the official account, 9 for harmony, and 63 for 'philosophy of the East'. Yoshimi Shun'ya, *Banpaku Gensō: Sengo Seiji no Jubaku* (Chikuma Shobō, 2005), 45–53, for a critique. Katō, *Nihon Bunka no Tenbō*, 12–15 and 19–20 for claims that futurology was a science. See also Gardner, 'The 1970 Osaka Expo', 32–34.

own discrete web. Distributed labour had been integrated by a 'centre brain' (sic). Systems thinking would allow Japan to make the micro changes needed to navigate the new landscape and ensure the macro stability of the emerging 'soft society', no longer tied to industrial production, but enabled by information technology.[22]

By the same token, vision and expo were anathema for the Japanese left. Having mobilized vast protests against the renewal of the US–Japan Security Treaty in 1960, it had since been in retreat, its purchase on the body politic defanged by income-doubling and consumer goods. The escalation of the Vietnam War provided an opportunity to regroup and recruit, not least at universities, and the expo a tempting target, given that it had been planned in part to distract from the second renewal of the treaty. But it proved hard for its opponents either to get their act together or to pin the expo down. A five-day event in Tokyo in 1968 set the terms for ongoing sniping at its artistic and intellectual collaborators: they were producing fake art, using their gifts to serve and disguise the state; they were simply worker ants, 'shrine maidens' at a festival of capital. The year 1969 saw a series of attempts to produce alternatives: performance art, teach-ins, and an Anti-Expo (*Hanpaku*) in the grounds of Osaka Castle. It included a handful of exhibits, some desultory panels, and fervent discussions about the possibility of 'peace and liberation' trumping 'progress and harmony', but soon descended into squabbling. The local Christian churches in Osaka were more organized. One-third of local priests were opposed. By early 1970, though, the prospect of mass protests had fizzled out. A small-circulation, short-lived journal used photomontage to counterpose the hazy expo promises to the hard-edged reality of atomic destruction, homeless workers, and American occupation. But, the critics admitted, the possibility of meaningful opposition had been foreclosed. The expo provided a model of social control, by which the illusion of progress and harmony might be sustained. The expo as a system had rendered revolutionary action impossible and itself invisible.[23]

[22] Ikeguchi Kotarō, '"Nihon" o Hyōgen Shiyō: Bankokuhaku e no Kitai to Fuan', *Asahi Jaanaru* 9.44 (1967), 103–6, for an early statement; *Nihon no Bankoku Hakurankai*, 31–52 for the expo's significance and 235–38 for the post-industrial age; *Bankokuhaku to Mirai Senryaku* (Daiyamondosha, 1970), for the strategy of the future, including 192 for labour, 314 for the producer system, and 338 for systems thinking.

[23] Koschmann, 'Intellectuals', 414–18 for an overview. Yoshimi, *Banpaku Gensō*, 11–16, for the ideology of income-doubling. *Yomiuri Shimbun* 1970/2/23, pm, 1, for the expo as distraction. Yasufumi Nakamori, 'Criticism of Expo '70 in Print: Journals Ken, Bijutsu

For someone like Ikeguchi, this was precisely the point. Systems thinking allowed bureaucrats to step back, analyse, and adjust the 'optimum conditions for society's well-being'.[24] Expos, cities, and states could afford to be indifferent to the objects, people, and information that passed through their buildings, networks, and systems. For most intellectuals, however, this was anathema. It was tempting to believe that they could participate on their own terms, if not to reform the system itself. As the expo processed their ideas, however, as the conceptual fat was rendered for corporate buy-in and public consumption, so they were exposed to charges of betrayal. The temptation and the dilemma were even more acute for creatives.

EXPO CREATIVES

The bureaucrats could play with numbers, the intellectuals with words, but the expo needed architects, artists, and designers to give them and it material form. Its preferred collaborators had made their reputations as standard-bearers of the avant-garde, but they too soon discovered how the expo was able to turn their work to its own account. The expo would change design more than design changed the expo.[25] The emblem provided an early indication of the limits of expression. The first committee chose a design consisting of 'three circles ... in the form of a pyramid', which symbolized people joining hands and expressed 'the desire to make this earth a splendid place to live in'. The Association determined that it

Techō, and Dezain Hihyō', *Review of Japanese Culture and Society* 23 (2011), 132–44, for an overview, including 133–35 for the short-lived journal, and 140–41 for the five-day event. Miyauchi Yoshihisa, 'Bankokuhaku: Geijutsu No Shisōteki Sekinin', *Gendai No Me* 9.9 (1968), 188–95, for fake art; Ichirō Haryū, 'Expo '70 as the Ruins of Culture (1970)', *Review of Japanese Culture and Society* 23 (2011), 47 for worker ants, and KuroDalaiJee, 'Performance Art and/as Activism: Expo '70 Destruction Joint-Struggle Group', *Review of Japanese Culture and Society* 23 (2011), 154–73, for performance art. *Asahi Shimbun,* 1969/3/18, pm, 7, and Hariu Ichirō, 'Hanpaku: Hansen Undō no Shikō Sakugo', *Gendai No Me* 10.10 (1970), 126–35, for the Anti-Expo. *Asahi Shimbun,* 1969/10/29, pm, 1 and *Yomiuri Shimbun* 1970/2/23, pm, 1, for the church. Awazu Kiyoshi and Hariu Ichirō, 'Genzō no Naka de', *Kikan Geijutsu* 4.2 (1970), 26–39, for social control, and Taki Kōji, 'Banpaku Hantairon: Dezain ni okeru Ningen Kaifuku o Mezasu', *Tenbō* 121 (1969), 172–77, and 'Kindai no Shūmatsu', *Ken* 1 (1969), 64–69, for the impossibility of revolutionary action.

24 Masukuni Kitazawa, 'Militarism in the Management Society', in J. Victor Koschmann (ed.), *Authority and the Individual in Japan: Citizen Protest in Historical Perspective* (University of Tokyo Press, 1978), 201, for optimum conditions.

25 Teasley, *Designing Modern Japan*, 263–82, for the impact of the expo on design.

was insufficiently graphic, the committee resigned en masse, and a new emblem was chosen. Based on the polysemic cherry, five petals represented the five continents, the circle in the centre the 'Rising Sun of the Japanese flag', the white part around it 'man's aspirations for development and progress', and the whole thing 'integrity and harmony'.[26] As the intellectuals had discovered, the expo proved adept at rendering radical experiment safe for popular consumption: cosmopolitan aspiration would be subordinated to corporate advertising, technological innovation to technocratic design.

Some insisted that it was possible to take the expo's yen without submitting to its demands. The most vocal was Okamoto Tarō, in his mid-fifties, the established embodiment of the Japanese avant-garde and so the natural choice to produce the theme exhibit. Okamoto insisted that he would liberate Japan from the 'heavy shell of the past' by creating something 'monstrous' (*berabō*), thereby turning the expo into a 'festival' (*matsuri*), which would consume surplus, prompt participation, and allow a new world to emerge. He was only allocated a billion yen, a third of what he wanted, but it was enough to pay for the Tower of the Sun, with a soaring white column, outstretched arms, and four contrasting faces, three above and one underground. Inside, the visitor went on a journey through time, from a subterranean realm of 'wisdom', 'prayer', and 'encounter', through a 'Tree of Life', leading up through the inside of the Tower, to a sample of futures suspended under the Festival Plaza's roof.[27]

In time, the whole thing would become an icon, but in 1970 it seems to have fallen considerably short of his claims. The Tower itself was a figure of fun, but the exhibit left the average visitor, not least the children, cold. It felt old-fashioned and didactic, when set against the attractions

[26] *Asahi Shimbun*, 1966/3/8, 1; Komatsu, 'Nippon 70 Nendai Zengo', 255; and *Kōshiki Kiroku*, vol. 1, 65–66, and vol. 3, 87.

[27] For Okamoto's selection and demands, Komatsu, 'Nippon 70 Nendai Zengo', 263–69. For his public statements, *Asahi Jaanaru* 22 (1967/10), 96–103, and 'Bankokuhaku ni Kaketa Mono', in Tange Kenzō and Okamoto Tarō (eds.), *Nihon Bankokuhaku: Kenchiku, Zōkei* (Kōbunsha, 1971). For the theme exhibit, *Japan World Exposition, Osaka, 1970: Official Report*, vol. 1, 477–95; Kawasaki-shi Okamoto Tarō Bijutsukan (ed.), *Okamoto Taro, EXPO '70: Taiyo no Tō kara no Messeji* (Kawasaki-shi Okamoto Tarō Bijutsukan, 2000); and Angus Lockyer, 'The Logic of Spectacle c. 1970', *Art History* 30.4 (2007), 571 and 578–82. Bert Winther-Tamaki, 'To Put on a Big Face: The Globalist Stance of Okamoto Tarō's Tower of the Sun for the Japan World Exposition', *Review of Japanese Culture and Society* 23 (2011), 81–101, for a less sceptical read. For a laudatory account of Okamoto's work in general, Sawaragi Noi, *Kuroi Taiyō to Akai Kani: Okamoto Tarō no Nihon* (Chūō Kōron Shinsha, 2003), which includes an analysis of the tower, 189–226.

available elsewhere: foreign hostesses! actual spacecraft!! a rock from the moon!!![28] It was also evident to his fellow artists and student activists that Okamoto had sold out. Their criticism pointed to the general dilemma, though. The expo mobilized a roll call of the anointed avant-garde, from architect Isozaki Arata to artist Yoshida Minoru, composer Takemitsu Tōru to choreographer Hijikata Tatsumi. Like graphic designer Yokoo Tadanori and filmmaker Matsumoto Toshio, the designers of the Textiles Pavilion, they may have assumed that they could use the opportunity to critique state, society, and capital from within. On site, however, their voices were inaudible amidst the cutting-edge technology and corporate bombast, their efforts simply one more attraction among the stream of images on the ubiquitous multiscreens. In retrospect, they could only distance themselves from their own efforts.[29]

The most original contribution, therefore, may have been that of the architects, who provided the frames within which objects and images, people and events were accommodated – both the pavilions, the most prominent eye-catchers on site, and the plan, which choreographed the whole. In particular, the expo was embraced, at least initially, by the Metabolists. Ten years earlier, the group had proposed understanding the city as a living form, rather than a fixed structure, so as to accommodate its ongoing growth and to liberate its population, from both the dead hand of the past and the chaos of the present. The scripture allowed for a broad church, ranging from the technocratic enthusiasm of Kurokawa Kishō to the technological scepticism of Isozaki Arata, but the expo had room for them all, with space for both megastructures and a vision of everyday life. Kurokawa provided various examples of modular construction: a 'capsule house' suspended from the plaza roof; a 'tetra-frame' for Toshiba's suspended 'dome theatre'; and the extensible steel frame of the Takara Beautilion. Isozaki was more cautious. He was suspicious that the 'invisible systems' of the city of the future would undermine its potential for liberation. Still, he collaborated. And sure enough, sitting inside one of

[28] Kushima Tsutomu, *Maboroshi Bankoku Hakurankai* (Shōgakkan, 1998), 91 for Apollo, 110 for foreign hostesses, and 188–90, for a lukewarm retrospective of the theme exhibit.

[29] *Asahi Shimbun*, 1970/4/2 and 3, pm, 7, for mobilization. Yoshimi, *Hakurankai No Seijigaku*, 236–38, for the dilemma. Midori Yoshimoto, 'Textiles Pavilion: An Anomaly and Critique of Expo '70', *Review of Japanese Culture and Society* 23 (2011), 113–31, for the Textiles Pavilion. Peter Eckersall, *Performativity and Event in 1960s Japan: City, Body, Memory* (Palgrave Macmillan, 2013), 116–17, for retrospective distancing.

his two robots, a control station for the plaza, at the opening ceremony, he slowly began to realize what he had wrought.[30]

Creative endeavour and social possibility was always subordinate to a master plan. Initially, this had been assigned to two architects. Nishiyama Uzō had proposed the Symbol Zone, subsequently delegated to Okamoto, and the Festival Plaza, to promote human interaction and give substance to the claims of progress and harmony. Both became core elements of the final product. But as visitor estimates were refined and transport links plotted, the bureaucrats turned exclusively to Tange Kenzō, who was experienced at both servicing the Japanese state and selling the product. Tange turned Nishiyama's ideas to useful rhetorical account: 'a metaphoric structure involving botanical trunk, branches, and flowers was employed; … the trunk … was deliberately referred to as the Symbol Zone'. But Nishiyama's draft had been no more than an 'image plan', which was refined as Tange and his team came up with the technical answer required by the question of how to accommodate the predicted number of visitors on site. The final solution was simple enough. The main entrance led to the 'trunk'. Four sub-gates added 'branches'. Moving walkways allowed visitors to pass through their 'arteries', which lead to seven sub-plazas, allowing the tree to further branch out and the visitors to find their way to 'the pavilions, which could be compared to flowers'. But function trumped metaphor. The whole was designed to ensure 'a smooth flow of spectators'. With such a masterful plan, moreover, there was no need for further control: '[we] let each pavilion be absolutely free'. Up to a point. The main attractions were disbursed, with corporate behemoths, Cold War warriors, and Japan itself pushed to the edges of the site. Tange also professed himself 'delighted' by Okamoto's cheek, puncturing the plaza roof with the golden face of the sun. 'Within the mechanical realization of the site as a whole, [here was] something

[30] For metabolism, Zhongjie Lin, *Kenzo Tange and the Metabolist Movement: Urban Utopias of Modern Japan* (Routledge, 2010), Rem Koolhaas, Hans-Ulrich Obrist, Kayoko Ota, and James Westcott, *Project Japan: Metabolism Talks* (Taschen, 2011), and William O. Gardner, *The Metabolist Imagination: Visions of the City in Postwar Japanese Architecture and Science Fiction* (University of Minnesota Press, 2020). For Kurokawa and capsules, Thomas Leslie, 'Just What Is It That Makes Capsule Homes so Different, so Appealing? Domesticity and the Technological Sublime, 1945 to 1975', *Space and Culture* 9.2 (2006), 180–94. For his projects, 'Works and Projects', www.kisho.co.jp/page/211.html. For Isozaki's doubts, 'Mienai toshi ni idomu', *Tenbō* 107 (1967), 49–62, and for his subsequent horror, 'Recalling the Days of Expo Art (?) (2001)', *Review of Japanese Culture and Society* 23 (2011), 72–80.

human'.[31] Tange, however, could confine his attention to the core structures and invisible systems that allowed exhibitors to exhibit and crowds to move.

EXPO MEDIA

And so, more or less, they did. It is hard, when contemplating the deluge of visitors on site, as well as the flood of material that encouraged them to go there, not to be reminded of other times and places. In Toyama, more than 50,000, half the total agricultural population, made their way to the expo, courtesy of local cooperatives. In one town, the station master arranged a party to wave goodbye. 'It was just like sending off troops.' But there were other echoes, too, of campaigns early in the twentieth century to improve conditions in the countryside and, through them, of older habits of hierarchy and deference. The expo could also exploit a century of compulsory education, through the institution of the school trip. The Ministry of Education had taken control of these following the war and distributed 200,000 copies of a handbook, so that teachers might ensure 'correct understanding' of the expo among their charges. It was met by a predictable response from the indomitably left-wing Japan Teacher's Association, which produced its own reader, highlighting the potential risks for children and the exclusion of darker stories on site. It is not hard to hear the echo of earlier mobilizations during wars.[32]

[31] For Nishiyama, Andrea Urushima, 'Genesis and Culmination of Uzô Nishiyama's Proposal of a "Model Core of a Future City" for the Expo 70 Site (1960–73)', *Planning Perspectives* 22.4 (2007), 391–416. Tange Kenzō, 'Ikutsuka No Keiken', *SD* 184 (1980/1), 187, for metaphoric structure; 'Bankokuhaku No Keikaku to Mirai Toshi', *Kenchiku Bunka* 282.4 (1970), 71–72 for image plan, branches, arteries, and free pavilions and 75 for something human; and 'Sakuhin', *SD* 184 (1980/1), 60 for smooth flow. Lin, *Kenzo Tange*, and Seng Kuan and Yukio Lippit, *Kenzo Tange: Architecture for the World* (Lars Müller, 2012), for overviews of his work.

[32] *Asahi Shimbun*, 1970/9/9, pm, 10, for Toyama. Gluck, *Japan's Modern Myths*, 186–204 for local improvement. Tatsuo Kasama, 'A Century of School Excursions', *Japan Quarterly* 34.3 (1987), 287, and Markus Oedewald, 'Meanings of Tradition in Contemporary Japanese Domestic Tourism', in Sylvie Guichard-Anguis and Okpyo Moon (eds.), *Japanese Tourism and Travel Culture* (Routledge, 2009), 115–24, for the school trip. *Expo '70: Nihon Bankokuhaku Shidō no tame ni* (Monbushō, 1970) for the Ministry's handbook. *Asahi Shimbun*, 1970/2/24, 14 for the teachers' union response. Awazu and Hariu, 'Genzō no Naka de', and Sakazaki Otsurō, 'Hanpaku No Shisō', *Kikan Geijutsu* 4.2 (1970), 58–61, for the claims of state mobilization, echoed in Yoshimi, *Hakurankai No Seijigaku*, 227–34.

There were differences, however, which suggest that the mix of compulsion and persuasion was new. The novelty was evident not least in the use of the media to assemble an audience – and public relations firms to coordinate the message. Much of the work was done by Dentsu, already an advertising monolith. Founded at the turn of the century, by the 1960s it was taking home a quarter of all advertising revenue, courtesy of what it called 'total communication', which combined a rationalized management system and comprehensive production facilities. Starting in September 1965, it produced 5,000 mini-exhibits in banks, companies, and shops in 6 major cities to kick things off; the rules, regulations, and general classification; posters, slides, and lectures; a touring exhibition that visited 29 cities, a special feature for 61 domestic newspapers, and a film for the overseas market; 46 issues of a glossy monthly magazine, with English and French editions, to document the bulldozing of the Senri Hills; 3 corporate pavilions and the PR for 6 more; over 2.5 million copies of the Official Guide; and over 30,000 of the Official Report. What became, for some, 'Dentsu's Expo' netted the company nearly 7 billion yen, diversified its portfolio, and made it the third largest advertising company in the world.[33]

Dentsu's genius lay in exploiting the symbiosis between big events and the increasingly differentiated media system. Media make their money by processing noise into information, but they do it in different ways. Television was already in nearly 90 per cent of households by 1963, ready for the Olympics, which had provided a laboratory for colour, satellite, and real-time broadcasts. The expo offered even more telegenic content: pavilions and exhibits, events and national days, winsome hostesses and curious foreigners. Eternal recurrence in the guise of breathless novelty. And TV responded, with an 'epoch-making joint broadcast' of the opening ceremony, followed by a daily expo slot and multiple special programmes on all the major channels. Three times a week, NHK, the public broadcaster, presented national days. Once a week, TBS, a commercial channel, got stars to evaluate the expo. Ten days after the opening

[33] *Kōshiki Kiroku*, vol. 3, 84–112, for the official account of publicity activities. Dentsū hyakunen shi henshū iinkai, *Dentsū Hyakunen Shi* (Dentsū, 2001), for the dream of total communication, 240–42 for the Olympics, and 259–63 for the expo. Ino Kenji, *Dentsū Kōgairon* (Nisshin Hōdō, 1971), 1–3 for Dentsū's size, 175–78 for the Olympics, and 191–95 for the expo. The monthly magazine was *Nihon Bankoku Haku Nyūsu* (1966–70). Bulldozers featured in issues 10/11 (February 1967) and 20 (November 1967). See also Yoshimi, *Hakurankai No Seijigaku*, 238.

ceremony, a Tokyo housewife confessed that seeing the expo every day on TV had made her want to go in person.[34]

She had also likely read about it in a newspaper. The press was now using its own new technologies – full-colour, high-speed printing – to mimic the immediacy of TV. The expo had burst onto the front pages in 1964, establishing a pattern that would persist for the next six years, with important, eye-catching material on page one, and less newsworthy, more predictable items somewhere inside. Newspapers also retained the advantage of being able to provide more than one take, covering what was expected while also distinguishing themselves from their rivals. The *Asahi Shimbun*, for example, the standard-bearer for the mainstream left, quickly deployed a tripartite division of labour, between reporting, opinion, and critique. The BIE decision to award the expo to Osaka spawned a page of factual reporting and background, a supportive editorial about its historical significance, and a sceptical commentary on the likely financial burden. Both-sidesism soon became a recurrent feature. Landmarks prompted the editorial staff to opine, for example, on the need for nationwide collaboration. Criticism congealed in a regular front-page snippet, 'the issue of the day', which highlighted ongoing difficulties, for example, on the rising prices of cabbage, onion, and daikon. By 1969, they had been joined by longer, probing specials. How was the expo going to cope with 50 million visitors? The scepticism culminated two weeks before the opening ceremonies in a seven-part series, which collated the many reasons to be 'cautious': packed trains, a crowded site, a heavy police presence, inadequate lodging, rising prices, the shoddy treatment of developing countries, and the heavy hand of the state. Not that the *Asahi* wanted out. It had organized a competition for expo-related ideas back in January 1966, attracting over 10,000 entries and doling out prizes.[35]

[34] Niklas Luhmann, *The Reality of the Mass Media* (Stanford University Press, 2000), and Gotthard Bechmann and Nico Stehr, 'Niklas Luhmann's Theory of the Mass Media', *Society* 48.2 (2011), 142–47, for the media as a system. Partner, *Assembled in Japan*, 247, for television ownership. Dentsū hyakunen shi henshū iinkai, *Dentsū hyakunen shi*, 241, for the Olympics as lab. *Kōshiki Kiroku*, vol. 3, 54–83, for press coverage. Yoshimi, *Banpaku Gensō*, 89–90, for saturation coverage on TV, and *Hakurankai No Seijigaku*, 230, for the housewife. Andrew Painter, 'Japanese Daytime Television, Popular Culture, and Ideology', *Journal of Japanese Studies* 19.2 (1993), 295–325, for Japanese television's bias towards predictability.

[35] *Asahi Shimbun*, 1964/6/9, 1, for the first front page. 1964/6/15, 1964/7/27, 1965/1/7, 8, and 9, all 1, for early important news. 1964/8/15, 5, 1964/11/5, 7, and 1964/2/4, 7, for a few of the many less newsworthy items. 1965/9/14, 1–2, for the approval of the bid by the BIE. 1965/8/19, 2, for nationwide collaboration. 1965/11/3, 1966/3/8, 1966/5/13, 1966/

The pattern continued as the expo got under way. Opening day saw full-spectrum coverage. The morning edition included a preview of the attractions and of the day itself, full of 'ethnic colour'; another supportive editorial; poems by sceptical readers – 'income doesn't bring "harmony" / to the "progress" of prices, / does it, mum?'; and a catalogue of last-minute confusion, missing exhibits, and the thirty people who had already been hurt. The evening edition repeated the mix: three pages of illustrated reporting at the front, set off against a double-page spread further in, which asserted that Japan's 'façade' as an economic power was merely covering up its 'suffering citizens'. The first few weeks brought a serialized guide to the attractions, set against reports of teething problems and lengthening queues, which suggested that the expo was going to be a 'forced march'. There were disappointing early numbers and mixed reviews from overseas – laudatory French and Russians, but sniffy British and Americans. By mid-April, the critics were passing judgement, probing behind the expo's 'mask' to reveal the 'true face' of the city of the future: empty space, virtual worlds, visual overload, discriminatory access, technological standardization, and electronic surveillance.[36]

From mid-April, however, the expo became thin gruel. Whereas Japanese TV thrives on scheduled predictability, newspapers need stories with twists and turns. The *Asahi* continued to publish a weekly guide, so that visitors knew what to expect. It was quick to seize on any and all incidents that disturbed the regular, less-than-newsworthy operation of the site: officials continuing to censor material about protest and war, removing 10 out of nearly 200 photographs in a montage in the Local Government Pavilion; a young man occupying one of the eyes in the Tower of the Sun for a week. A couple of months in, though, the news had slowed to a trickle. The two-month mark produced a balance sheet, with the continuous stream of visitors (and ringing of tills) set against the vacuous theme and the familiar litany of a crowded site, high prices, and traffic accidents. The early summer brought a focus on labour conditions,

6/27, 1967/8/28, 1969/10/29, 1969/12/15, 1970/2/14, all 1, for various issues of the day, and 1970/3/6, 1, for vegetable prices. 1969/10/29, 4, for the special on visitor numbers, 1969/12/26, 4, for another on whether it would be ready on time. 1970/2/26–28, 14, 3/1–2, 1, and 3/3–4, 14, for the seven-part series. 1966/5/5, 1, for the competition.

[36] *Asahi Shimbun*, 1970/3/12, 17–32 for the special edition. 3/14, 1, 5, 8, and 15, and 3/14, pm, 1–3, 10–11 for opening day coverage. 3/15, 17–27, for a comprehensive guide. 3/17, pm, 2, saw the beginning of a series on 'unusual companies', 4/20, pm, 9, on performers. 3/15, 15, and 3/16, 1, 7, and 14 for teething problems, queues, and 'forced march'. 3/21, 11, and 3/22, 12, for the foreign reviews. 4/2–17, pm, 7, for mask and true face.

FIGURE 6.2 The site ground to a halt in September, but already in early August visitors had to queue to get in.
Source: Keystone-France/Gamma-Keystone via Getty Images.

with a newly formed expo union facing off against a security-obsessed Association. Their standoff was still wending its way through the legal system when the expo closed, having seen eighty-eight disputes and thirty-four full-blown strikes. After the first flurry, the *Asahi* had not been that interested.[37] By September, there were other tales to tell.

The story, this time, was visitor numbers, which would determine, in official eyes, the expo's success. The early returns had not been promising. The numbers had ticked up during the spring holidays, but not enough to meet official estimates, and had tailed off during the rainy season. Come the summer, though, they started to climb. And as records started to fall, so the expo became newsworthy again (Figure 6.2). On 7 June, 620,000 flocked to Senri, surpassing the daily record of its predecessor, Montreal. On 14 August, that record fell in turn, in the midst of a typhoon, with

[37] *Asahi Shimbun*, 1970/3/24, 14, for the first weekly guide. Censorship was a long-running theme: 2/6, 10; 2/24, 14; 3/4, 14; 3/14, 1 and 10; 4/12, 14; 5/16, 4; 9/10, 4; and 9/11, 10. For the protestor 4/27, 5/3 and 5/4, all 15, and Lockyer, 'The Logic of Spectacle', 571–89, 576–78. For the two-month balance sheet, *Asahi Shimbun*, 1970/5/16, 4. For union activity, *Kōshiki Kiroku*, vol. 3, 21 and 315–16, and *Asahi Shimbun*, 1970/6/4, 23, 7/13, 19, and 8/26, 19.

680,000 squelching around the site. It was the penultimate weekend in early September, however, that brought the expo to its knees and back to the front page. Saturday saw 835,832 visitors, easily topping the previous world record of 703,664, at the Brussels Expo in 1958, with 5,000 sleeping on site. Sunday was almost as bad. Despite emergency measures, some 783,682 visitors managed to make their way to the site, this time with 2,000 staying overnight. The *Asahi* was captivated, documenting the situation on the front cover and elaborating its critique inside. The expo had catalysed the kind of 'general mobilization' (*sōdōin*) required by war. It needed rechristening. The world exhibition (*bankokuhaku*) was in fact a brutal expo (*zankokuhaku*), which had blown a tire (*panku shita*).[38]

One weekend was not enough, however, to derail the achievements of the whole. As the expo wound down, the *Asahi* made a valiant attempt to hold the party line, with three days each of critical reflection and editorial balance sheet. On the one hand, the expo was a 'magnificent zero', underlining the 'terminal condition of machine civilization'. On the other, the infrastructure investment, together with visitor spending, had catalysed the first 'world renewal' (*yonaoshi*) in Kansai in nearly four centuries, with 400 kilometres of new roads, a doubling of local hotel capacity, and a quadrupling in the takings of a local bentō company. The same ambivalence was evident when they turned to the visitors themselves. Like foreign observers, they were puzzled by the 'buffalo stampede' when the gates opened in the morning, and the strange patience with which visitors had queued for four of the six hours they would spend on site. Behold an 'administered crowd' of mobilized sheep, driven to the site by the corporatist state. A closer look, however, suggested a curious audience, which had streamed past the avant-garde art, and towards the four pavilions, the US, the USSR, Japan, and Mitsubishi, roundly condemned by the progressive press. The US pavilion, with its lump of rock from the moon, may have been a portable shrine. But visitors were friendly and confident as they waited, for hours, to hear its message from a world beyond their own.[39]

The *Asahi* provides one example of the way in which the expo fed the media system, creating facts on the ground that evaded both the designs of the technocrats and the charges of the left. This was clearest, ironically,

[38] *Kōshiki Kiroku*, vol. 2, 362–67, for visitor numbers. *Asahi Shimbun*, 1970/3/16, 1; 3/21, pm, 11; and 3/22, 12 for early returns; 6/8, 19, for Montreal; 8/15, 22, for the mudfest; 9/6 and 7, 1 and 3, for the penultimate weekend.

[39] *Asahi Shimbun*, 1970/9/7–9, pm, 7, for critics, and 9/9–11, pm, 10, for balance sheet. 9/10, 4, for foreign impressions.

the further left you went. *Akahata*, the in-house paper of the Japanese Communist Party, rehearsed the obvious failings of the expo: the vast expense, the hollow corporate exhibits, and the omnipresent multiscreens; the poor working conditions, the disregard for visitors, and the exclusion of Asian socialist representation. The popular interest in the expo, however, meant it would have been a tactical mistake to try to 'crush' it. It was possible, moreover, to see the expo and the world in a different way. On the opening day, union members expressed pride in the collaboration that had produced such a vast event. One visitor's contrasting experiences in the 'beautiful but cold' US pavilion and the informative Soviet pavilion made them think it was 'pitiful that we trail along after America, destroying nature and cultural heritage'. In the Nigeria pavilion, another visitor was underwhelmed by poverty, but his friend was struck by progress. A twenty-three-year-old female worker from Matsue underlined the point. 'I understood that the world is one ... that people are essentially the same. The Expo ... [made] the barriers disappear'.[40]

Japan's first international exhibition fell short, therefore, either as the paean to progress promised by its organizers or as the authoritarian spectacle feared by its critics. True enough, it was impossible to stop, as implacable as the developmentalist agenda of which it was a part. The combined forces of government, business, academy, and media, together with the manifold interests of an invested public, easily outweighed the anti-expo antics of the fractious left. Also true, it was impossible to ignore, hogging the limelight for most of 1970, if not half a decade. Rather than festival or fascism, however, the expo turned out to be what expos always had been: planned and chaotic, revelatory and humdrum. It provided space for First, Second, and Third Worlds; for dignitaries, visitors, and protestors; for journalists, critics, and writers to make of it what they would.

The possibilities were exploited best, I would argue, neither by its boosters, prescribing the society of the future, nor its critics, bemoaning the state of the present, but by the pedlars of popular culture, who grasped that it was possible to turn the expo inside out. Tsutsui Yasutaka, an emerging god of science fiction, kept returning to Senri.

[40] *Akahata* 1970/3/16, 2 for working conditions, and 11, for expense; 4/30, 6, for cramped site, and 11 for labour struggles; 5/5, 6, for multiscreens, 5/9, 6, for hollow exhibits. 9/14, 1, for the complete charge sheet and for crushing the expo as a mistake. 3/16, 11, for union members. 9/15, 10, for trailing along after America. 9/14, 10, for Nigeria and the female worker from Matsue.

FIGURE 6.3 In the real world, the expo was heavily policed, but in a virtual one, rival spies could make hay.
Source: Minamoto Tarō, *Homo Homo 7* (1970), Gurūpu Zero.

First, it provided a killing ground for spies after dark, then a place where the 'great disharmony of mankind' was put on show, with American soldiers massacring Vietnamese villagers, and hungry refugees emerging from the Biafra Mansion. Minamoto Tarō also used the expo as a battlefield for spies, but this time in a gag manga, which saw the hapless agent Homo Homo 7 trying to protect the rights of men in a parallel Japanese world where feminism was on the march. The site provided a distinctive stage for one in a series of showdowns with his nemesis in the Lez Lez Block, albeit at the cost of most of its distinctive architecture (Figure 6.3).[41] Three decades later, the expo would be refigured again, as other creatives explored what else it might have to offer.

[41] Gardner, 'The 1970 Osaka Expo', 28 and 31, for Tsutsui. Minamoto Tarō, *Homo Homo 7* (Wakagi Shobō, 1971). For gag manga in general, Mio Bryce and Jason Davis, 'An Overview of Manga Genres', in Toni Johnson-Woods (ed.), *Manga: An Anthology of Global and Cultural Perspectives* (Bloomsbury Academic, 2010), 34–61. For the turn to the visceral in the 1960s, Angela Drummond-Mathews, 'What Boys Will Be: A Study of Shonen Manga', in Toni Johnson-Woods (ed.), *Manga: An Anthology of Global and Cultural Perspectives* (Bloomsbury Academic, 2010), 63–64.

At the time, though, the success of the expo was measured not by its reception but by its numbers: its ability to attract the crowds, to monitor them while on site, and to anticipate the contingencies that would inevitably result. In the lead-up to the show, the bureaucrats had surveyed public opinion repeatedly, to make sure they were going to make the trek to the site. Once things were under way, they only surveyed visitor attitudes three times, revealing that a bare majority had been impressed. The Association was obsessed with movement, however, boasting about the information technology that enabled real-time visitor tracking, and discomforted when more than half of the 120 female 'data swallows' went on strike. When all was said and done, the average visitor had spent six-and-a-half hours on site and managed to see eight pavilions; families with children eight hours, which allowed them to add one more; and agricultural cooperatives only five hours, running between fifteen. Nearly half a visitor's time was spent resting, walking, and queuing.[42]

The Association's focus was shared by its collaborators. The Osaka Prefectural Police had been on the case from October 1965. Early on, they had grasped the 'true character of this 'huge event': the need not only for prophylactic patrols (*keiei*), emergency response (*keibi*), and VIP protection (*keigo*) but also for congestion prevention (*zattō keibi*). The complexity of the problem required, so they claimed, 'total war' (*sōryokusen*). They documented their victory at length, in number, word, and image, in a commemorative report. The expo had seen 6.42 million visitors and 2.8 million vehicles, but only 1,713 criminal incidents. Photographs at the front showed the simultaneous transformation of the Senri Hills into the expo, and the creation of their own 'perfect system'. There were nods to the moments when it had trembled: accidents and protests, billed as 'happenings' (*hapuningu*); the human tsunami on the penultimate weekend. But the system had held, the moment had passed.[43]

As did the expo itself, leaving at least a couple of traces behind. First, there was the 20-billion-yen profit. Of this, 15 billion went into a fund, which for forty years provided a useful sinecure for bureaucrats descending from the clouds following their exertions in the Ministry of Finance and the National Police Agency, as well as smallish grants to

[42] *Kōshiki Kiroku*, vol. 2, 373–77, for visitor surveys; 478–99 for comms and data processing; and 371–72 for visitor movements. *Yomiuri Shimbun* 1969/9/15, 3, and 1969/12/10, 18 for 'expo flowers'. *Asahi Shimbun*, 1970/7/13, 19, for the strike.

[43] *Nihon Bankoku Hakurankai no Keisatsu Kiroku* (Ōsakafu Keisatsu Honbu, 1971), 15 for the scale of the problem, 21–23 for an outline of the system, 52 for total war, and 164–68 for the statistics.

some 4,500 projects that have been deemed to 'contribute to international understanding'.[44] Then there was the site. The set was being struck by December 1970 and the site was more or less cleared by summer 1971, but it took another decade to transform the pseudo-city into a commemorative park. A few iconic remnants survive in the northern half: the Tower of the Sun; a fragment of the festival plaza roof next to the pond; the Japanese Garden. To the east, windswept plazas are all that is left of official Japan and its corporate titans. To the west, the information overload of the city of the future has given way to the older idea of museums in a garden. A plum orchard, maple valley, and camellia wood swaddle the Japan Folk Crafts Museum and the National Museum of Ethnology, the latter incorporating the objects collected by Umesao for Okamoto's subterranean world. South of the expressway, the amusement park muddled on until a fatal accident in 2007 spelled its end. It was replaced by EXPOCITY, a shopping and entertainment complex, which provides a faint echo of the expo's promise to combine 'the joy of playing, learning, and discovering in one place'.[45]

Again, though, the visible, material remnants of the expo are only the tip of the iceberg that was conjured into being. The expo had never been intended to describe the city of the future, or to convince the general public that they lived in the best of all possible worlds. Rather, the expo served to canalize investment and catalyse spending in the service of specific official ends. It confirmed the lesson of the Olympics, that big international events could advance the national interest, as conceived by bureaucrats, businessmen, and their intellectual and creative consiglieres. Spectacle has continued to foster development in Japan ever since. It was not long before the planning for the next expo was under way.

[44] *Asahi Shimbun*, 1970/9/27, 3, for the debate. *Kōshiki Kiroku*, vol. 3, 124, for the establishment of the Commemorative Association, and 'What Is the Japan World Exposition 1970 Commemorative Fund?', for an account of the fund. Kawamura Masayo and Shin'ichi Mishima, 'Konna Dame Hōjin ga Nihon o Horobosu', *Shūkan Asahi*, 2010/4/30, for sinecure. *Yomiuri Shimbun* 2010/4/26, Osaka pm, 13, for discussions about the abolition of the first Association, and 2014/4/6, Osaka, 8, for the establishment of the second.

[45] 'Expo '70 Commemorative Park', www.expo70-park.jp/languages/english/ and 'EXPOCITY', www.expocity-mf.com/expo/, for their public faces. Umesao, *Chosakushū*, vol. 21, 13–18, for the birth of the ethnology museum. Yomiuri 2007/5/6, Osaka, 39, for the accident, and 2015/11/26, Osaka, 10, on EXPOCITY.

7

Boom Time

By the end of the 1980s, expos were everywhere: commemorating an undersea tunnel up north and a big bridge down south; serving up food in Hokkaido, celebrating greenery in Hyōgo, and combining them in the Japanese Alps; trumpeting Nara's status as the eastern terminus of the Silk Road; reinventing Nagoya from Toyota town to design destination, Osaka from domestic industrial heartland to international cultural metropolis, and Hiroshima from peace industry central to the hub of an 'island resort paradise'; smoothing a path for Japanese investment and tourism in Brisbane, in the face of Australian distrust. In the fevered imagination of one wide-eyed, right-wing commentator, when set alongside the construction of infrastructure and the rise in living standards, the staging of spectacle echoed developments in the UK in the mid nineteenth and the US in the early twentieth centuries. Given the building of bridges, the holding of exhibitions, and the consumption of shrimp (her preferred index for quality of life), might Japan too be on the verge of global hegemony?[1]

It was not to be, as soon became apparent. She was not alone, however, in her preoccupation with the wave of expos, which had built throughout the decade. In part, this was driven from the top. Expo 70 in Osaka had proved that large international events could catalyse regional development. Expo 75 in Okinawa would close the gap between

[1] David Tracey, 'Japan's Astonishing Expo Boom', *World's Fair* 9.2 (1989), 8, for island resort paradise; Rachel Sanderson, 'Queensland Shows the World: Regionalism and Modernity at Brisbane's World Expo 88', *Journal of Australian Studies* 27.79 (2003), 70–74, for Brisbane; and Arakawa Chisato, 'Ebi to Hashi to Hakurankai', *Shokun* 20.12 (1988), 72–79, for shrimp, bridges, and expos.

the islands and the mainland. Expo 85 in Tsukuba would 'image up' the nation's first science city. Five years later, Osaka would use a horticultural expo to rebrand itself. Olympics were useful too, as Sapporo showed in 1992. Nor did the bursting of the bubble in the early 1990s undermine the utility of both. The 1998 Winter Olympics in Nagano, Expo 2005 in Aichi, the 2020 Summer Olympics in Tokyo (in 2021), and the plans for Expo 2025 in Osaka suggest that the official imagination remains captivated by the possibilities of big events, as will become apparent.

It would be a mistake, however, to linger too long on national plans and the international scale. The government provided a framework within which local authorities could find their place. But cities and regions had to do their bit, too, as the economy pivoted away from heavy industry and the young left town for the bright lights of big cities. So they too turned to expos, as they had fifty years before. Many of the initiatives were predictable, rehashing local history, borrowing corporate exhibits, and promising new dawns. Many of them shared the assumptions of the plans being drawn up in Tokyo, about the desirability of growth, the possibility of planning, and the shape of the urban future. Here and there, though, the expos of the 1980s revealed fault lines, too. They acknowledged that growth had its costs, not least 'public hazards' (*kōgai*), that is, environmental pollution. By the end of the decade, some were even suggesting that the future could be crafted from the bottom-up, by local citizens.

These were only small sprouts. It would take the bursting of the bubble in the early 1990s for expos to become a site of confrontation, rather than consensus, and for the spectacle to acknowledge constraint, rather than encourage excess. In the round, the expos of the 1970s and 1980s, at home, had more than a little in common with those of the 1920s and 1930s, in the colonies. In ministerial offices and corporate headquarters, an expo was an opportunity to find some nominally virgin territory where control could be extended, capital invested, and returns secured. For their municipal compradors, willing collaboration and local initiative were the way to show their eagerness to play developmentalist ball. The authorities may not have had everything their own way. For the time being, though, they could pretend that everything was under control and that the best use of the earth was to reclaim more land.

REMODELLING THE ARCHIPELAGO

At first glance, Expo 75 in Okinawa and Expo 85 in Tsukuba looked rather different. They were held to address nominally different concerns, in very different places, with seemingly different sites, themes, and exhibits. Beneath the surface, though, they exhibited a consistency of purpose and practice, foreshadowed by Expo 70 in Osaka, which makes them a compelling diptych. Through the uncertainty of the early 1970s and the bubble of the 1980s, big events continued to reconcile bureaucratic agenda, corporate appetite, regional aspiration, and creative ambition; to accommodate interpretive diversity, market disruption, local protest, and intellectual dissent; and to ready the regional landscape for its prescribed future, when the circus left town.

The initial idea for the Okinawa Ocean Expo emerged in 1969. Following the savage Battle of Okinawa in 1945, which killed a third of the civilian population, the islands had passed into US military 'trusteeship' at the end of the war, transforming Okinawa into a staging area for American activity in the Korean and Vietnam Wars. Now, discussions were under way to transfer the islands back to Japanese control in 1972. An expo would not only serve as the third and final 'jewelled product' (*medama shōhin*), with which to commemorate reversion. It might also begin to wean the local economy from its dependency on the US military, and to 'rectify disparities' (*kakusa zesei*) with the mainland. There were differing visions of what this meant. Okinawan leaders imagined a future of tourism, with one suggesting the Kerama Islands, twenty miles west of the main island, as an expo site. Mainland companies were set on industrial development, however, and Mitsui eventually carried the day, with a proposal for a site in Motobu, on the untouched northwest coast. Come reversion in May 1972, the expo had been approved in Paris and its Association was in business. It would focus on the bountiful resources of, contemporary threats to, and technological possibilities for the ocean. More important, it would justify a gush of spending, on airports, roads, and more. Come December, the site had been identified as a future resort zone.[2]

[2] John Michael Purves, *Island of Military Bases: A Contemporary Political and Economic History of Okinawa* (2001), https://cdn.preterhuman.net/, 50–122, for an overview of reversion. Takara Yūsei, *Jiritsu no Keizaigaku: Fukki to Kaihatsu no Seisaku Dokuhon* (Roman Shobō, 1988), 43, for 'jewelled product'. *Expo '75: Official Report* (Japan Association for the International Ocean Exposition, 1976), 26–29, 37, and 68–71, for chronology, theme, and public works. *Kaiyōhaku Handobukku* (Nihon Kōgyō Shinbunsha, 1973), for its significance for development. Tada, *Okinawa Imēji no Tanjō*, 37–40 and 140–42, and Yoshimi, *Banpaku Gensō*, 109–10 and 136–39, for the differing visions. Takara, *Jiritsu no Keizaigaku*, 1–5, for contemporary comment. Sakaiya Taichi, 'Banpaku ni Sengoshi o Yomu: Okinawa Kaiyōhaku (1975 nen) o Chūshin ni', in Sano Mayuko (ed.), *Banpakugaku: Bankoku Hakurankai toiu, Sekai o Haakusuru Hōhō* (Shibunkaku shuppan, 2020), 369–86, for retrospective self-aggrandizement.

With backers, organization, and theme in place, the expo's progress was inexorable, though hardly smooth. As ever, the prospect of an expo prompted the speculators to move in. By the end of the year, 30 million square metres in the north of the island, much of it agricultural land, had changed hands. Small businesses began to go under, the prices of vegetables, fresh fish, and building materials to jump, and workers to leave agriculture for expo construction. By early 1973, 'expo inflation' was a thing. The unions began to mobilize in January. The local intellectuals got organized by June. 'Indiscriminate development' was destroying Okinawan culture and nature. The expo was meant to have been a 'catalyst' (*kibakuzai*), but it was turning into a 'suicide bomb' (*jibakuzai*). Soon, however, the oil shock, a fiscal crisis, and the possibility of cancellation moved public opinion back on side. Expo funding was the only item maintained at existing levels in the government's budget for 1974. The sniping continued, but the expo helped the new prefecture ride out the economic uncertainty. It opened in July 1975.[3]

The site sprawled for four kilometres along the Motobu peninsula, with a thin strip of pavilions, in four clusters, sloping down the hill to a subtropical park by the shore. There was space for Okinawa's own complicated three-act history in the 'Peoples and History' cluster: a golden age, when the islands had lived in harmony with the gods, the sea, and their neighbours; the closure of the sea, with invasion by Satsuma in the early seventeenth century, annexation by the Japanese state in the late nineteenth, and the violence of the Pacific War; and the possibility of now recovering Okinawa's true identity. Elsewhere on site, however, the local story was subsumed in larger claims. Just up the slope, in the vast Oceanic Culture Pavilion, Okinawa was put in its regional place, as the penultimate staging post of the Black Current (*kuroshio*), which emerged from the Pacific Islands, swept past Southeast Asia – and Okinawa – and culminated in Japan. Six Satawalese from Micronesia arrived at the expo in December, underlining the point, having made the journey in a canoe, using only the wind and the stars (Figure 7.1).[4]

[3] Takara, *Jiritsu no Keizaigaku*, 44–46, for speculation and inflation. Tada, *Okinawa Imēji no Tanjō*, 118–28, and Yoshimi, *Banpaku Gensō*, 115–23, for unions, intellectuals, and shifting public opinion.

[4] *The Sea We Would Like to See: The International Ocean Exposition, Okinawa, 1975* (Japan Association for International Ocean Exposition, 1975), 46–59, for the Peoples and History cluster. Tada, *Okinawa Imēji no Tanjō*, 96–100 and 104–11 for the two pavilions. Yoshimi, *Banpaku Gensō*, 124–28 for the intention behind the Okinawa pavilion.

FIGURE 7.1 Two views of the ocean. The Satawalese had crossed the Pacific. Aquapolis had been towed from Hiroshima and would end its life in Shanghai. *Source*: Image N-2638.05, Trust Territory Photo Archives, Pacific Collection, University of Hawaii-Manoa Library.

The past was merely prelude, however. In the present, the expo suggested wiping away the tears of subjection with a tourist-friendly vision of blue sea and sky, which was embodied in three parks on site: an amusement park for the kids, a marine animals park, close to an aquarium, and a Shore Park next to the water, planted like the roads leading to the expo to make Okinawa over into a simulacrum of Hawaii. It was the future, however, and the technology that would bring it into being, which was the main focus of the state and its corporate compradors. The government had paid to make it so, spending four times as much as the other participants combined, and forking out 32 billion yen for the site as a whole as well as 28 billion for five official exhibits. The jewel in the crown, costing 13 billion, and the long-held dream of its architect, Kikutake Kiyonori, loomed a few hundred metres offshore. Aquapolis was the world's largest

Expo '75: Official Report, 374, for a contemporary impression, and 354 for the Satawalese.

semi-submersible structure, generating its own power and incinerating its own trash: 30 metres high, with a 100-square-metre deck and a population of 40 resident-employees. Visitors walked across a plastic bridge, suspended above an 'Ocean Ranch', where 50,000 fish ranged and sea 'cowboys' rounded them up to be fed; descended via escalator beneath the waves, to a re-creation of the primeval ocean; then ascended back into mid-air, to 'experience' the sea today. The exhibits were underwhelming, but the claim was clear. It was finally possible to live a settled life on the ocean wave.[5]

Back on land, however, the discrepancy between expo vision and reality was clear. The *Asahi* was outspoken but not out of line when it condemned Aquapolis as a 'sideshow' (*misemono koya*) and the expo for its 'futility' (*munashisa*). Where was the war? The American bases? The economy or the environment? Radicals turned words into action, throwing a firebomb in the Crown Prince's direction at the opening ceremony, and following it up with a waterborne attack on a Chilean training ship in the expo port.[6] The official results were more measured, but still mixed. Only 3 million rather than the predicted 4.5 million visitors made it through the gates, to battle exhaustion, dehydration, and inflated prices. Still, three-quarters had a good time and the whole thing had ended up in the black. The local economy was not so lucky. Most visitors had used mainland companies for transport and lodging, producing a string of local expo bankruptcies, which kept coming. Tourist numbers collapsed in 1976. When they recovered, the trend of steady growth of mainland tourism to Okinawa resumed, which had increased even under American trusteeship. The expo may have caused considerable temporary local disruption, it turned out, for little permanent gain.[7]

[5] *The Sea*, 88–92, 96–100, and 106–8 for parks, 113–21 for Aquapolis and ranch, and 60–62 for Mitsubishi. Tada, *Okinawa Imēji no Tanjō*, 48–54 for the roads. Gerald Figal, *Beachheads: War, Peace, and Tourism in Postwar Okinawa* (Rowman & Littlefield, 2012), 101–19 for the long history of tropical planting. *Expo '75: Official Report*, 358 for government expenditure. Lin, *Kenzo Tange*, 111, and Koolhaas et al., *Project Japan*, 18 and 129–55, for Kikutake. *Expo '75 Bulletin*, for official publicity. Blaxell, 'Preparing Okinawa', 10–16 for another account.

[6] *Expo '75: Official Report*, 359 for the Asahi commentary, and 14–15 for radical activity, for which see also Tada, *Okinawa Imēji no Tanjō*, 128–31.

[7] *Expo '75: Official Report*, 16 for mixed results, 356 for visitor reactions, and 358–60 for media reports. *Yomiuri Shimbun* 1976/1/16, 3, for bankruptcies, 1976/1/19 for visitor numbers and satisfaction. Tada, *Okinawa Imēji no Tanjō*, 134–38 and Figal, *Beachheads*, 211–16, for the impact on tourism.

It had cemented a particular image of the islands, however, distilling the tourist elixir of blue sea and blue sky, condensed in the three waves of the expo mark. It had also put in place some infrastructure required to accommodate and disperse the tourists – when they eventually came back. Airlines and others began to build up their Okinawa operations in the late 1970s, but it was only in the 1980s that Motobu began to come into its own. First came additional infrastructure: a national highway in 1982 and a bridge to an offshore island in 1985. Then further incentive: a national resort law in 1987 and a prefectural master plan in 1990. But the site was showing its age. Aquapolis closed in 1993, sold off in 2000, and towed to Shanghai, to be taken apart for scrap. The aquarium shut down in 2002, and a massive replacement was built. Finally, too, there were places to stay. Small resort hotels had started to appear at the end of the 1980s. In 2014, a brewery company opened a sprawling resort and spa at the north end of the expo site.[8] It had taken four decades, more or less, before the vision of sun, sand, and sea had been realized.

Where Expo 75 was an attempt by state and corporate actors to bring a newly reacquired periphery to heel, Expo 85 was an attempt by a peripheral state agency to get in on the action. Both Tsukuba the city and Tsukuba the expo, however, were dogged by the questionable attractions of a big project in a faraway place, as well as local opposition. The city had started up in 1963, as somewhere to put bureaucrats and relieve overcrowding in Tokyo. Soon, the Science and Technology Agency (STA) took over, aiming to consolidate the nation's advanced research capacity at a single site. But the going was slow. The budget was limited, locals suspicious, academics unenthusiastic, and transport links non-existent. In the mid-1970s, it was still a lonely place. So, in the wake of Expo 75, the STA began to wonder if an expo might help. The turn from heavy industry to information society was no longer academic prediction, but economic fact. Why not show how science could advance 'the health of humanity and ... the planet', how Japan could base itself anew on science and technology (*kagaku gijutsu rikkoku*)? Why not show off Japan's new 'science mecca'? By the end of 1978, the proposal for a third

8 *Expo '75: Official Report*, 357, for the original plans for the site. Tada, *Okinawa Imēji no Tanjō*, 150–59, and Figal, *Beachheads*, 215–26, for Okinawa resorts. David Nguyen, 'Tourism Development in Okinawa: Spatial and Temporal Patterns' (MA thesis, University of Hawaii, 2012), 45–51, for the official plans, 78–88 and 112–14 for the actual resorts. 'Tōsha no Rekishi', www.maedasangyo.net/history/, for local resort hotels. *Okinawa Times* 2000/10/28, for the end of Aquapolis.

expo 'bullet' (*tama*) had been endorsed by local authorities, cabinet, and Diet.[9]

The organizers still had to navigate some quagmires. First, in Tokyo, the initial ask was slashed in half, courtesy of the Ministry of Finance. The expo would have to rely even more than its predecessors on the gambling industry but would still not be accessible by rail. Then, in Tsukuba, there was a squabble between the six municipalities into which the city was divided. Five made their case, but Yatabe, the cheapest choice for the site, was determined not to share the spoils, and the status quo remained. During the expo, foreign dignitaries would arrive on site to be greeted by six local mayors. It was unclear until the last minute, however, if the expo would attract either foreign participation or domestic visitors. In the end, there were forty-seven governments on site, but two years ahead of time only twenty countries had agreed to participate. One year later, only one in ten members of the public were planning to go, with women especially reluctant. The advertising professionals went to work. PR shows were staffed by female 'companions', who promised that it would be possible to touch your 'dreams'. A seven-carriage train, Ekisupo-go, went on a six-month tour, stuffed with models of the pavilions, satellite comms, and more. The press pitched in to big the expo up. Three months before the opening ceremonies, things had improved slightly. Just over five in ten were now interested in going, but four were still determined to stay away.[10]

They had their reasons. Lodging was limited, despite the biggest capsule hotel in Japan (eight kilometres from the site), and access was

[9] Yoshimi, *Banpaku Gensō*, 151–58. For Tsukuba, James Dearing, *Growing a Japanese Science City* (Routledge, 1995), 45–56; Hiroshi Nishimaki, 'The Making of Garden Cities: Milton Keynes, Irvine and Tsukuba' (PhD thesis, University of California Berkeley, 2001), 21–25 and 36–43. For the shift from *jū-kō-chō-dai* (heavy, thick, long, large) to *kei-haku-tan-shō* (light, thin, short, small) products, *Yomiuri Shimbun* 1982/12/26, 8, and *Ekonomisuto*, 1985/2/4, 26–33. For the expo as a bullet, *Yomiuri Shimbun* 1978/07/16, 1.

[10] Yoshimi, *Banpaku Gensō*, 158–62. For gambling, *Yomiuri Shimbun* 1982/4/27, pm, 2, and *Ekonomisuto*, 1985/02/04, 52–55. For the local squabble, *Yomiuri Shimbun* 1982/10/01, pm, 1 and 1985/2/18, 22; and Nishimaki, 'The Making of Garden Cities', 88–93. For foreign participation, *Yomiuri Shimbun* 1981/9/25, pm, 2; 1983/9/8, pm, 1; 1984/3/17, 1; 1985/3/16, 1; and 1985/4/29, 23. For prospective visitor numbers and publicity campaigns, *Yomiuri Shimbun* 1981/4/13, 4; 1983/1/23, 3; 1983/8/30, 22; 1984/6/4, 3; 1984/9/17, 22; 1984/12/10, 1; 1985/1/19, 1. For fluffing by the press, 1984/7/18–20, 6; 1984/8/27–31, pm, 3; 1984/9/10–17, 22; and 1985/2/13–18, 22. For companions, Takai Masashi, 'Hakurankai to Jendā, Kaisō: "Konpanion" o Jirei Toshite', in Fukuma Yoshiaki, Nanba Kōji, and Tanimoto Naho (eds.), *Hakuran no Seiki: Shōhi Nashonariti Mejia* (Azusa Shuppansha, 2009), 161–81.

difficult. A day trip from Tokyo meant two hours each way by train and shuttle bus, or three there and back by car.[11] Once through the gate, moreover, the pickings were thin. Budgetary constraints had meant a bare-bones site, divided into functional blocks, distinguished by colour rather than design, which reminded the press of a hutment, like those at a building site, a circus, or a barracks.[12] Nor were the exhibits too compelling. A few tried to follow the official brief and display the benefits of science and technology for 'humanity, dwellings, and environment'. The Japanese government used a story from the early chronicles about the 'giant snake' (*orochi*) of Yamata to bloviate about 'man's longing for iron and rice'. It exhibited a tomato plant with 10,000 fruits. Most exhibits focused, instead, on the wonder of technology, but tech proved hard to display. A maglev train trundled along a test track at one-tenth of the speed promised in Osaka fifteen years earlier. The American introduction to artificial intelligence was underwhelming, if not incomprehensible. Twenty-six out of the twenty-eight corporate pavilions therefore resorted to the familiar massive screens, including Sony's débute of the Jumbotron, prompting the press to characterize it as an 'image expo' (*eizō hakurankai*), and one novelist to suggest that it was like walking along a street of cinemas. Given the lack of competition, it was easy enough for the robots to steal the show, lifting a 440-pound barbell for Fujitsu, drawing portraits for Matsushita, playing the organ and performing (simulated) surgery for Waseda, sculpting ice for Hitachi, and spinning tops for Toshiba. In the Fuyo Robot Theatre, fifty smiling, curvy, companion bots on wheels tried to demonstrate the future of care. For one editorialist, they were notable mainly for breaking down.[13]

11 For lodging, *Yomiuri Shimbun* 1984/7/3, 3; 1985/3/16, 37. For transport, ibid. 1984/9/17, 22; 1984/12/7, pm, 1; 1985/2/18, 22; and 1985/3/16, 29–30. For problematic access, *Expo '85: Dwellings and Surrounding – Science and Technology for Man at Home* (Japan Association for the International Exposition, Tsukuba, 1985, 1981), 8–9.

12 For hutment, Yoshimi, *Banpaku Gensō*, 170. For more positive spins on the site, *Yomiuri Shimbun* 1981/11/27, pm, 1; 1984/3/17, 1 and 12–13; and 1984/12/7, 22.

13 For an overview, *Ekonomisuto*, 1985/2/4, 15–33; Osamu Senna, 'The Top Ten Pavilions at Tsukuba Expo '85', *Japan Quarterly* 32.3 (1985), 304–7; and Yoshimi, *Banpaku Gensō*, 163–67. For Japan, *Tsukuba Expo '85: Official Guide Book* (Japan Association for the International Exposition, 1985), 140–43. For tomatoes, *Yomiuri Shimbun* 1984/5/24, pm, 9; 1985/7/23, pm, 6; and 1985/9/17, pm, p. 1. For maglev, *Globe and Mail* 1985/3/1, 8. For AI and IT, *Ekonomisuto*, 1985/2/4, 56–59. For multi-screens and image expo, *Yomiuri Shimbun* 1982/2/26, pm, 15, 1985/3/14, pm, 7, and 1985/9/17, 3; and *Ekonomisuto*, 1985/2/4, 51. For robots, *Yomiuri Shimbun* 1985/9/17, 3; *Sunset*, 174 (1985/3), 72–73; Reuben Hoggett, '1985 – Marco and the Fuyo Robot Theater Expo '85' (2011), http://cyberneticzoo.com/robots/1985-marco-and-the-fuyo-robot-theater-expo85-

The critics, too, were unimpressed. Hariu Ichirō, doyen of leftist expo sceptics, insisted that the whole thing was 'belchingly boring' (*geppu ga deru hodo*). Sakaiya Taichi, who'd made his reputation at Osaka in 1970 and helped plan Okinawa in 1975, dismissed it as a mah-jong tile one could safely discard (*anzenpaishugi toiu inshō*); the conservative *Yomiuri* newspaper as a 'Disneyland' of images and robots.[14] The press was also quick to seize on evidence that the expo was falling short. Visitors were few on the ground at first and businesses soon suffering. The first bankruptcy came at the end of April, of a business that had supplied beds to the capsule hotel, and continued throughout the run. By September, twelve companies had gone bust. By Christmas, the merchants were mobilizing for over 1 billion yen in compensation.[15] The expo itself, however, was in the black. Visitor numbers had hit their target a few days before it closed, in part thanks to the 30 per cent of middle schoolers who had trooped around the site, and the Association ended up 15 billion yen to the good. There were complaints about long lines, by now de rigueur, but one in four children had become more interested in science and technology; nine out of ten adults felt that the expo had 'helped familiarize them with new ... developments'.[16]

As in Osaka and Okinawa, though, the expo itself had only been part of the point. The hutment disappeared quickly, with 300,000 tons of expo trash dumped in what had once been a river, as well as some flooded gravel pits.[17] The site was transformed into a research park, which quickly filled up, corporate demand spilling over to all six municipalities, which finally merged in 1987 to manage the growth (Figure 7.2). Land prices tell the story. One tsubo (roughly three and a third square metres) cost 100 yen in 1963, 40,000 in 1973, 200,000 in 1985, 2,000,000 in 1987 (thanks in part to the nationwide asset price bubble), and 600,000

automax-japanese/; and Yuji Sone, *Japanese Robot Culture* (Palgrave Macmillan, 2016), 61–71.

14 Yoshimi, *Banpaku Gensō*, 167; Sakaiya Taichi, 'Tsukuba Kagakuhaku de Nihon wa dō Kawaru', *Chūō Kōron* 100.4 (1985), 112–17; and *Yomiuri Shimbun* 1985/3/16, pm, 1; 1985/3/17, 3.

15 Yoshimi, *Banpaku Gensō*, 173–77, and *Yomiuri Shimbun* 1985/5/1, 23, and 1985/8/14, 6, for bankruptcies; 1985/12/25, 22, for merchants demanding compensation.

16 *Yomiuri Shimbun* 1985/8/16, 22, 1985/9/2, 22, 1985/9/9, 23; and 1985/9/13, pm, 1, for breathless commentary on visitor numbers; 1986/6/25, 2, for the profit; 1985/9/17, 3, for general reactions; 1985/9/26, 8, for long lines. Claude Servant and Ippei Takeda, *Study on the Impact of International Expositions* (Bureau International des Expositions, 1996), 106 and 110–11, for public opinion surveys.

17 *Yomiuri Shimbun* 1985/9/28, 9; 1985/10/7, 22.

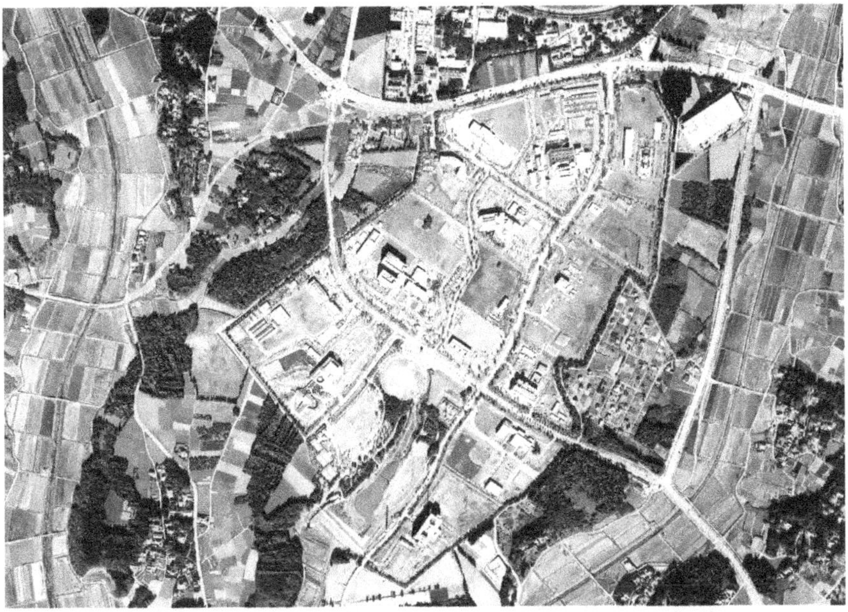

FIGURE 7.2 Five years after Expo 85, its outline was still visible, but the site had been turned into a research park.
Source: Geospatial Information Authority of Japan.

when things settled back down.[18] The expo was a watershed for Tsukuba, as the STA had hoped, though it took another twenty years before it finally got a high-speed rail link to Tokyo. The expo's effects were also not limited to the town. It had cost 118 billion yen to stage, but this was dwarfed by the 1.7 trillion yen of indirect expenditure it prompted between 1980 and 1987, the 2.7 trillion yen of production this stimulated during the same period, and the additional 3.3 trillion in the succeeding eight years. By 1995, one study, commissioned by the Japanese government, used macroeconometric models of prefectural and national economies to conclude that every yen spent in Ibaraki in 1985 had generated three-and-a-half somewhere else in Japan.[19]

At first sight, Tsukuba and Motobu do not have much in common, beyond the parks commemorating the expo sites. In Okinawa, subtropical flora provides a frame for beach, blue sea and sky. In Tsukuba,

[18] Dearing, *Growing a Japanese Science City*, 56–58; Nishimaki, 'The Making of Garden Cities', 132–38.
[19] Servant and Takeda, *Study*, 120–25.

greenery sits placidly among the corporate research facilities. On the surface, too, the two expos seem rather different. The one in 1975 celebrated the reunification of orphan Okinawa with proud parent Japan, while 1985 provided toddler Tsukuba with some much needed economic stimulus. Under the surface, though, the continuity is clear. In each case, the expo was an opportunity to channel public funds and catalyse private investment to parts of the archipelago that had previously been off the map. In each case, the expo provided the opportunity to subject a local place to central design, bulldozing the environment to produce a blank slate on which producers could project their visions, companies their pitches, and consumers their dreams. In neither case did state or capital get things entirely their own way. Metropolitan investment prompted local interest, intellectual reaction, and the occasional competing story on the ground. The logic of development, however, continued to compel. In Okinawa and Ibaraki, the state had to step in to make it so. In other places, locals were increasingly doing it to themselves.

REBRANDING THE CITY IN THE EARLY 1980S

Come the 1970s, an intermittent trickle of regional expos was becoming a steady stream. Some were testament to state-driven development. Smaller cities had long used expos to celebrate their incorporation within the national rail network. Okayama in 1972 and Fukuoka in 1975 commemorated the arrival of the bullet train, the latter with a 'symphony of humanity, nature, and science', blissfully unaware that it would likely facilitate the flight of their young out of town.[20] Four years later, Okayama was at it again, this time celebrating the beginning of construction on the world's largest double-decker bridge with an expo dedicated to the future of mobility, which ranged from hydrogen cars, through a virtual flight through the Grand Canyon, to a high-speed rollercoaster, which had debuted in Washington DC three years before. Larger places made the link between national interest and Cold War collaboration explicit. Summer 1978 saw the opening of a Space Science Expo in Tokyo, the brainchild of Sasakawa Ryōichi, tech-washing his ascent from wartime profiteer, through gambling monopolist, to well-connected pillar of the establishment. Four years earlier, his foundation had opened

[20] For the mixed blessing of the shinkansen, Komei Sasaki, Tadahiro Ohashi, and Asao Ando, 'High-Speed Rail Transit Impact on Regional Systems: Does the Shinkansen Contribute to Dispersion?' *Annals of Regional Science* 31.1 (1997), 77–98.

a ship-shaped Museum of Maritime Science on some reclaimed but unused land in the bay. Now it was surrounded by twenty-five items of 'national-treasure-level kit', sent over by the National Aeronautics and Space Administration (NASA), including the seventy-metre-long, eighteen-ton Saturn 1B rocket. The combination was catnip to the elementary-school science fan, and by mid-January 1979 well over 5 million had come through the gates, prompting a repeat performance from March to September, which brought in another 6 million. The progressive governor of Tokyo had been opposed, but Sasakawa's cronies in the ruling party thought it was the ideal Japanese contribution to the International Year of the Child. The expo did little to transform a strange, lonely place, however, until plans for a 'teleport' emerged in the mid-1980s.[21]

Other cities had to think a bit harder about where they belonged in Japan's evolving economic geography, as companies moved industrial production offshore and people flocked to the big cities. As in the 1930s, local authorities not only had to plug into the circuits of capital and culture. They also had to highlight their distinction, promoting their potential and persuading their people to stay.[22] The size of the problem, the scale of the ambition, and the shape of one potential solution became clear in Kobe in 1981. The planning for what would become Port Island had begun in 1966. 'Against the geographic odds', the city had 'bulldozed off the craggy ends of some ... hills', moving 80 million cubic metres of earth via an overhead conveyor belt to 'reclaim the vast expanse of the sea' (Figure 7.3). The operation made space for Kobe's first new town on the flattened mountain and a 'new culture city' on the 436-hectare artificial island a kilometre offshore, with a 'pentagonal division of functions': an 'international square' in the middle of the island for work and leisure;

[21] 'Uchū Kagaku Hakurankai', https://nippon.zaidan.info/kinenkan/history30/2/2531.html, for the official account. *Yomiuri Shimbun* 1978/2/17, 21; 1978/2/25, 22; 1978/5/1, pm, 9; 1978/6/3, pm, 11; 1978/8/20, 28; 1979/1/15, 22; and 1979/1/31, 22 for contemporary promotion and comment. For Sasakawa, Richard J. Samuels, *Machiavelli's Children: Leaders and Their Legacies in Italy and Japan* (Cornell University Press, 2005), 242–46. For gambling, Tom Gill, 'Those Restless Little Boats: On the Uneasiness of Japanese Power-Boat Gambling', *The Asia-Pacific Journal: Japan Focus* 11.42.2 (2013). For reclamation, 'Tokyo Minato Umetate no Ayumi', www.tokyoport.or.jp/43pdf_01.pdf, and *Yomiuri Shimbun* 1978/2/19, 21. For the teleport Asato Saito, 'Global City Formation in a Capitalist Developmental State: Tokyo and the Waterfront Sub-centre Project', *Urban Studies* 40.2 (2003), 293–96, and the following chapter.

[22] Louise Young, *Beyond the Metropolis: Second Cities and Modern Life in Interwar Japan* (University of California Press, 2013), for the interwar dilemma and attempted resolution.

FIGURE 7.3 In 1976, Port Island was under construction. Five years later, it would become Portopia.
Source: Geospatial Information Authority of Japan.

a 'community square', with 6,500 units for 20,000 residents, for living and learning; and a port with berths for 16 liners and 12 container ships, equal to nearly half of Kobe's exports and four-fifths of its container traffic. There were three parks, a green belt, and a new computer-operated transport system, the Portliner, which would minimize sound pollution.[23] What better way to announce the post-industrial vision to the

[23] *New Cultural City on the Sea: Port Island, Kobe City* (Kōbe-shi Kaihatsu-kyoku, 1979), for promotion. *Yomiuri Shimbun* 1980/12/2, 5 and 1981/3/19, 4 for reporting. See also *Kenchiku to Shakai* 70.6 (1989), 52. For the new town, Hisako Koura, 'Old "New Towns" in Kobe and Future Perspectives of Inhabitants in Shrinking Detached Housing Areas: Case of Takakuradai in Kobe City', in Y. Kadono, A. Beilein, J. Polívka, and C. Reicher (eds.), *Maturity and Regeneration of Residential Areas in Metropolitan*

world than an expo? What became known as Portopia gestured both back to the piers, from which the city had connected to the world in the late nineteenth century, and forward, to the utopia that would emerge on the island in its wake.

And who better to get the show on the road than Kobayashi Kōhei, the adopted, third-generation heir to the railway, retail, and entertainment conglomerate founded by grandpa Ichizō – and inaugurated with expos – in the early twentieth century? Kōhei deployed the full genealogical arsenal to get the word out: a Takarazuka festival; a 'culture train', which made its way down from Tokyo, with 545 students being tutored on the meaning of life (*ikigai*); and an expo song, which hymned the transformation of sea into land, 'solid as a rock', as the realization of a dream, 'sail[ing] into dock'. Also, crucially, the crown jewel (*medama*), which any expo needed in order to attract the crowds. Two one-year-old pandas, Long Long and Zhai Zhai, luxuriated in 'deluxe VIP quarters … air-conditioned to simulate … the Chinese climate', captivating the crowds and drawing 10 out of the 16 million who visited the expo. They were not the only thing to see. By the end of the summer, people were queuing for eight hours to get into the last showing of the day at the Omnimax theatre staged by Daiei. They were also spending. Suntory sold more than four times as much product as expected, including 10,000 souvenir bottles of whisky, at 12,000 yen each. Most important for Kobe, they were experiencing what it might be like to live and work and relax and even learn in the new culture city by the sea.[24]

Not everyone was convinced, of course. For some, island and event were simply the last fruitless flower (*adabana*) of high economic growth. For one Osaka professor, the expo was little more than a 'laboratory for calculating the patience of the Japanese'. 'I went, I queued, I was knackered', went one headline on the closing day.[25] But this was what expos did. And Kobe provided a model, showing how they could not only lustrate new infrastructure and attract the crowds but also rebrand the

Regions: Trends, Interpretations and Strategies in Japan and Germany (TU Dortmund, 2017), 130–45.

[24] *Kobe Port Island Exposition: Official Guide* (Kobe Port Island Exposition Association, 1981). *Yomiuri Shimbun* 1978/8/15, 5; 1981/2/14, pm, 8; 1981/4/27, pm, 10; and 1981/9/15, 3. Nanba Kōji, 'Hachijū Nendai no Chihōhaku "Ransai"', 10.

[25] *Yomiuri Shimbun* 1980/12/2, 5 and 1981/9/15, 3.

city itself. The next few years, leading up to Tsukuba, saw a regional expo boom, which crested in 1984 with six major expos and 11 million visitors.[26] Some tried to ride its utopian coattails. There was a Green-pia in 1982 (Hokkaido), a Marine-pia in 1983 (Fukuoka), and two pias in 1985 (Iwate and Naruto, which simply tagged the two syllables onto their own names). There would be six more in the second half of the decade. Some turned instead to the past, riding the burgeoning confidence in Japan's pre-industrial achievements, the media's mining of early modern history, and the consequent, emerging Edo boom. Osaka kicked things off with a Castle Expo in 1983, building a re-creation of the castle town on the north side of the ferro-concrete keep. Nagoya responded the following year, with its gold *shachi* in the starring role, and a dramatic mash-up of Edo-period automata (*karakuri ningyō*) and fifteen-metre-long, late twentieth-century robot dragons.[27]

More important than plaster and pastiche, though, were the increasing awareness of and emphasis on an expo's economic effects. Portopia had cost just over 27 billion yen and made just under 34, largely from entrance fees. But the 16 million visitors had also spent 100 billion on site and injected two-and-a-half times that amount into the local economy. Putting it all together, the expo prompted 500 billion yen of demand locally and nearly 2 trillion nationwide. Kobe benefited from a relatively low-cost structure, some genuine novelties (island, Portliner, pandas), and a vast potential audience in the immediate region. Expos in more remote places had a harder time attracting the crowds. In 1984, the small city of Otaru, an hour west of Sapporo in Hokkaido, considerably undershot its 2-million visitor target but still generated well over 20 billion yen of demand in the region. Tsukuba, the following year, would drive the point home. Expos were a game-changer, even when they failed to convince. They were becoming less important for what they said about a place than for what they could do to transform its fortunes and place in the world. Already in 1985, Japan was looking like an 'expo archipelago'.[28] By the end of the decade, their promise was accepted as gospel, producing an expo frenzy that exceeded the enthusiasm fifty years before.

[26] Nanba, 'Hachijū Nendai no Chihōhaku "Ransai"', 13.

[27] Carol Gluck, 'The Invention of Edo', in Stephen Vlastos (ed.), *Mirror of Modernity: Invented Traditions of Modern Japan* (University of California Press, 1998), 273–75 for the Edo boom. *Yomiuri Shimbun* 1983/10/1, pm, 14 and 1983/11/5, pm, 9, for Osaka.

[28] *Yomiuri Shimbun* 1981/12/8, 8 and 1985/8/29, 3; and *Ekonomisuto*, 1985/2/4, 60–65.

REVITALIZING THE REGIONS IN THE LATE 1980S

The wave of regional expos that crested in 1989, with the municipal system and therefore many cities turning a hundred, exceeded anything the archipelago had seen. How you measure it depends on your definitions and instruments. The contemporary media counted seventeen major expos in 1988, which attracted nearly 29 million visitors, and 15 in 1989, which pulled in roughly 50 million. But this was only the tip of the iceberg – and the municipal anniversary may have been merely a convenient excuse. By one count, by 1989, Japan was seeing 2,000 local government 'expos' a year and 7,000 in total, including privately organized exhibitions. Those numbers likely stretched the definition, to include quite a few 'events' (*ibento*), whose number had also been growing throughout the decade. The Regional Revitalization Centre had already counted considerably more than 8,000 between September 1987 and August 1988.[29]

At first glance, again, it is tempting to dismiss this rush of expos as yet another thread in the tapestry of development being woven by the state, its corporate collaborators, and cultural enablers. The 1970s had started with a government campaign to 'discover Japan', as a way of mitigating the lull in domestic consumption following the Osaka Expo, which had fertilized an ongoing metropolitan discussion about the need to reconnect with one's hometown (*furusato*, literally 'old village'), back in the countryside. The Fourth Comprehensive National Development Plan in 1986 focused on the regions, in part as a stage for leisure, which was outlined in yet another plan, for regional resorts, in 1987. But the government understood that it also had to fertilize the local roots, as expressed by the determination of Takeshita Noboru, the new premier, to create a 'wonderful country' (*subarashii kuni*), where everyone could enjoy a 'safe, rich, and convenient' life 'in their region'. A project to stimulate 'hometown creativity' (*furusato sōsei*) took concrete form in 1988, with grants of 100 million yen to municipalities nationwide. None of that went to expos. Even the smallest one worthy of the name required 1 billion yen. In 1989, though, together with an announcement of yet more funding for 'hometown municipal zones', the government approved

[29] *Yomiuri Shimbun* 1986/1/26, 29, for a premonition; 1988/5/1, 11, for the beginning of the flood; and 1989/1/30, 12–13, and 1989/2/11, 2, for a forum on its significance. See also *Kenchiku to Shakai* 70.6 (1989), 73, and Nanba, 'Hachijū Nendai no Chihōhaku "Ransai"'.

lotteries as a way of funding expos and other events with an expected attendance of over 700,000.[30]

Seen in the aggregate, too, the newly prominent regions produced familiar, predictable shows. Some turned to the past, celebrating its national and/or global dimensions. The year 1987 saw three more castle-focused expos in central Japan, including Okazaki, which trumpeted its status as the birthplace of Tokugawa Ieyasu. A one-tenth model of the Tōshōgū, his commemorative shrine, had been exhibited at the 1964 New York World's Fair and spent the time since resting in Hawaii, but was now welcomed 'home'. The following year brought a Silk Road Expo and 6 million visitors in Nara. Following a decade of Silk Road broadcasting by NHK, it rebranded its own Tang-period national treasures as world heritage, courtesy of 650 exhibits from 10 countries, a slew of replicas, and 30 tons of sand from the Taklamakan Desert in China's far west.[31] 'Dahha Rando' in Sakai in 1989 celebrated the brief encounter the town had enjoyed with the Netherlands four centuries earlier.

Others celebrated their incorporation in the national network. The simultaneous inauguration in 1988 of the huge Seto Ōhashi Bridge, which connected southwest Honshu to northern Shikoku, and the Seikan tunnel, which did the same for northern Honshu and Hokkaido, brought two sets of paired expos, in Kurashiki and Sakaide, at either end of the bridge, and Aomori and Hakodate, up north, seeking to exploit the increase in traffic.[32] Whatever they called themselves, though, most expos combined local past and connected present in a story about where the region was going. Sendai in 1987 had provided a template. It capitalized on rail privatization, the surging nostalgia for the rural past, and the ongoing

[30] For Discover Japan, Marilyn Ivy, *Discourses of the Vanishing: Modernity, Phantasm, Japan* (University of Chicago Press, 1995), 29–59. For the resort law, Peter Rimmer, 'Japan's "Resort Archipelago": Creating Regions of Fun, Pleasure, Relaxation, and Recreation', *Environment and Planning A: Economy and Space* 24.11 (1992), 1599–625. For hometown creativity, *Yomiuri Shimbun* 1988/9/8, 11; 1988/12/19, 3; and 1989/1/22, 3. For regional revitalization, *Kenchiku to Shakai* 70.6 (1989), 78; John Knight, 'Rural Revitalization in Japan: Spirit of the Village and Taste of the Country', *Asian Survey* 34.7 (1994), 634–46; and Sam K. Steffensen, 'Ideals versus Realities in the Japanese Periphery: The Case of Endogenous Development', *The Copenhagen Journal of Asian Studies* 9 (1994), 54–79.

[31] For Nara, *Yomiuri Shimbun* 1988/5/1, 11; 1988/8/23, 7; and 1989/1/30, 12–13; and *Kenchiku to Shakai* 70.6 (1989), 67 and 71. Tim Winter, 'Silk Road Diplomacy: Geopolitics and Histories of Connectivity', *International Journal of Cultural Policy* 26.7 (2020), 898–912, for context.

[32] *Yomiuri Shimbun* 1988/5/1, 11, for big projects; 1988/9/9, 13, and *Kenchiku to Shakai* 70.6 (1989), 65, for the two ends of the bridge.

dramatization of the story of its founder, Date Masamune, the 'one-eyed dragon', on NHK. Nearly 3 million flocked north, to 'rediscover Japan's hometown' and to find a 'magic vision' of life there fifty years hence: parents and children up in space, grandparents on holiday in Fiji, while robots ran the show back home.[33]

The volume and predictability of expos, predictably, made them catnip for critics and the press. An expo, so the experts claimed, needed a crown jewel (*medama*) – cute pandas! big rockets! – with which to attract the crowds. But most cities lacked the reputation or connection to land the biggest foreign fish. Too often, organizers were content to outsource the creative work; too often, the PR companies simply recycled their back catalogue; and too often, this meant a predictable corporate line-up, in the form of the familiar multiscreens. Before long, there was a metropolitan consensus: expos were all the same. Like Kintarō candy (or Brighton rock), wherever you cut into the archipelago, you would see the same thing.[34] Even better for the media vultures was an expo that proved a drain on the public purse. The World Food Festival tried to show how 'eating is good' in Sapporo in 1988, but it was one of four concurrent expos in Hokkaido alone, the exhibits were little more than text panels, the menus were unappetizing, and everything cost too much. Then, in September, the emperor started dying, the country began to restrain itself (*jishuku*), and the visitors dried up. The eventual loss of nearly 9 billion yen would take ten years to pay off.[35] The journalists made hay and the critics went to town, taking a sample of one for the truth of the lot and setting course for much academic commentary ever since. November saw a psychologist and a novelist lumping expos together with financial recession and feral children as evidence of a fin de siècle. By the following

[33] *Yomiuri Shimbun* 1987/7/2, Katei Keizai, 8; and 2003/9/4, pm, 4. For nostalgia, William W. Kelly, 'Rationalization and Nostalgia: Cultural Dynamics of New Middle-Class Japan', *American Ethnologist* 13.4 (1986), 603–18, and Jennifer Robertson, 'Furusato Japan: The Culture and Politics of Nostalgia', *International Journal of Politics, Culture, and Society* 1.4 (1988), 494–518.

[34] *Yomiuri Shimbun* 1986/1/26, 29; 1988/5/1, 11; and 1988/12/19, 3. Manita Yukio, *Chiiki o Tsukuru Yume Sōchi: Hakurankai kara Chiiki o Miru* (Seibundō Shinkōsha, 1991) and Nanba, 'Hachijū Nendai no Chihōhaku "Ransai"'. See also Hirano Shigeomi and Hirano Akiomi, *Ibento Fukokuron: Sofuto-ka Jidai no Chiiki Kaseika Senryaku* (Tōkyū Ējenshī Shuppan Jigyōbu, 1987).

[35] *Yomiuri Shimbun* 1988/7/16, 9; 1988/10/5, 30; 1988/10/7, 31; 1988/10/11, 9; and 1989/6/30, 31. For *Jishuku*, Osamu Watanabe, 'The Sociology of *Jishuku* and *Kichō*: The Death of the Shōwa Tennō as a Reflection of the Structure of Contemporary Japanese Society', *Japan Forum* 1.2 (1989), 275–89.

June, an architect was insisting that there was 'nothing in the world as annoying as an expo'.[36]

Before following their lead, though, it is worth trying to flip the lens. Looking harder at expos in the regions suggests that the gaze from the metropolis obscures more than it reveals. First, by the late 1980s, it was clear that an expo was as important for what it brought about as for what it put on show. The World Food Festival was an exception in ending up badly in the red. Otherwise, a small loss on the expo account was easily offset by the increase in demand it prompted. As a rule of thumb, spending 5 billion yen on expo construction produced 10–20 billion yen of direct stimulus, even before consumer spending was taken into account. Sure enough, in 1988, Gifu, Kagawa and Saitama ended up with three to five times their direct investment in induced demand. In 1989, Nara made a small loss on its 15 billion yen of direct expenditure but generated ten times that in terms of demand in the prefecture, well over 5 per cent of its total economic activity a few years earlier.[37] Second, the primary audience for a regional expo was not establishment intellectuals, but local people, above all children and students. For many schools and parents, pressed for time and short of money, somewhere like Tsukuba was far too far away. What was wrong with being a mini-Osaka?[38]

Third, even if one restricts one's focus to what was happening on and with the sites, it was clear that expos were beginning to provide a way for cities, and regions, to redescribe their past, reorient their present, and so reimagine their future, not always in the way prescribed in Tokyo. This is easiest to see in the larger expos in the larger cities, three of which followed Kobe's lead in reclaiming land from the sea. Hiroshima was the least successful, pulling in a less-than-expected 3 million to the Sea and Islands Expo, and turning the site into a sewage treatment plant rather than a future city. Yokohama did better. Its 'exotic showcase' was underwhelming, with more NASA kit for the kids, and a mix of nineteenth-century nostalgia and futuristic transport for the adults, but 13 million came to Minato Mirai 21, another 'intelligent', multifunctional city, which now linked the formerly separated commuter hub and government centre. Fukuoka, too, did well. The Asia-Pacific Expo, or Yokatopia, revealed the city's history and potential as an intra-Asian

[36] For the psychologist and novelist, *Yomiuri Shimbun* 1988/11/29, pm, 15. For the architect, *Kenchiku to Shakai* 70.6 (1989), 67.

[37] *Ekonomisuto*, 1985/2/4, 62–63; *Yomiuri Shimbun* 1988/8/23, 7; 1989/1/30, 12–13.

[38] *Yomiuri Shimbun* 1988/5/1, 11.

hub, but also inaugurated a new eighty-hectare urban neighbourhood, bequeathing a communication hub and museum to the city as a whole. It was Nagoya, however, that topped the visitor charts in 1989, 15 million, rather than the expected 6, flocking to a World Design Expo on existing land. Part of its plan to make over the 'dependable' but 'dowdy' industrial city for a post-industrial age, the expo spread over three sites. A new bridge linked the amusement park to a new aquarium down by the port. The history of design, East and West, could be found next to the castle. Design present and future sat between the two, in the neglected southern half of the city, including not only a predictable celebration of space-age materials but also a 'design Olympics', with twenty-eight countries, in the largest wooden structure in Japan, which would become an international convention centre the following year.[39]

By the late 1980s, in other words, cities and expos were no longer simply following dictates from Tokyo, if they ever had, but trying to prescribe the future, at least somewhat on their own terms. They were also reimagining how it might be brought into being. The need had long been recognized. Already in the late 1960s a handful of municipal pioneers, notably in Oita in the southwest, had begun to think about how to revitalize the local community and resist externally imposed models of development. By the late 1970s, the prefecture was leading a 'one-village, one-product' movement, albeit with mixed results.[40] It would be too

[39] For Hiroshima, *Sea and Islands Expo Hiroshima 1989* (Sea and Islands Expo Association, 1989); *Yomiuri Shimbun* 1989/1/30, 12–13; *Kenchiku to Shakai*, 70.6 (1989), 84; and Tracey, 'Expo Boom', 8. For Yokohama, *Uchū to Kodomotachi: Yokohama Hakurankai* (Yokohama Hakurankai Kyōkai, 1989); *Yomiuri Shimbun* 1988/4/4, pm, 19, 1989/3/24, pm, 11 and 13, 1989/4/11, p. 31, and 1989/10/1, 1; *Kenchiku to Shakai* 70.6 (1989), 72 and 76; and Tracey, 'Expo Boom', 3 and 6. For Minato Mirai, Takayuki Suzuki, 'Downtown Revitalization through Public Private Partnerships: The Case of Minato-Mirai 21, Yokohama, Japan' (MA thesis, Tufts University, 2013), 1–4 and 21–23. For Fukuoka, *Asian-Pacific Exposition Fukuoka '89* (Association for the Asian-Pacific Exposition, 1987); *Yomiuri Shimbun* 1989/3/14, 3 and 1989/4/11, 31; Mizutani Eisuke, 'Hakurankai o Kikai ni', and Tanaka Naoto, 'Toshi no Shīn: Fukuoka no baai', *Kenchiku to Shakai* 70.6 (1989), 71 and 75; and Tracey, 'Expo Boom', 5–6. For the aftermath, *Fukuoka Japan: Helping to Create the 21st Century* (Tourism Section, Fukuoka City Hall, 1993), 23; and 'Hito ga Atsumaru "Nyū Taun" Shīsaido Momochi', *Shishi Dayori Fukuoka* 22 (2016), 2–5. And for Nagoya, *Yomiuri Shimbun* 1988/8/25, 13; 1989/3/14, 3; 1989/7/10, 24; 1989/11/18, 9; 1989/11/19, 30; *Kenchiku to Shakai* 70.6 (1989), 64 and 75. Tracey, 'Expo Boom', 7–8.

[40] Jeffrey Broadbent, *Environmental Politics in Japan: Networks of Power and Protest* (Cambridge University Press, 1998), 62–83; and Knight, 'Rural Revitalization', 638–40. For the intellectual backing, Tamanoi Yoshirō, Kiyonari Tadao, and Nakamura Hisashi, *Chiiki Shugi: Atarashii Shichō e no Riron to Jissen no Kokoromi*

much to claim that expos, too, were part of the resistance, but some local authorities were beginning to branch out. An expo was simultaneously a way to 'image up' the region on the national and international stage, a reason for a far-flung son or daughter to return and reconnect, and a place where the community itself could begin to rethink what it was and might be. A region did not have to rely on Tokyo-based corporations. An expo could focus on things other than high-tech. In 1989, Yamagata went with souvenirs from around the world, Tottori with 40,000 toys, and Kurume with 200,000 azaleas. An expo should appeal as much to people's hearts as their minds or eyes. Above all, an expo had to include a region's citizens, not simply as passive spectators but as active participants. Again, Nagoya led the way, insisting that the World Design Expo should stimulate 'citizen solidarity', and mobilizing 50,000 residents as volunteers. But the trend was general, and Kagoshima took the prize. Sazanpia 21 focused on volcanoes and featured a rocket, but the retrospective theme pavilion had been created by local citizens, the twenty-first-century Wonderland Pavilion was full of 3D animations of pictures by local students, and the organizers had mobilized 1,500 groups in support of the expo.[41] Citizen participation was a long way from civic activism, of course. It would take a little while before a citizen movement was powerful enough to transform the landscape.[42] Even in the late 1980s, though, at the height of the economic bubble and expo boom, there was beginning to be space, even at expos, for something new to emerge.

EXPO GREENWASHING, CIRCA 1990

By 1990, then, expos in Japan were beginning to do some useful work, allowing cities and regions to reinvent themselves and find a new place on the national map. They suffered, in some cases, from the predictability of the story – a rich past, a challenging present, a shining future; the familiarity of its tropes – pliant nature and active culture, local distinction

(Gakuyō Shobō, 1978); and Yoshihiro Nakano, 'Postdevelopment in Japan: Revisiting Yoshirō Tamanoi's Theory of Regionalism', in Elise Klein and Carlos Eduardo Morreo (eds.), *Postdevelopment in Practice: Alternatives, Economies, Ontologies* (Routledge, 2019), 37–51.

41 *Yomiuri Shimbun* 1986/1/26, 29; 1988/11/10, 9; 1989/1/30, 12–13; 1989/2/20, 12; and 1989/3/14, 3. *Kenchiku to Shakai* 70.6 (1989), 64. *Ekonomisuto*, 1989/12/5, 10–15.

42 Simon Avenell, *Making Japanese Citizens: Civil Society and the Mythology of the Shimin in Postwar Japan* (University of California Press, 2010), 153–85.

and global connection; and their continuing dependence on corporate exhibits. But they also had appetite and space for the new.

One of the new things was a focus on the environment. It was about time. Already, back in 1970, the ground had seemed to be shifting. An international symposium in Tokyo had asserted the right to an environment 'free of elements which infringe on human health and well-being and nature's endowment, including its beauty'. The 'Pollution Diet' had determined that environmental protection was not to be subordinate to 'sound economic development'. Even the government exhibit at the Osaka Expo had admitted that industrial development produced 'environmental disruption' (*kōgai*).[43] Slowly, environmental concerns made their way into national planning. How might nature be repurposed for the post-industrial future? One obvious way was to 'green' the cities in which the majority of the Japanese population lived. The Ministry of Construction began to monitor such efforts in the 1970s and to reward them in the 1980s. The first Prime Ministerial Green City award went to Kobe in 1981, with the parks and tree-lined streets of Port Island as evidence, and Portopia as climax, of a ten-year campaign to green the city as a whole.[44] An annual National Urban Greening Fair soon followed, endorsed by the new cabinet headed by Nakasone Yasuhiro, which came up with a plan to triple the amount of greenery in Japanese cities, even as it was beginning to let the market rip through urban real estate.[45] Soon, the Ministry of Agriculture, Forestry, and Fisheries was getting interested, advocating expos showing 'the importance and wonder of the relationship between food and the environment that brings it forth'. Four 'food and greenery' expos duly followed in 1988, two in 1989, and three in 1990. At Eatopia Tochigi, in 1988, twenty-three-year-old Miss Eatopia insisted that Tochigi's strawberries were the best in Japan; the theme pavilion traced the Japanese diet back 2,000 years; and freshwater

[43] Shigeto Tsuru, *The Political Economy of the Environment: The Case of Japan* (University of British Columbia Press, 1999), 67–68, for an overview. For the symposium, William Kapp, 'Environmental Disruption: General Issues and Methodological Problems', *Social Science Information* 9.4 (1970), 15–32, and Shigeto Tsuru, *Environmental Disruption: Proceedings of [an] International Symposium* (International Social Science Council, 1970). For Expo 70, *Official Report*, vol. 1, 183; *Asahi Shimbun*, 1970/3/1, 1; 3/12, 18; and 3/14, 10; and Yoshimi, *Banpaku Gensō*, 48–53 and 71–76.

[44] *Yomiuri Shimbun* 1974/12/30, 2; 1980/6/16, 2; and 1981/10/30, 1. For the winners, 'Midori no Toshi-shō: Jushō Ichiran', https://urbangreen.or.jp/grant/3hyosho/green-city/greencity_list.

[45] For privatization, Jordan Sand, *Tokyo Vernacular: Common Spaces, Local Histories, Found Objects* (University of California Press, 2013), 29–30.

and saltwater fish mixed in Za Sakana Pavilion.[46] Nature could be engineered to provide the food of tomorrow.

Both story and moral were underlined two years later, when 23 million flocked to Osaka for the International Garden and Greenery Expo. The city had decided to reinvent itself for the twenty-first century as a green, 'international cultural metropolis', with smaller expos in 1983 and 1987 to get the process under way.[47] Following Nakasone's announcement of his greenery-tripling initiative at a party with the city's elite, they started to think about turning a flower expo, already being mulled for 1989, into an international extravaganza. The city had the ideal site. Tsurumi Ryokuchi ('crane view green space'), a few miles northeast of the castle, had been paddy fields and lotus ponds at the beginning of the century; acquired by the city during the war, to preserve green space for the rapidly expanding metropolis; but transformed in the 1950s into a dumping ground for the waste produced by a new subway construction. The city had begun to clear the waste in the 1970s, with pipes to vent the off-gassing of the trash underneath. Where better to demonstrate Osaka's green credentials to the world? By 1985, the Ministries of Construction and of Agriculture were shepherding the local authorities through the bureaucratic maze required to stage an international horticultural exhibition. As ever, however, they needed creative talent to attract the crowds, turning among others to Komatsu Sakyō, local boy, science fiction author, and Expo 70 veteran.[48]

Komatsu did his best. His career had been devoted to imagining the future, mainly in science fiction but also through expos. In the mid-1960s, he had hoped that an expo might do for Osaka what the Olympics had done for Tokyo. By the mid-1980s, he was struck by the fragility of the blue planet, revealed in the view from space. Perhaps, he mused, the botanical superpowers of the global south might prompt a shift of focus from industrial competition to ecological coexistence (*kyōzon*), which became the expo theme. The vision needed selling, however, if potential visitors were going to 'participate emotionally'. So, for the intellectuals, he staged six symposia, with 'big brains' ruminating on the global

46 *Yomiuri Shimbun* 1986/1/26, 29; 1988/7/24, 6; and 1988/10/4, pm, 7.

47 *Yomiuri Shimbun* 1983/10/1, pm, 14. The plan has gone through various iterations since then. For the latest, 'Shin Kankyō Sōgō Keikaku: "Ōsaka 21 Seiki no Shin Kankyō Sōgō Keikaku"', www.pref.osaka.lg.jp/kannosomu/shin-kankyousoukei/.

48 Komatsu Sakyō, *Kyodai Purojekuto Ugoku* (Kōsaidō, 1994), 17–21. For the history, 'Tsurumi Ryokuchi History', www.tsurumi-ryokuchi.jp/history/index.html.

environmental moment. For the public and the media, rather than *idobata kaigi*, the traditional opportunity to gather and gossip at the village well, he laid on six *hanabata kaigi*, where fans sat and gawped at youngish celebrities talking about their love for flowers. The *Yomiuri* newspaper sponsored a nationwide search for a hundred famous trees. Nippon TV aired a documentary on the world's biggest on New Year's Day 1990. Komatsu's efforts climaxed during the expo itself with a two-hour-long 'landscape opera', *Gaia*, on a massive stage constructed on and around the 'Sea of Life' pond in the middle of the site. The story went from the birth of the planet, continents, forests, and civilizations, through conflict, industry, and destruction, to a view of earth from space and a hymn of praise. Komatsu had been inspired in part by the memory a review of the Imperial Fleet in Kobe Bay, which he had seen from the deck of a warship as an elementary school student in 1940.[49]

Komatsu's message was a powerful one, but frustrated as eco-logic became expo reality. There was the familiar quagmire of expo bureaucracy. Group think plus organizational silos made a novel theme and innovative programming a tough sell. There were the inevitable compromises over the budget. Nakasone's neoliberal insistence that urban development should be driven by private capital (*minkatsu*) meant that the organizers had to turn to the corporate sector and the gambling industry. Predictably, too, the lead-up to the expo was marked by a commercial bunfight, 'the riotous blooming of a hundred flowers' (*hyakka ryōran*), as department stores and others sought to exploit the opportunity. Tokyo-based Mitsukoshi stole a march by establishing a project office three years before the expo, sponsoring the on-site monorail, and setting up thirteen outlets on the site itself. Hanazukin-chan, the mascot, got sent to space, and appeared on over a thousand products, ranging from beer and candy to towels and fans.[50] The retail extravaganza was only a prelude to the site itself, however.

[49] Komatsu, *Kyodai Purojekuto*, 6–7 for the full list of projects, 22–25 and 32–35 for the vision, 50 for emotional participation, 62–64 for big trees and media tie-ups, 69–83 for *Gaia*, 84–90 for the symposia, and 90–94 for the *hanabata kaigi*. *Expo '90, Osaka, Japan: Official Guidebook* (Japan Association for the International Garden and Greenery Exposition, 1990), 15–21, for the theme. See also *Yomiuri Shimbun*, 1989/4/29, 14 and 1989/11/2, 43.

[50] Komatsu, *Kyodai Purojekuto*, 36–44 for financing, 56–57 for bureaucracy, and 131–34 for Hanazukin-chan in space. 'Kokusai Hana to Midori no Hakurankai', https://nippon.zaidan.info/kinenkan/history30/4/4541.html, for support from gambling. *Yomiuri Shimbun* 1990/2/12, 13, for the 100 commercial flowers.

Isozaki Arata, another Expo 70 veteran, had been charged with the plan and had divided the site into three. The 'mountain' area was the place for highbrow musings, with the Japanese government garden, including terraced rice paddies and imperial bonsai, and five pavilions riffing on the theme. Even here, all was not well. Senegal illustrated the dangers of desertification with ninety tons of sand from the Sahara, which invaded the Israeli garden, making for a 'dreary' (*samuzamu*) scene. The 'Golden Bell Pavilion', a re-creation of a hole at the Augusta National Golf Club, required four different pesticides to be applied for a year ahead of time, producing a run-off that flowed downhill into the fish-rich Sea of Life. The pond, in turn, was at the south of the 'field' area, with two hectares of flower terraces and 2 million blossoms making good on the Deputy Mayor's promise to produce a '*hana*-ful', or flower-filled, Osaka. The rigours of an urban summer, however, meant that some had to be replaced after only a week. The salvia, licianthus, and petunia of June quickly became the hibiscus and zinnia of July.[51] They were also overshadowed by the corporate pavilions in the 'city' area, to the south and west, which occupied half the site. The Bureau Internationale des Expositions (BIE) had raised concerns, since horticultural exhibitions were not meant to include built environments, but the Association argued that the expo needed more than 'plain flowers' (*jimi-na hana*). The corporations tried. The Mitsubishi Pavilion represented 'a huge sprout bursting from the earth', the Hitachi one a chestnut tree. But technology still ruled the show. In the Matsushita pavilion, a room of animatronic creatures and electronically controlled flowers provided a prelude to eight of Henri Rousseau's jungle fantasies, reproduced on a nearly 200-inch, high-definition screen.[52]

Not everyone was convinced. Already during the planning stages, some committee members had wondered if a 'dialogue with nature' would be possible. As the summer went on and the crowds mounted, the question recurred. There was not enough signage to find the greenery. Too many of the exhibits felt like souvenir shops. With one week to go and one person

[51] *Kenchiku to Shakai* 70.6 (1989), 72, for 'hana-ful Osaka'. *Yomiuri Shimbun* 1990/7/8, 3, and Komatsu, *Kyodai Purojekuto*, 119, for replacing flowers.

[52] *Expo '90, Osaka, Japan: Official Guidebook*, 22–25, for an overview of the site. Komatsu, *Kyodai Purojekuto*, 114–18 for his take and BIE concerns. *Yomiuri Shimbun* 1990/2/12, 12, for Matsushita; 1990/4/5, pm, 19, and 1990/4/12, 12, for Golden Bell pesticides, and 1990/3/15, pm, 18, for previous suffering by the fish; 1990/7/8, 3, for Saharan sand; and 1990/9/30, 11, for plain flowers. *New York Times* 1990/4/15, 18, for Mitsubishi and Hitachi.

on every square metre of the site, people were asking 'Where are the flowers?' Still, the expo hit its numbers. It attracted over 23 million visitors and made 5 billion yen in profit, which funded a commemorative association and the International Cosmos prize, for contributions to the 'harmonious coexistence of nature and humankind.' There was also nearly 2 trillion yen of economic stimulus.[53] The last word, however, should go to eleven-year-old Okabe Masato in Tokyo, who read along as the crowds swarmed during the last days and the crews began to clean up afterwards. The flowers were still blooming, but the stage was being cleared, so they disappeared into the back of a garbage truck. He felt sorry for them. If he had been one, he would have wanted a visitor to take him home.[54]

In 1990 in Osaka, Komatsu's view from space and Okabe's plant's eye view of the world still seemed a long way off. Expo 90 was an artefact of the imagination that had spawned its ancestor, Expo 70, its predecessors in Okinawa and Tsukuba, and the swarm of municipal offspring in the years since. Development, it was assumed, could be planned, with expos helping to realize the plan and put the future on show. Already, though, it was unclear if the good times would continue to roll. By the time the last visitor squeezed through the gates, the stock market had plummeted to less than half its peak at the beginning of the year. Other asset prices soon began to fall and the economy to stagnate, opening up space in which to question the assumptions underwriting the expo boom. This did not mean an end to expos. Over two decades, they had spawned an infrastructure and industry, institutions and individuals, who were invested in their survival. In order to endure, however, they would have to change: to provide space for citizens and even opponents to contribute to their planning; to produce sites and exhibits that acknowledged the limits of nature; and thereby, perhaps, to begin to redefine the nature of development itself.

[53] *Yomiuri Shimbun* 1990/7/8, 3, for midterm report and reservations; 1990/9/30, 3, for 'dialogue with nature' and the final week. Ibid. 1990/10/1, 3, and Komatsu, *Kyodai Purojekuto*, 42, for economic effects. Ibid., 46, for the prize.

[54] *Yomiuri Shimbun* 1990/10/9, 12.

8

Beyond Development?

For the last half century, in East Asia and beyond, development has taken a binary visual form. On the one hand, mega-events – expos, Olympics, and World Cups – have provided opportunities for countries to showcase their progress to the world, whitewash the costs of growth, and justify spending on the infrastructure required when the circus comes to town.[1] On the other, the infrastructure itself, embedded in national plans, for planes, trains, and automobiles, as well as container ships, has enabled city to connect to country and world. It is only fitting, therefore, that the transport of the future has featured at expos. Pride of place has long gone to maglev rail. It was previewed in Osaka in 1970, desired in Nagoya in the early 1980s, and demonstrated in Tsukuba in 1985 and Yokohama in 1989. By then, though, the vision of the future that it embodied was coming into question. Land prices in Tokyo started to collapse the same year and stock prices the following one, with nominal gross domestic product (GDP) and real wages following from the middle of the 1990s. Mired in what seemed, after four decades of fetishizing growth, as a lost decade (or three), it may have seemed that both future and technology had been put on hold.[2]

[1] According to Google's Ngram Viewer, the use of 'mega-event' began in the 1980s, accelerated in the late 1990s, and took off in the late 2000s, likely prompted by sports mega-events in the formerly developing world. Maurice Roche, *Mega-Events and Modernity: Olympics and Expos in the Growth of Global Culture* (Routledge, 2020), and Graeme Evans (ed.), *Mega-Events: Placemaking, Regeneration and City-Regional Development* (Routledge, 2019), for more historically informed, less myopic takes.

[2] O'Bryan, *The Growth Idea*, 1–4 and 177–80, for loss as an artefact of the growth fetish.

It was also tempting to assume that expos no longer had much point. With capital in short supply, how could the authorities justify the funds for a discretionary good? With growth in the rear-view mirror, how to make a spectacle of the development on which they had been premised? As the 1990s unfolded, expos started generating headlines that seemed to support the claim. Tokyo's world city expo ran into trouble in early 1991, was postponed at the end of the year, and a few years later cancelled for good. Soon, the plans for an expo in Aichi in 2005 were also running into quicksand. By mid-decade, there was a wave of nostalgia for the promised land of Expo 70: a lost generation of middle-aged 'expo boys' revisited their elementary-school playground to find consolation for their vanished dreams.[3] More generally, the intellectuals were suggesting, the time for expos had passed. 'Spectacle space' had moved from the real to virtual worlds online. Expos were dead.[4]

But they were and are not. It took a while, admittedly, for them to become something different. While the problems in Tokyo and Aichi were grabbing column inches, the Ministry of International Trade and Industry (MITI) was inaugurating its Japan Expo system. Initially planned as a way of bringing order to the expo frenzy that had broken out at the end of the 1980s, eleven expos over the following decade held out the possibility of reversing decline. Others were beginning to repurpose the expo for Japan as it was, with slow growth and an ageing society, which needed local initiative and ecological awareness. Both strands were visible in Aichi in 2005, which went ahead, despite it all. There were robots and there was maglev rail, gesturing to the development that had been planned, promised, and more or less delivered, for almost fifty years. But there were other things to see and stories to hear, too, which suggested that development might proceed otherwise, and even that the future could be different.

Besides, even if developmentalism 1.0 had run its course in Japan, it was still a compelling export good. As Korea and then China followed Japan, inheriting its industrial mantle and moving up the value chain, so they too turned to expos, as a way of mitigating the costs of growth and

[3] Kikuchi Fumihiko, '1970 Osaka Banpaku: Banpaku Shōnen-Tachi', *Ronza*, 2016/11/16, https://webronza.asahi.com/culture/articles/2016100900001.html, for expo boys.

[4] Mizuno Ami, 'Sayonara! Bankoku Hakurankai', *Gijutsu to Ningen* 21.3 (1992), 38–45; 'Tokushū: 20-Seiki no Supekutakuru Kūkan', *InterCommunication* 20 (1997), 64–164; Oda Masanori, 'Sayonara Banpaku, San Tabi / Nippon Banpaku 70 Nyūsu', *10+1* 36 (2004), 112–15, and the other articles in this special issue, 63–192, which served as an extended critique of Aichi and farewell to expos; and Yoshimi, *Banpaku Gensō*, 273–87.

recalibrating its direction.[5] Two decades after Osaka established the template, five years after the Seoul Olympics, South Korea held its first expo in 1993 in Daejeon, insisting that science and technology remained a way for the state to restructure industry and balance regional development. Two decades later, two years after the Beijing Olympics, Shanghai would host China's first in 2010, to trumpet the country's arrival on the global stage, the city's status as a social hub, and the arrival of middle-class consumers, finally outdoing Osaka in terms of area, participants, and visitors.[6]

Already in 2005, though, back on the outskirts of Nagoya, the writing had been on the wall. Shanghai had opened Asia's first maglev the year before, covering the 30 kilometres between the airport and the financial district in under 10 minutes at speeds of over 400 kmh. Japan's first, Linimo, opened fourteen months later, in time for the Aichi Expo, but took over fifteen minutes to cover nine kilometres, with a stop along the way at the expo site (Figure 8.1). Nearly twenty years later, there are still only six lines in operation anywhere in the world – three in China, two in South Korea, and lonely Linimo in Japan – but the technology continues to progress. A new maglev bullet train will open in 2027, to whisk passengers from Tokyo to Nagoya in 40 minutes at speeds topping 500 kmh. Nowadays, however, this is not the only story to tell about development. Aichi provided space not only for those who believed in the old dream but also for those who wanted to change minds and maybe even course.

TROUBLE IN TOKYO

Tokyo has never been that lucky in expos. It was the preferred site for state-sponsored industrial exhibitions in the late nineteenth century, and took the lead with municipal and private initiatives in the early twentieth, but attempts to do something bigger had repeatedly foundered. Plans for

[5] Kaname Akamatsu, 'A Historical Pattern of Economic Growth in Developing Countries', *The Developing Economies* 1 (1962), 3–25, and Bruce Cumings, 'The Origins and Development of the Northeast Asian Political Economy: Industrial Sectors, Product Cycles, and Political Consequences', *International Organization* 38.1 (1984), 1–40, for East Asian development.

[6] For Daejeon 1993, *Expo '93: Official Guide* (Taejon Expo Organizing Committee, 1993); and Yu-Min Joo, Yooil Bae, and Eva Kassens-Noor, *Mega-Events and Mega-Ambitions: South Korea's Rise and the Strategic Use of the Big Four Events* (Palgrave Macmillan, 2017), 47–67. For Shanghai 2010, see the Conclusion.

FIGURE 8.1 Maglev has long been an attraction. Come Aichi in 2005, there was a working line to the expo, which maxed out at 100 kmh, and a test car on display, which had hit nearly 600 kmh.
Source: Yuji Ohsugi / Stringer via Getty Images.

international exhibitions in 1912 and 1940 collapsed thanks to fiscal uncertainty and total war. There was a stream of smaller events in the post-war period: atoms for peace in 1955 led to space (1960 and 1966) and aviation (1963 and 1964); a Canada expo in 1955 had paved the way for the UK (1965), the US (1968), Siberia (1973), and the USSR (1977). Cold War enthusiasms converged in a Space Expo in 1978–79 on some reclaimed land in the bay. Come the consumerist 1980s, promoters were catering to the whims of middle-class children with manga and anime (1983) and their parents with the lure of travel (1984 and 1986). It was hardly enough for the expo faithful, however, who had never stopped believing in the promise of the form.

The Tokyo World City Expo was the spawn of Suzuki Shun'ichi, who had cut his teeth on the cancelled World's Fair of 1940 and served as Secretary-General of the Expo 70 Association.[7] By 1980, he was Governor of Tokyo, hosting a tenth anniversary gathering of his old expo buddies. A few years later, following his re-election, he was thinking

[7] Hatanaka, *Gorin to Banpaku*, 150–212, for a detailed account.

bigger: six sub-centres would relieve concentration in the urban core; a seventh on the reclaimed land in the bay would serve as a 'teleport', underwriting a new vision of 'World City Tokyo'; though both would rely on private capital, as the neoliberal Prime Minister insisted. The plan emerged in March 1988. The seventh waterfront sub-centre would have space for over 100,000 to work and 60,000 to live. It would be inaugurated with a 'World City Expo' in 1994, over 300 days, with 30 million visitors, which would 'build social capital', announce Tokyo's ascension to the first rank of global cities, and commemorate the 400th anniversary of Tokugawa Ieyasu's arrival in town.[8] December 1988 saw the Expo 70 old boys get to work at an inaugural roundtable in Osaka. Suzuki kicked things off by asking that the expo should be something possible 'only in Tokyo' (*Tokyo de koso*). Sakaiya Taichi suggested that it might 'renovate' (*isshin*) the idea of what an expo might be. By March 1989, they had decided that this would be the third in the series of international mega-events (*dai kibo kokusai gyōji*), following the Tokyo Olympics and Expo 70. In their final report, in July, they ditched the word for 'expo' (*hakurankai*) entirely, proposing instead that they call it 'Tokyo Frontier'. By April 1990, the Association was at work and the plan was being fleshed out.[9]

The timing was not great, however. The commercial real estate market in Tokyo had already been stagnant for a couple of years. Come the autumn, the stock index had halved. By March 1991, with another election in the offing, the media were asking questions and discovering that the work on sub-centre and expo had been crammed in any old how (*muriyari oshikonda*). The other candidates for governor urged reconsidering the plans, if not postponing the expo. However, Suzuki stuck to his guns, with just enough bluster to get re-elected. But by the end of the summer, the timeline had been pushed. The expo would open after the end of his term, depriving him of a garlanded exit (*hanamichi*), with duration and visitor numbers cut by a third.[10] Still, the planning proceeded. The site plan, published in November 1992, made space for

[8] For the vision, Saito, 'Global City Formation', 292–96; Sand, *Tokyo Vernacular*, 32–36; and Kawashima Yūsuke, 'Hitsuzen demo Gūzen demo Naku: 1995-nen "Sekai Toshi Hakurankai" Chūshi no Seijigakuteki Bunseki', *Nagoya Daigaku Hōsei Ronshū* 269 (2017), 310–12.

[9] *Yomiuri Shimbun* 1988/12/2, 25; 1988/12/29, 11; 1995/5/9, 9. *Shin Kenchiku* 70.5 (1995), 195.

[10] *Yomiuri Shimbun* 1991/3/19, 30; 1991/6/11, 25; 1991/8/29, 2; 1991/11/28, 13; and 1995/5/9, 9.

twenty-five company pavilions, but ten months later only eleven had signed up. The Association tried to pump up the volume in August 1994. There would be a full-scale model of the international space station! The expo would create 625 billion yen of direct demand, 90,000 jobs, and 1.7 trillion yen of indirect effects!! Two months later, however, it emerged that corporate participants were expected not only to build their own pavilions but also to buy 100,000 or more advance tickets, ten times as many as Tsukuba in 1985 or Yokohama in 1989.[11]

Squaring the budget was one thing. Designing the site was another. Rather than a corporate-powered celebration of urban potential in the information age, the architectural team wanted the expo to return its visitors to nature, community, and the past. Information technology would be there, for example in the 150-metre-diameter, doughnut-shaped centrepiece for 'Media Land'. Itō Toyō, its architect, and his colleagues were insistent, however, that the buildings were subordinate to their vision of the expo as a liveable, lively, and enriching city, with water and greenery, roads to walk on and places to meet. People would mingle, on two eighty-metre-wide promenades and multiple plazas, two of them in 'Community Land', with its recycling studio and citizen stage, to foster civic participation. A huge fountain filled with rainwater would form the centrepiece of 'North Intercity Land'. Entering the expo from the south, they would walk straight into an urban forest with a stream running through 2,000 trees, and a wooden teahouse suspended overhead, with space for 10,000 attendees and the heads of a hundred different tea schools. Until now, expos had taken the pavilions for the figure (*zu*) and the space between them simply as ground (*tsuchi*). This time, green space would be the main event.[12]

The contrast between the two visions was striking, but in the end neither was meant to be. By the end of 1994, the bells of the press were tolling for Suzuki, announcing that he had tripped on the bubble.[13] The following spring, even in the new forty-eight-storey Tokyo Metropolitan Government building, fewer and fewer staff were wearing the expo badge. Four out of the six candidates in the upcoming election were pledging to cancel it outright, as a woeful misuse of taxes. The most striking was sixty-two-year-old Aoshima Yukio, actor, writer, and

[11] *Yomiuri Shimbun* 1991/11/19, 25; 1993/9/15, 25; 1994/3/15, 27; 1994/8/19, 24 and 30; and 1994/10/8, 33.

[12] *Shin Kenchiku* 70.1 (1995), 257–62; 70.5 (1995), 194–97; and 70.9 (1995), 190–97.

[13] *Yomiuri Shimbun* 1994/12/2, 3.

comedian, who had sat in the House of Peers for thirty years, and who insisted that the expo should be cancelled and the sub-centre redesigned, as an urban forest or waterfront park.[14] On 9 April 1995, Aoshima was elected and reiterated his line. The sub-centre was a 'dream-like mega-project, planned at the very height of the bubble', imposed on the citizenry by a 'governor sitting above the clouds' (*kumo no ue no chiji*). Admittedly, cancelling it would cost something. But proceeding would cost eight times as much. Like Okinawa 75, the expo had become a suicide bomb (*jibakuzai*) rather than the intended trigger (*kihakuzai*) for development.[15] Even so, taking the decision took some time. Aoshima was immediately besieged by outraged local politicians, corporate executives, and citizen groups. Older voters and younger folk wanted the expo, but the majority remained against. A day before the deadline, he pulled the plug. And the media made predictable hay.[16]

On the streets, opinion was split. A forty-four-year-old banker thought the decision was foolish (*baka-na koto*). A fifty-two-year-old, employed by an electrical equipment subcontractor, was worried about his job. But a forty-seven-year-old housewife was happy that Aoshima had 'kept his promise'. The expo would have been just another jamboree. On the thirty-sixth floor of the new government building, the promotion office was turned into a liquidation agency.[17] Even as its accounts were being wound up, however, it became clear that the expo had had its effects.[18] First, like its hapless predecessor in 1940, cancelling the expo may have publicized the new sub-centre as effectively as holding it would have been. The administrator had to halve the rents on its office buildings to attract tenants, but Hotel Nikko opened in March 1996, Tokyo Big Sight in April, and by May people were flocking to what had once been reclaimed land (Figure 8.2). By the end of the decade, Odaiba was being trumpeted by Ishihara Shintarō, the new right-wing governor, as an

[14] *Yomiuri Shimbun* 1995/3/29, 33; 1995/4/5, 29; 1995/4/7, 33. For the corporate pavilion announcements, ibid. 1995/2/4, 33; 1995/2/12, 29; 1995/2/21, 29; 1995/3/3, 33; 1995/3/4, 34; and 1995/3/14, 29.

[15] *Yomiuri Shimbun* 1995/4/10, 2; 1995/5/9, 9.

[16] *Yomiuri Shimbun* 1995/5/31, pm, and 1995/6/1, passim. Itō Hisao, 'Sekai Toshi Haku Chūshi no Konnichi-teki Igi', *Gijutsu to Ningen* 24.5 (1995), 16–21; and Kawashima, 'Hitsuzen demo Gūzen demo Naku', 314–19.

[17] *Yomiuri Shimbun* 1995/5/31, pm, 14–15.

[18] Kawashima, 'Hitsuzen demo Gūzen demo Naku', 318. See also Yukio Aoshima, 'Introduction: Cities and the Environment – Towards Eco-Partnerships', in Takashi Inoguchi, Edward Newman, and Glen Paoletto (eds.), *Cities and the Environment: New Approaches for Eco-Societies* (United Nations University Press, 1999), xv–xxiii.

FIGURE 8.2 Three years earlier, the expo had been cancelled. The Theme Pavilion became the centrepiece of Tokyo Big Sight.
Source: ullstein bild via Getty Images.

urban resort.[19] Second, the sacrifice of Tokyo Frontier suggested that expos were changing. Suzuki, of course, continued to believe that it should have gone ahead.[20] It would take a while for the developmentalist fantasy to lose its grip. It was becoming clear, however, that expos had to begin with their environment: both the local community, to which they had to be accountable, and whose participation was required; and nature, no longer as mute resource, but as living partner, in collaboration with which an expo had to take its place.

PROGRESS IN THE PROVINCES

Even before the collapse of the bubble, in the late 1980s, it had been clear that something had to give. On the one hand, there was a swarm of regional expos calling out for bureaucratic oversight. On the other, there

[19] *Yomiuri Shimbun* 1996/5/28, am, p. 6; 1996/5/29, am, p. 11; 1997/6/1, p. 29; Saito, 'Global City Formation', 299. See also James O'Leary and Machimura Takashi, 'Between State and Capital: Third Sector Organizational Development in Tokyo', *Comparative Politics* 27.3 (1995), 317–37.
[20] *Yomiuri* 2004/4/5, 11.

were regions suffering from deindustrialization and depopulation. Perhaps expos could be disciplined to build social capital and stimulate consumption. Still in its planning pomp, MITI decided that a new expo system would 'create hometowns' (*furusato sōsei*), with two officially designated 'Japan Expos' each year. MITI would provide a subsidy, the Ministry of Home Affairs the sign-off for the necessary municipal bonds, and the government as a whole a pavilion and publicity. The bursting of the bubble in 1991 prompted a delay and some downsizing. Expos would now last sixty days, rather than three to four months, and aim to attract 1 million, rather than 2 to 10 million visitors.[21] Still, the system kicked off in 1992, with both expos easily exceeding expectations. The 'hand-made' Sanriku Sea Expo addressed the decimation caused by the rationalization of the local steel industry in Kamaishi, turning the population of just over 50,000, 500 kilometres north of Tokyo, towards the 'sparkling future' and the 'shining sea' (*hikaru umi, kagayaku mirai*). The organizers had hoped for 700,000 visitors, while a snubbed, sniffy advertising agency suggested under 300,000 was realistic. But 2 million passed through the gates, generating a surplus of 400 million yen. Another 2 million came to Toyama to revel in life, love, laughing, and light, embodied in a wonderland, a 'star corridor', a robot utopia, and an eerie, empty cube that still serves as an observation tower.[22]

MITI insisted that things were now different. Previously, regional expos had been copycats, outsourced to the major advertising agencies. Now they were 'by regions and for regions' (*chihō no chihō no te ni yoru chihōhaku*).[23] As the decade unwound, however, it was clear that not everything had changed. Many expos fell back on the tried and tested strategies of the 1970s and 1980s. Some drew on malleable local heritage. The third Japan Expo in Matsumoto in 1993 ran alongside a festival commemorating the 300th anniversary of the castle's construction. The fifth in Ise in 1994 used the eponymous shrine as the starting point for a celebration of festivals around the world, culminating in the immolation of a huge puppet (*falla*) from its sister city, Valencia in Spain.[24] Some boasted their suitability for middle-class recreation. The fourth, also in 1994, on the other side of the peninsula from Ise, inaugurated Wakayama

[21] *Yomiuri Shimbun* 1987/3/26, 7; and 1988/12/29, 7. Fujita Yoshihisa, '"Japan Ekisupo" ni yoru Chihō Hakurankai o Megutte', *Aidai Shigaku: Nihonshi, Ajiashi, Chirigaku* 11 (2002), 66–68.

[22] *Yomiuri Shimbun* 1992/7/4, pm, 18; and 1992/9/22, 17.

[23] *Yomiuri Shimbun* 1992/9/22, 17.

[24] *Yomiuri Shimbun* 1994/7/23, Chubu, 25, and 1994/11/7, Chubu, 27.

Marina City with a World Resort Expo, adding a West Samoan village to the permanent attractions: a marina, a Mediterranean-port-themed theme park, and a fisherman's-wharf-style market.[25] Finally, some followed prescriptions for 'internationalization' and 'exchange'. In 1997, a 'Dream Port Expo' in Tottori used local boy and manga artist Mizuki Shigeru to trumpet the region's abundance of ghosts (*yōkai*). Its location in a 'Foreign Access Zone', however, made clear that the region's future would be one of servicing international trade. In the same year, a budget-strapped 'International Dream Exchange' in Sendai used site and facilities from an earlier 1987 expo, which Tokyo-based companies filled with hackneyed, tech exhibits.[26] There was also the question of what to do after the expo. Fifteen years after its Japan Expo in 1992, Toyama still had nearly fifteen hectares of unused land, with a book value of nearly 2 billion yen and weeding costs of over 5 million a year. After its one in 1994, Ise had eighty-seven empty hectares and a 16-billion-yen 'Sun Arena' in a place without a regular bus line.[27]

Through the familiar weave of local dependence, Tokyo stories, recycled exhibits, and wasted land, though, other threads were beginning to emerge. The 1980s had already seen local citizens not only as volunteers but also as performers and producers. Citizen participation (*shimin sanka*) was an explicit priority in Kamaishi in 1992 and soon became a staple, part of a larger rethinking of what an expo might be and do. Rather than short-term economic effects, why not aim at long-term community development? Rather than passive gawping on a crowded site, why not provide a participatory experience in the local landscape? Kobe showed the way in 1993, pulling in 16 million to a six-month-long Urban Resort Fair, with 400 events spread out through the city, which ranged from Yukio Ninagawa's staging of 'Midsummer Night's Dream' to a 200-metre-long *sōmen* noodle run.[28] Three years later, the sixth Japan Expo, a World Ceramics Expo in Saga supplemented a conventional, corporation-heavy main site in Arita with a number of satellite venues. In Imari, a group of potters threw the world's largest platter, decorated it with local specialities, including pears and horseshoe crabs, repaired the crack from firing, and then paraded the 500-kilo monster

[25] *Yomiuri Shimbun* 1993/5/7, Osaka, 19; 1994/6/11, Osaka pm, 5; 1994/7/16, Osaka, 15; 1994/9/26, Osaka, 28.

[26] *Yomiuri Shimbun* 1997/8/9, 1; Fujita, '"Japan Ekisupo"', 72 and 75.

[27] *Yomiuri Shimbun* 2007/3/29, 35, and 1994/11/7, Chubu, 27.

[28] *Yomiuri Shimbun* 1993/4/11, 30, and 1993/9/2, 1; *Rejā Sangyō Shiryō* 34.7 (2001), 40–45.

around town during the summer festival on a four-and-a-half-ton float in the shape of a cow.[29] Finally in 1999, the ninth Japan Expo back in Wakayama abandoned the notion of site and admission fee entirely. The 'site' for the Nanki Kumano Experience Expo was the 'pavilion of nature' provided by the region as a whole, centred on the old pilgrimage route linking the three Kumano shrines, along which 360,000 now walked and which was supplemented by 700 events, ranging from a fire festival to beach volleyball. The expo attracted 3 million visitors and 100 interested local governments, though a twenty-nine-year-old housewife was a bit puzzled, and a Tokyo-based professor felt compelled to wonder if it was an expo at all.[30] The same year 1,000 events and 10 million visitors commemorated the opening of the Nishi Seto Expressway between Honshu and Shikoku at the Shimanami Kaido 99. You did not need 'magnificent pavilions like a normal expo' to attract the crowds.[31]

'Normal' expos were adapting too. Wakayama's invocation of nature had been a long time coming. The National Urban Greening Fair had been held annually since the early 1980s; Osaka had proved that flowers were a dependable attraction in 1990; and 'gardening' had been one of ten words of the year in 1997. Three years later, Hyōgo doubled down, attracting 7 million to Awaji Island for Japan Flora 2000. From one angle, it was yet another attempt to promote tourism, consumerism, and so development. The 100-hectare site sat 17 kilometres above the epicentre of the Great Hanshin Earthquake of 1995. The Akashi-Kaikyō Bridge had opened in 1998, linking island to mainland and bringing a 300 per cent increase in tourists. The expo was a useful way to consolidate the island and prefecture's dominant position in the domestic cut flower market, and to counter the 2,000 tons arriving at Kansai International Airport every year, which included cheap Korean chrysanthemums and expensive Taiwanese orchids. Stretocarpus, a promising entry in the potted plant market, became the official expo souvenir. Rafflesia, the world's largest flower, returned to delight the crowds. But the expo, like its predecessors, also suggested that a different relationship

[29] *Yomiuri Shimbun* 1996/6/23, West, 27; 2001/9/25, 11.

[30] *Yomiuri Shimbun* 1999/4/28, Osaka pm, 18; 1999/4/29, Osaka, 27; 1999/5/5, Osaka, 27; 1999/9/12, Osaka, 9; 1999/9/22, Osaka, 26; 1999/10/1, Osaka, 35; 2002/11/25, Osaka, 36. See also Kanda Kōji, 'Nanki Kumano Taikenhaku to Kumano no hyōshō', in Sano Mayuko (ed.), *Bankoku Hakurankai to Ningen no Rekishi* (Shibunkaku Shuppan, 2015), 321–47.

[31] *Yomiuri Shimbun* 1999/1/1, Osaka, supplement, 7; 1999/7/28, Osaka, 28; 1999/10/17, Osaka, 23.

with nature might be possible. The need to reduce, reuse, and recycle had been evident at expos for a while. Awaji went further, restoring a degraded environment, which had provided 100 million cubic metres of earth for construction projects in the region, not least the new airport. A 12-hectare woodland appeared on the slope above the site, courtesy of 550,000 acorns, collected by local residents, and 240,000 native seedlings. Ando Tadao, the expo architect, had prioritized the 'restoration of lost nature', using recycled materials, designing permanent facilities, rather than temporary structures, and burying many of them underground. It was clear to observers that the rules were changing.[32]

By the new millennium, then, expos were in transition. For local authorities, it was no longer enough to raze a site, outsource the contents, and wait for the visitors to arrive, in line with commands from Tokyo (*chiiki o kanri*). Instead, they had to start actively managing the region (*chiiki o keiei*). For state bureaucrats, too, it was time for a rethink. The Japan Expo system had got off to a decent start, but come 1998 a MITI report suggested it be wound down.[33] The last three, in 2001, indicated both where expos had been and where they might be going. All three 'wrestled with environmental issues' and did so by involving the community. But they did this in different ways. The Kitakyushu Expo Festival was an attempt to turn the town from heavy industry to a zero-emission, advanced-technology town. Site and exhibits were familiar, however. Even though the organizers let in 600,000 visitors for free in order to hit their 2 million target, they ended up 2 billion yen in the red. The Utsukushima Future Expo inaugurated a virgin site in the hills of Fukushima, with recycled materials, self-contained toilets, and trash separated into eighteen different categories, but here the organizers fell well short of their 2 million target and struggled to sell off the site. The Yamaguchi Kirara Expo, however, managed to thread the needle between industrial deadweight and ecotopia, attracting 2.5 million visitors to what had been an unused piece of reclaimed land. There were 20,000 volunteers, elementary school kids separating garbage, and 2,000 local

32 *Yomiuri Shimbun* 1999/9/22, Osaka, 26; 1999/9/23, Osaka, 31; 1999/9/25, Osaka, 27; 2000/1/1, Osaka, 3; 2003/3/17, special section, 1–2, 4–6; 2000/8/17, Osaka, 25; and 2000/9/16, Osaka, 30; and *Rejā Sangyō Shiryō* 34.7 (2001), 46–51. For earlier recycling efforts, *Yomiuri Shimbun* 1993/4/11, 30, and 1993/9/2, 1.

33 *Yomiuri Shimbun* 1992/10/27, 14; *Rejā Sangyō Shiryō* 34.7 (2001), 42–45. A new Japan Expo started up in 2000, celebrating Japanese pop culture in Paris: Japan Expo, 'The History of Japan Expo' (2020), www.japan-expo-paris.com/en/menu/history_100541/info/the-history-of-japan-expo_474.htm.

performers in a piece of community theatre orchestrated by fashion designer Yamamoto Kansai.[34]

THE LOST BOYS OF EXPO 70

The end of the Japan Expo system did not mean the end of expos in Japan, although they were fewer and different than they once had been. There were regular festivals of food in Osaka (2005, 2009) and confectionery nationwide (Kumamoto 2002, Himeji 2008), as well as annual celebrations of greenery and gardening (Yamagata 2002, Oita 2003, et al.). History remained a reliable go-to, whether for anniversaries (Yokohama 2009, Nara 2010) or castles (Nagoya 2005, Kumamoto 2008), particularly when the local past had been recently televised as a morning drama by the national broadcaster, NHK (Kanazawa 2002). For some, though, this was a strange landscape given the promises of their youth. By the late 1990s, someone who had been in elementary school in 1970, captivated by the future on display at the Osaka Expo, was on the cusp of their forties, struggling not only with the standard disappointments of middlescence but also a specific betrayal of their childhood. Where was the future they had been promised in the Senri Hills? Why were they still waiting to travel into space? They were not alone in their wistful looking back. Freed from the futurist certainties of the Cold War, nostalgia was a 'global epidemic' by the end of the century.[35]

It also had a long track record in modern Japan. In the late nineteenth century, amidst the onrush of industry and empire, aesthetes had begun to bemoan the heritage that was being lost. In the early twentieth, novelists had captured the dislocation of those forced to live between two worlds and folklorists had excavated places on the periphery where they imagined the past to live on. In the 1970s, government and business got involved, encouraging urban consumers to discover a bowdlerized rural past. Come the 1980s, urban history too provided consolation in the vacuous post-industrial present, with a revaluation of both early modern Edo and early twentieth century Tokyo. The death of the emperor and the doldrums of the economy upped the ante, providing further impetus.

[34] *Yomiuri Shimbun* 1998/10/31, West pm, p. 11; 2000/7/11, West am, p 36; 2000/11/30 West am, p. 37; 2001/9/25, am, p. 11; 2001/10/1, West am, p. 39; 2001/10/2, am, p. 28; 2001/10/2, West am, p. 37; 2001/10/3, am, p. 32; 2001/10/3, West am, p. 31; 2001/10/5, am, p. 32; and 2001/10/7, am, p. 30; *Rejā Sangyō Shiryō* 34.7 (2001), 56–67; and Fujita, '"Japan Ekisupo"', 73–81.

[35] Svetlana Boym, *The Future of Nostalgia* (Basic Books, 2001), xiv.

By the late 1990s, a wholesale 'Shōwa boom' was under way, focusing on the decade leading to the 1964 Tokyo Olympics, when families were moving into cities and the future had still been bright.[36]

Osaka was also part of the story, with Expo 70 animating the archaeology and Okamoto Tarō's Tower of the Sun as the grail on which the digs converged. They took various forms, however. Some were charmingly well-intentioned, providing a place to rummage through the past. 'EXPO', a small bric-a-brac shop, opened in 1994 (and burned down twenty years later) just down the hill from the Tokyo National Museum, selling Expo 70 goods and allowing its customers to recreate a 'Shōwa lifestyle'. Some were baldly opportunist, producing little more than ersatz schmaltz. One of many retro food courts opened in 2002 opposite the Universal Studios theme park on the Osaka waterfront, promising that the customer could 'time slip' back to 1965, experiencing 'nostalgia' for the 'deliciousness' of the moment when expo excitement began to build.[37] The event provided more than period flavour and plastic memorabilia, however. By the turn of the millennium, both expo and Tower were becoming potent sources with which to redescribe and renavigate the present.

For some, going back to 1970 was a way to revisit old battles, lambasting the expo as a state-sponsored distraction from the dark side of progress and rehabilitating the reputation of those now seen to have been on the right side of history. Okamoto Tarō had died in 1996 but did well from the rewrite, with his name on two museums, a series of special exhibitions, and a stream of commentary, which endorsed his persona as prescient critic rather than opportunist showman.[38] Living artists, too, could get mileage from going back to the show and its barker. The most

[36] Ivy, *Discourses of the Vanishing*, for nostalgia in general. Kelly, 'Rationalization and Nostalgia', and Robertson, 'Furusato Japan', for rural nostalgia. Gluck, 'The Invention of Edo', for the Edo boom, and Sand, *Tokyo Vernacular*, 110–42, for Shōwa.

[37] For 'EXPO', https://konoike.org/expo/. For ersatz schmaltz, Margaret Hillenbrand, 'Nostalgia, Place, and Making Peace with Modernity in East Asia', *Postcolonial Studies* 13.4 (2010), 383–401. For the food court, *Yomiuri Shimbun* 2002/5/27, Osaka pm, 13; 2002/7/13, Osaka pm, 13; 2002/12/19, Osaka, 29; and 2003/4/18, Osaka, 23.

[38] For the museums: Taro Okamoto Memorial Museum, https://taro-okamoto.or.jp/; and Taro Okamoto Museum of Art, Kawasaki, www.taromuseum.jp/. For a special exhibition, National Museum of Ethnology, 'Kaikan 40 Shūnen Kinen Tokubetsuten "Taiyō no Tō kara Minpaku e: 70-nen Banpaku Shūshū Shiryō" | Tenji Gaiyō' (2018), https://older.minpaku.ac.jp/museum/exhibition/special/20180308taiyou/outline. For commentary, Sawaragi, *Kuroi Taiyō* and *Sensō to Banpaku* (Bijutsu Shuppansha, 2005); and Winther-Tamaki, 'To Put on a Big Face'.

prominent was the contemporary artist Yanobe Kenji, who had spent his childhood playing on the abandoned site. The expo figured large in his subsequent work, much of which investigated the 'ruins of the future' (*mirai no haikyō*) that had been promised in 1970. Some pieces were sombre. Wearing a radiation suit and standing in what had been the Festival Plaza, Yanobe pointed to the desolation left behind when the circus left town. Some were more optimistic. In 2003, still wearing the radiation suit, he recreated the 'eye-jacking' of the Tower of the Sun by a student radical, climbing up the inside to occupy one of the eyes.[39]

For the producers of more popular culture, too, the expo served as a stage for battles between good and evil, but ones that were playing out in the present. Released in 2001, the ninth anime movie in the hugely popular, frequently scatological Crayon Shin-chan manga-and-more franchise earned 1.4 billion yen at the box office by skewering the expo's continued hold on the present. Disgusted by the early twenty-first century, an evil genius creates a twentieth-century 'expo' in a department store surrounding Tokyo Tower, which includes a reproduction of the Expo 70 site, so that grown-ups can rediscover their childhood. Shin-chan's father gets caught up fighting a Godzilla lookalike, but Shin-chan saves the day, by making his dad smell his own feet, bringing him back to the present, and so inspiring other adults to turn their backs on the past.[40] A couple of years earlier, Urasawa Naoki had used Expo 70 as inciting incident for *20th Century Boys*. Neither the hero, Kenji, nor his nemesis, Friend, had gone to the expo as children, but both studied it obsessively and pretended that they had. For Friend, the disappointment, one of many slights in childhood, builds to a series of atrocities. They climax in a recreation of the expo site in Tokyo, with his organization's symbol replacing the golden face of the Tower of the Sun, providing a set on which to stage his resurrection from a faked assassination. For Kenji, this is a betrayal, a 'knockoff' – 'When you arrive in Osaka on the bullet train, that's the

[39] Yanobe Kenji, https://yanobe.com/; *Yomiuri Shimbun* 2018/3/19, Osaka pm, 10; and Gunhild Borggreen, 'Ruins of the Future: Yanobe Kenji Revisits Expo 70', *Performance Paradigm* 2 (2006), 123–36. For context, Yoko Hayashi, 'Les Enfants de l'Exposition Universelle d'Osaka: La Génération-symbole de la Scène Artistique Japonaise des Années 90', in Éric Mézil (ed.), *Donai yanen! Et maintenant!* (École nationale supérieure des beaux-arts, 1998), 39–59.

[40] Hara Keiichi, *Kureon Shinchan – Arashi o Yobu Mōretsu! Otona Teikoku No Gyakushū* (Tōei, 2021). See also Makino Osamu, *Banpaku Seisen* (Hayakawa Shobō, 2020), which moves from a fictionalized 1970 to a second Osaka Expo in 2037, exploring what might be possible if the younger generation were allowed to have its day.

world of tomorrow' – but the fake expo turns out to provide the only place to shelter from Friend's plan to destroy the world. The story ends with Kenji's old band reassembling on a stage at the fake Festival Plaza, allowing good to triumph over evil and city and country to move on.[41]

In the alternate worlds of manga and science fiction, it was possible to render the ambivalence of the expo as binary plot, but in Shigematsu Kiyoshi's *Twilight* (2002), it plays out as midlife crisis. A group of forty-something classmates returns to their elementary school, in a lightly disguised Tama New Town, west of Tokyo, to dig up a time capsule they had buried in their final year. In real life, Katsuya is a middle manager, facing redundancy, and Mariko, his childhood crush, is married to Tetsuo, his childhood tormentor, though their marriage is on the rocks. In the capsule, though, plastic bags reveal their younger fears and dreams. Katusya's includes a small replica of the Tower of the Sun. 'The expo? That brings back memories, doesn't it?' (*banpaku kā, natsukashi nā*). He tutors a twenty-year-old colleague on the way the expo had promised a future of global peace, space travel, and medical advance. 'See the expo and you could believe.' He goes back to Osaka with Mariko, but everything has changed. The expo is a park, Expoland seems like a dinosaur, and the Expo Tower resembles frog spawn. Lying in bed with her, he confesses that, like Kenji and Friend, he had never actually been. Back in Tokyo, with the memory of the expo in ruins, real life also falls apart. But the group meets again, to bury another time capsule. Tetsuo and Mariko include their divorce certificate, Katsuya a robot and a new replica of the Tower of the Sun. The important face is not the golden one at the very top, gazing at a fantasy future; it is the one in the middle, between the arms, staring down a difficult present.[42]

Come the turn of the century, creatives and historians were beginning to understand that expos remained protean and present because they never delivered simple stories. Expo 1970 had been a watershed, but not the one described at the time. For a few, returning to the archive became a key with which to extricate themselves from its lingering hold and recalibrate their place in the world. Kushima Tsutomu spent his Sundays combing through documents, artefacts, and the responses to a

[41] Urasawa Naoki, *Nijū Seiki Shōnen* (Shōgakkan, 2000), especially vols. 15, 21, and 22. See also Ōsawa Masachi, *Fukanōsei no Jidai* (Iwanami Shoten, 2008), 10–11; and Marc Yamada, 'Trauma and Historical Referentiality in Post-Aum Manga', *Japanese Studies* 34.2 (2014), 153–68.

[42] Shigematsu Kiyoshi, *Towairaito* (Bungei Shunjū, 2002).

questionnaire, to reconstruct the expo and to work out why it continued to haunt the present.[43] Hashizume Shin'ya had been in the fourth grade in 1970, had visited the expo eighteen times, and has returned repeatedly to expos to understand how the modern city, not least Osaka, became what it is.[44] For many, the Tower of the Sun remains an icon, which promises to provide a new lease on life. In 2018, it reopened to the public, with the tree of life restored. Shigematsu was one of the first inside. The tower retained a 'fundamental vitality', though its lesson had changed. When he wrote *Twilight* in his late thirties, he had been struggling to bridge the gap between the future he had been promised and the present in which he was living. Now in his fifties, he realised that he could still be nostalgic, but he could move on (*ore mo natsukashiminagara, mae ni susumu zo*).[45] It was a lesson that had also been on display during the complicated genesis and unexpected triumph of the Aichi Expo in 2005.

LOVING THE EARTH IN AICHI

Aichi 2005 was born in grief, which took at least two decades to overcome. Nagoya, the prefectural capital, had been a confident candidate for the 1988 Olympics, but in 1981 they were awarded to Seoul. The denial and the anger took a while to subside, but by the second half of the decade, the authorities were bargaining that an expo would 'fill the hole' the loss left in their plans for the region. Nagoya had always been overshadowed by its larger neighbours, too easily dismissed as 'splendid sticks' (*ōinaru inaka*), stuck between Tokyo and Osaka. Aichi prefecture was also known for making things (*monozukuri*), but demand was shifting. Seto, on the outskirts of the city, had been a famous centre for ceramics, but was struggling with technological change and global competition. Toyota dominated the local economy, but had offshored much production and was pondering the turn away from oil. How to navigate the shift from industry to information? How to get the region to 'stand up' again after losing the Olympics? In good high-growth fashion, new infrastructure was imagined as the way to save the day. A new governor was elected in 1983, in part thanks to his promise of three big

43 Kushima, *Maboroshi Bankoku Hakurankai.*

44 Kikuchi, 'Banpaku Shōnen-Tachi'. Hashizume Shinya, 'Expo 1970 Osaka: The Story of Japan's First World Expo', www.bie-paris.org/site/en/focus/entry/expo-1970-osaka-the-story-of-japan-s-first-world-expo (2020).

45 *Yomiuri Shimbun* 2018/3/19, Osaka, 36; 2018/3/20, pm, 1.

projects: new expressways, an international airport, and maglev rail. Then, in October 1988, as the regional expo boom crested and the emperor began to die, an expo at the beginning of the new millennium joined the 'big three' as his preferred 'plus one', promising to catalyse a 'flow of people' through the exciting new connections.[46]

The expo soon started taking predictable form, with a faint green wash. By 1991, the tired initial theme, 'Peace and Civilization in the Twenty-First Century', had become 'Technology, Culture, and Exchange: Creation of a New Earth'. By mid-1994, a plan was taking shape. The site would be 250 hectares in the hills southeast of Seto. Like Osaka in 1990, it would have three zones: a 'mountain village' for smaller pavilions, a 'pastoral' one for foreign participants, and an 'urban' one for corporations. More important for its boosters, though, was the way it would spur a transformation of the local landscape. By 1995, the site was being imagined as the germ of a 2,000-hectare 'future exchange city', the core of a 12,000-hectare 'Aichi Academic Research Development Zone', which would be linked to research clusters in Gifu and Mie. The whole would instantiate a 'next-generation urban concept', which would help the region transition from an industrial to a knowledge economy. The expo would attract 40 million visitors, cost 500 billion yen, unlock 1 trillion of infrastructure spending, create 2.5 trillion in demand, and prime the national economy to the tune of over 8 trillion. Its successor city would accommodate 100,000 during the day and over 30,000 at night. Already, though, it was clear there were issues. The site was home to over fifty species of endangered plants and animals, with the status of twelve acknowledged by the government to be critical.[47]

The twists and turns (*kyokusetsu*) the expo took from the mid-1990s have been well-described elsewhere.[48] Briefly, at each stage, it found itself

[46] *Yomiuri Shimbun* 1988/10/20, 2; 1988/12/11, 2; 1996/3/30, pm, 11; 1996/9/20, Chūbu, 34; 1996/9/28, Chūbu, 30; 1996/10/2, special feature, 26; 1997/1/3, Chūbu, 30; 1997/7/8, Chūbu, 30; 1997/7/21, Chūbu, 26; 2004/11/12, Chūbu, 27. Yoshimi, *Banpaku Gensō*, 193–95.

[47] *Yomiuri Shimbun* 1991/5/9, Chūbu, 26; 1991/10/17, Chūbu, 1; 1994/6/15, Chūbu, 1; 1996/6/5, Chūbu, 30; 1995/6/16, Chūbu, 26; 1995/6/20, Chūbu, 20; 1997/1/11, Chūbu, 5. Yoshimi, *Banpaku Gensō*, 195–98, and 2006, 398–99.

[48] Yoshimi, *Banpaku Gensō*, 200–51, and 'A Drifting World Fair: Cultural Politics of Environment in the Local/Global Context of Contemporary Japan', in Tomiko Yoda and Harry Harootunian (eds.), *Japan after Japan: Social and Cultural Life from the Recessionary 1990s to the Present* (Duke University Press, 2006), 399–412. For *kyokusetsu*, see the ten-part series in the *Yomiuri Shimbun*, 2004/11/12–23, Chūbu. For Aichi's planning in historical perspective, Sawada Yūji, 'Aichi Banpaku Zen'ya: Hakurankai no

blocked by a series of opponents. The earliest and most persistent were local citizens, with housewives often taking the lead, who observed, photographed, studied, documented, and proposed alternative futures for what became identified as 'Kaishō Forest'.[49] The government was also changing tack, compiling a 'red list' of endangered species in 1991 and passing a new law in 1993, which transformed the environment from an object of human activity to a subject in its own right.[50] Similarly, in Paris, in 1994, just as the organizers were releasing their first plan, the Bureau Internationale des Expositions (BIE) stipulated that expos needed a theme that could still showcase progress but had to demonstrate 'a commitment to the supreme importance for Humanity of due respect for Nature and the environment'.[51] The plans for the expo were thus assailed from multiple sides. Local groups coordinated with their national organizations, who appealed to international organizations for support, who put pressure on the BIE to insist that the expo authorities should change their plans. And the media, of course, revelled in the conflict.[52] By the time the plans were finalized in the new millennium, they revealed a new kind of expo – and maybe even a new understanding of the environment, too. The story can be told briefly in three simple acts.

The first saw a rearguard action by MITI, in cahoots with Tokyo-based intellectuals and Toyota, to ensure that Nagoya beat out Calgary and the expo was approved in Paris. By the end of 1995, the expo had been rebranded as 'Beyond Development', with two-fifths of a shrunken site devoted to ecological co-existence.[53] One year later, however, when a BIE

Kikaku Seisaku Genba kara', in Sano Mayuko (ed.), *Bankoku Hakurankai to Ningen no Rekishi* (Shibunkaku Shuppan, 2015), 475–503.

49 Yoshimi, *Banpaku Gensō*, 200–3. *Yomiuri Shimbun* 1992/1/13, Chūbu am, 31, for birdwatching, and Sophie Houdart, 'L'Image ou Sa Dissolution au Moment de la Préparation de l'Exposition Internationale Japonaise de 2005', *Autrepart* 4 (2002), 154–58, for photo books.

50 *Yomiuri Shimbun* 1996/09/21, Chūbu, 30; and Ministry of the Environment, 'The Basic Environment Law and Basic Environment Plan' (1994), www.env.go.jp/en/laws/policy/basic_lp.html.

51 Bureau International des Expositions, 'A Short History of Expos', www.bie-paris.org/site/en/expos/about-expos/history-of-expos. See also Yamada Munenori, 'Kokusai Hakurankai no Hen'yō: Ai•Chikyūhaku no Seika to Hyōka', *Nihon Gomu Kyōkai Shi* 80.3 (2007), 112–18.

52 Yoshimi, 'A Drifting World Fair', 410–11. As for the press, the regional edition of the *Yomiuri Shimbun* ran 1,662 articles on the expo, roughly two every three days, between June 1994 and June 2001.

53 *Yomiuri Shimbun* 1995/12/6, 3; 1996/5/29, Chūbu, 30; 2004/11/13, Chūbu, 31; Yoshimi, *Banpaku Gensō*, 205–7; Sophie Houdart, '"Un Chaos Savamment

team visited, they found unpersuaded opponents lying in the road and yet another theme, 'Nature's Wisdom: Creating a New Earth'.[54] Meanwhile, Toyota was dangling prospective investment in front of potential voters. When Aichi won the BIE vote, in mid-1997, parties celebrated the 'expo that Toyota won' (*Toyota ga yūchi shita banpaku*).[55] MITI reassembled its fractious squad of intellectual 'green berets' and was soon producing promotional material featuring nature as protagonist: technology was simply a 'branch', through which to access 'the limitless information contained within life'.[56]

In April 1999, however, the discovery of a pair of goshawks, nesting within the proposed site, threw the plans into reverse, inaugurating a second act.[57] The main site was moved to the Aichi Youth Park, a mile to the southwest, but 200 hectares of forest were still going to be used for expo and housing.[58] The BIE visited again and, the leaked minutes subsequently revealed, characterized the plans as a 'land mine', which persisted with 'large-scale, twentieth-century-style land development', and ran the risk of international opprobrium and boycott.[59] The authorities scrambled, abandoning the plans for housing and roads, setting up a Review Council with local citizens, and asking creatives to provide some cover: an architectural rendition of a four-kilometre-long walkway through the forest, suspended in mid-air; pamphlet, posters, and packing

Ordonné…": L'élaboration Conceptuelle de l'Exposition Internationale Japonaise En 2005', *Ateliers d'anthropologie* 30 (2006), 35–62.

[54] *Yomiuri Shimbun* 1996/1/1, Chūbu, 1; 1996/6/3, Chūbu, 26; 1996/6/5, Chūbu, 30; 1996/6/7, Chūbu, 30; 1996/6/8, Chūbu, 30; 1996/08/10, Chūbu, 30; 1996/10/2, special feature, 26; 1996/11/21, Chūbu, 30; 2004/11/14, Chūbu, 34. For the MITI position in mid-1996, 'Tokushū: 21 Seiki Kokusai Hakurankai No Kaisai Ni Mukete', *Tsūsan Jānaru* 29.7 (1996), 8–21. Yoshimi, *Banpaku Gensō*, 213–16, for more on local opposition.

[55] *Yomiuri Shimbun* 1996/6/3, Chūbu am, 26; 1997/6/4, Chūbu am, 30; 1997/6/5, Chūbu am, 30; 1997/6/14, Chūbu am, 1; 1997/7/18, Chūbu am, 30; 1997/7/21, Chūbu am, 26; 2004/11/14, Chūbu am, 34; Yoshimi, *Banpaku Gensō*, 256–58.

[56] For the intellectuals, Yoshimi, *Banpaku Gensō*, 216; *Yomiuri Shimbun* 1997/7/13, Chūbu, 30; and 1997/12/30, Chūbu, 22. For the promotional materials, Yoshimi, *Banpaku Gensō*, 209–12, and 'Tokushū: 21 Seiki Kokusai Hakurankai No Reinen to Kadai', *Tsūsan Jānaru* 31.1 (1998), 12–25.

[57] 1999/5/1, Chūbu, 27. For goshawk prehistory, *Yomiuri Shimbun* 1994/3/6, 1; 1998/7/28, Chūbu, 31; and 1999/3/21, Chūbu, 38.

[58] Yoshimi, *Banpaku Gensō*, 221–27; *Yomiuri Shimbun* 1999/6/18, Chūbu, 27; 1999/10/15, special feature, 29; and 2004/11/16, Chūbu, 30.

[59] Yoshimi, *Banpaku Gensō*, 227–30; *Yomiuri Shimbun* 1999/11/18, Chūbu, 34; 2000/1/15, Chūbu, 1 and 38; 2000/1/21, 38, and Chūbu, 35; 2000/2/9, Chūbu, 11; and 2004/11/17, Chūbu, 34.

tape promoting 'forest thinking'.[60] The business community was not happy, however, and turned to Sakaiya Taichi. The Expo 70 propagandist and Expo 75 facilitator insisted that the expo needed 50 million visitors, and therefore a bigger draw than the environment, if it were to be a 'holy singularity' (*sei-naru ikkaisei*), with which civilization would be catapulted into a new era. Why not a hundred-metre-tall 'World Tower'? A one-hectare screen? An outdoor theatre for a million? Tanioka Ikuko, the chair of the Review Council, pointed out that times had changed. 'What's needed isn't the wisdom of a single big producer, but the diverse wisdom of ordinary people'. Soon enough, Sakaiya was gone.[61] He was not the only person to be chewed up and spat out by the expo.[62] By late 2001, however, it was gradually becoming apparent how to reconcile the demands of development and environment, which meant the third act could begin.

The nominal labour was divided between three septuagenarian 'producers', all old expo hands, including architect Kikutake Kiyonori, who had already indulged his long-standing predilection for mega-structures in Okinawa in 1975. But the work was delegated to and done by three chief producers in their fifties, together with a squad of even younger specialists, responsible for environmental management, citizen participation, and more.[63] By the end of 2021, the plan was to accommodate 15 million visitors on a split site. The Youth Park comprised a two-and-a-half-kilometre, gourd-shaped 'global loop', which linked a corporate pavilion zone next to the main gate; symbol and Japan zones in the centre; six 'global commons' for international participants; an interactive fun zone; and a sprawling forest experience zone,

[60] For the changing plans, Yoshimi, *Banpaku Gensō*, 236–43; *Yomiuri Shimbun* 2000/5/29, Chūbu, 31; 2000/7/25, 1; and 2004/11/20, Chūbu, 28. For the architect, *Yomiuri* 1997/7/13, Chūbu am, 30; 1999/10/15, am, 29; 2000/1/23, Chūbu am, 26; 2004/11/18, Chūbu am 30; Houdart, 'L'image'. For the pamphlet, Zaidan Hōjin 2005-nen Nihon Kokousai Hakurankai Kyōkai, *Expo Japan 2005* (2000); Hara Design Institute, 'EXPO 2005 AICHI Pamphlet & Posters', www.ndc.co.jp/hara/en/works/2014/08/expo.html.

[61] *Yomiuri Shimbun* 1997/6/14, special feature, 30; 1998/1/21, Chūbu, 34; 1999/10/15, special feature, 29; 2000/2/9, Chūbu, 11; 2000/12/5, Chūbu, 11; 2001/2/27, Chūbu, 34; 2001/2/28, Chūbu, 34; 2001/3/20, Chūbu, 1; 2001/3/23, Chūbu, 38; 2001/5/18, Chūbu, 1; 2001/6/2, Chūbu, 1 and 31; 2001/6/3, Chūbu, 29; 2001/6/9, Chūbu, 28; 2001/6/14, Chūbu, 31, and Chūbu pm, 18; 2001/6/29, Chūbu, 39. Yoshimi, *Banpaku Gensō*, 243–50. Sakaiya Taichi continued to insist that expos were 'the greatest events on earth' until the end of his life: *Chijō Saidai No Gyōji: Bankoku Hakurankai* (Kōbunsha, 2018).

[62] Kominami Yukihiro used the phrase *hito-kui banpaku* ('people-eating expo') in a conversation in November 2002.

[63] Yoshimi, *Banpaku Gensō*, 252, and *Yomiuri Shimbun* 2001/7/24, Chūbu, 34 and 37. For the producer system, *Banpaku o Tsukuru* (2005-nen Nihon Kokusai Hakurankai Kyōkai, 2005), 21 and passim.

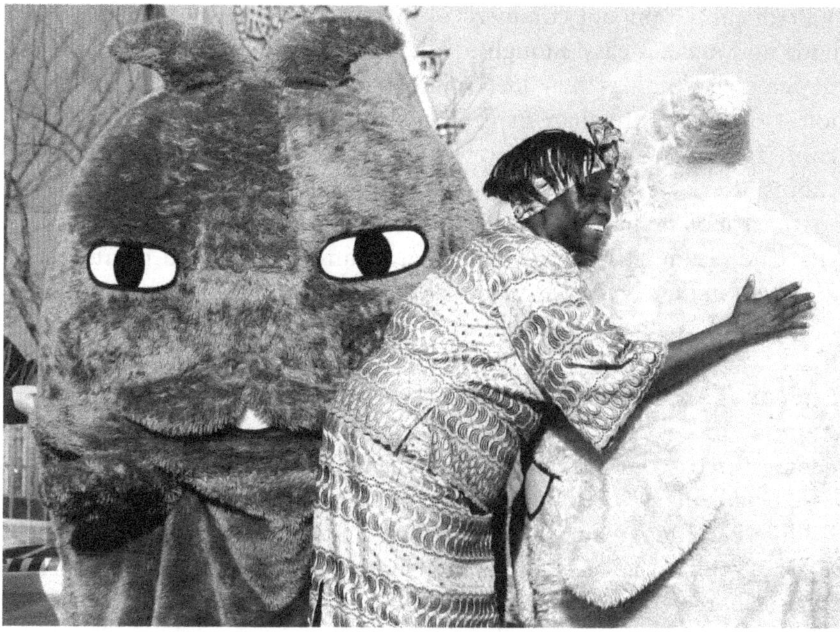

FIGURE 8.3 The stars of the show. Wangari Maathai was a constant presence in the early days of the Aichi Expo, Morizō and Kiccoro throughout.
Source: Toshifumi Kitamura/AFP via Getty Images.

accessed through a Japanese garden. A gondola from the main entrance took visitors to a much reduced second site, on the edge of the now sacrosanct forest, with pavilions for Aichi prefecture and the Japanese government, and a nature trail. Participation was the name of the game, 'with "nature's wisdom" as its warp and "great global exchange" as its weft'. Come the new year, the rebranding was taking off. A new nickname, 'Ai-chikyū-haku', suggested that the expo was going to 'love the earth' (and, homonymically, that Aichi might just save expos). Come spring 2002, two mascots showed how (Figure 8.3). Morizō was the old man of the woods, Kiccoro – tree plus piccolo – his youthful sidekick. Before long, they were starring in their own anime series and appearing in books, which three-year-old Princess Aiko pored over in a well-timed press shoot. By 2005, 200 companies had licence agreements for Morikoro goods, and the Association was expecting 68 billion yen in sales. In fact, they managed to shift 80 billion during the expo and another 20 in the next half year.[64]

[64] Yoshimi, *Banpaku Gensō*, 253. *Yomiuri Shimbun* 2001/10/16, am, 15; 2001/12/20, am, special feature, 25; 2005/1/10, Chūbu, 28; 2005/9/24, 3; 2005/9/26, Chūbu, 39; 2006/3/1, Chūbu, 7; and 2006/9/24, Chūbu, 1. Morizō and Kiccoro live on in the name of the

Predictably rampant commercialism was one of many indicators that made and make it easy enough to characterize the expo itself as more of the same. Technology remained omnipresent, both as spectacle and solution, nowhere more than in the American pavilion, which completely ignored the downsides of growth. Huge screens continued to loom. Robots were everywhere, from the female 'Actroids', through the Toyota robot orchestra, to their less-humanoid cousins, collecting garbage and cleaning floors. Advertising, meanwhile, focused on the main attraction in the 'Symbol Pavilion' at the centre of the site: a lab exhibited the partial remains of a four-to-five-ton Siberian mammoth, which had been excavated from the melting permafrost, while the earth's glories and its predicament were projected on either side in super high-definition.[65] It was easy to condemn the expo for treating nature as something to be memorialized, commodified, and betrayed, particularly given the perceived inadequacy of its environmental assessment efforts. The World Wildlife Fund Japan and two other environmental groups withdrew their participation at the last minute.[66] Given its troubled genesis, it was also easy to characterize the 'ostentatious' (*kyoshoku*) expo as the 'final festival' of the construction state.[67]

But these were partial takes. In the round, the expo suggested that, at least in Japan, things had started to change. Compared to the familiar offerings of the developmental state, the ideation and execution of the expo made space for many more than the usual suspects. It was the BIE, in Paris, that had struggled to understand nature's wisdom, and suggested that corporations needed double their allotted space. In Aichi itself, there were nearly 80 non-governmental organizations (NGOs), over 200 citizen projects, and an additional 30,000 volunteers, with input into 'what they wanted to do and how they were going to do it'.[68] At the expo, the environment was no longer a passive victim of human progress. The possibility of ecological co-existence was captured in the notion of *satoyama*. The space between the village (*sato*) and the mountain (*yama*),

expo memorial park: Ai•Chikyū Haku Kinen Kōen (Morikoro Pāku), 'Morikoro Pāku: Ai•Chikyū Haku Kinen Kōen', www.aichi-koen.com/moricoro/.

65 *Yomiuri Shimbun* 2005/2/25, 3; 2005/3/25, 3; 2005/9/24, 13; 2005/9/25, 24. Fred Nadis, 'Nature at Aichi World's Expo 2005', *Technology and Culture* 48.3 (2007), 577–80. For the Global House, Ai•Chikyū Haku, 'Gurōbaru Hausu', www.expo2005.or.jp/jp/C0/C3/C3.5/C3.5.4/index.html.

66 *Yomiuri Shimbun* 2005/3/8, Chūbu, 37.

67 Maeda Eisaku, *Kyoshoku No Aichi Banpaku* (Kōbunsha, 2005).

68 *Banpaku o Tsukuru*, 15, 61–63, 98–99, 170–81, and 191–201; *Yomiuri Shimbun* 1997/7/16, Chūbu, 26, 2001/12/20, am, 25, 2004/10/14, Chūbu, 30, and 2005/10/2, Chūbu, 34.

which had traditionally provided a community with fuel and food, emerged as a concern in the 1960s, under threat from suburban sprawl. It was easy enough to deploy the term in ponderous claims of cultural difference. But it was also possible to put it to work. The Kaishō Forest, it turned out, was one such site. The original woodland had been consumed as fuel for pottery. The existing forest, treasured by the expo's opponents, was the product of ecological destruction. Where better for a lab (*jikkenjō*) to experiment on a new relationship with nature?[69]

And experiment the expo did, even if some of its initiatives failed. Planning and operations were governed by environmental assessment standards developed through public hearings, as well as the triple imperative to reduce, reuse, and recycle. A 'new energy system', using restaurant waste, fuel cells, and solar panels, failed to produce zero emissions, but the expo only used three-quarters of the electricity and emitted three-quarters of the carbon dioxide predicted. The global loop, designed by Harada Shizuo for easy disassembly and made from recycled wood, allowed visitors barrier-free access, wildlife to pass under it, and trees to grow through it. The foreign pavilions, too, had been designed for reuse. Only one in four was, after severe discounting, but the steel frames of the rest were recycled and nearly three-quarters of the equipment was repurposed. The Japanese Government Pavilion was wrapped in a huge bamboo basket made from some 30,000 trees, protecting it from the elements, improving ventilation, and reversing 50 years of bamboo invasion in *satoyama* areas around Tokyo and down south. In the restaurants, the plastic, whose adoption in the 1960s had allowed the bamboo to spread, was abandoned in favour of biomass tableware, designed for composting and reuse. When the goshawks were surveyed a year after the expo, they were fine.[70]

[69] For *satoyama* at the Expo, *Yomiuri Shimbun* 1995/6/16, Chūbu, 26; 1996/6/4, Chūbu, 26; 1996/6/5, Chūbu, 30; 1996/10/2, special feature, 26; 1997/7/29, Chūbu, 30; 2001/12/20, special feature, 25; 2005/9/29, Chūbu, 34. For the context, Catherine Knight, 'The Discourse of "Encultured Nature" in Japan: The Concept of Satoyama and Its Role in 21st-Century Nature Conservation', *Asian Studies Review* 34.4 (2010), 421–41; and Yukihiro Morimoto, 'What Is Satoyama? Points for Discussion on Its Future Direction', *Landscape and Ecological Engineering* 7.2 (2011), 163–71. For the BIE reservations, Houdart, 'L'Image', 158. For ponderous claims, *Banpaku o Tsukuru*, 36–39.

[70] *Banpaku o Tsukuru*, 11–12, 31, 54–57, 68–69, 73–81, 157–61. *Yomiuri Shimbun* 1997/1/16, Chūbu, 5; 2001/6/9, Chūbu, 28; 2001/12/20, 25; 2002/1/1, Chūbu, 35; 2004/9/25, Chūbu, 1; 2004/9/28, Chūbu, 34; 2004/11/23, Chūbu, 30; 2005/9/28, Chūbu, 38; 2006/7/25, Chūbu, 32.

By conventional metrics, too, the expo had been a success. The 22 million visitors had produced a surplus of 14 billion yen, which was ploughed back into the commemorative park, the restoration of Nagoya castle, and more. The 3 trillion yen invested in expo and infrastructure had prompted close to 8 trillion yen of production and created nearly 500,000 jobs. But Aichi had also repurposed expos. A week after the closing ceremonies, the United Nations dubbed it the 'first Eco-Expo', a model for educating the world about sustainable development. Hot air, perhaps, but on the ground, the expo had made a difference. Being on site meant being educated: taught by volunteers to separate trash into nine different categories; nudged with eco-money to avoid plastic bags and spend on reforestation projects. Over 500,000 people attended an environmental education course during the expo, prompting even some of its harshest critics to change their tune. Repeated, in-depth surveys by the Association showed that most visitors were not just 'satisfied', as at expos past, but had learned something. Four out of five had got the message about the need for common purpose, new behaviours, and active participation. Three out of four wanted to come again. Almost nine out of ten wanted to get more involved.[71]

Not everything changed in 2005. There was still an appetite for old-style spectacle, whether horticulture, history, or food. The landscape was littered with purpose-built exhibition halls, which exploited the abbreviated term for expos (*haku*) to stage opportunistic extravaganzas. It was clear, however, that the environment had changed. It was no longer enough to level some land, concrete the site, throw up pavilions, and pack in the crowds. Yokohama tried and failed, in 2009, with a thin, familiar mash-up of futuristic technology, historical reminiscence, and ecological gesture, generating a 1-billion-yen deficit and a string of lawsuits.[72] Nara fared better, in 2010, by providing free admission to the main site and a diverse menu of events and exhibits, bringing nearly 5 million visitors to the rural areas of the prefecture.[73] But Aichi had done more even than this, redefining both spectacle and development, as

[71] *Yomiuri Shimbun* 2005/9/26, 1; 2005/9/29, Chūbu, 34; 2005/101/1, Chūbu, 32; 2005/11/10, 33; 2005/12/22, Chūbu, 33; 2006/1/24, Chūbu, 36; 2006/4/21, Chūbu, 31; 2006/9/23, Chūbu, 1; 2007/3/20, Chūbu, 32. 'Aichi 2005: The First Eco-Expo', https://sdgs.un.org/partnerships/aichi-2005-first-eco-expo. Yamada, 'Kokusai Hakurankai no Hen'yō', 113–16.

[72] *Yomiuri Shimbun* 2009/9/28, 38; 2009/9/30, 33; 2009/12/5, 33; 2010/1/20, 27; 2010/3/12, 33.

[73] *Yomiuri Shimbun* 2010/10/17, Osaka, 16; 2010/12/2, Osaka, 29.

the product of hesitant exploration, which went back to the past as well as forwards to the future; iterative testing, alive to failure; and ongoing negotiation, between those with authority and capital, and the communities – human, animal, vegetal, even mineral – they were charged to serve. One year after the expo, its lessons lived on. Local households were beginning to recycle food waste. Experiments were ongoing with the new energy system, relocated near the new airport. And 'dry mist', pioneered at the expo, was replacing traditional air conditioning in public spaces in Tokyo. Citizen activities were 'more widespread than expected'. Ten years later, the volunteer centre continued as a non-profit organization.[74]

So did one of the major inspirations and attractions of the show. The public awareness of *satoyama* had taken off at the end of the 1980s, in part due to its animation by Miyazaki Hayao. In *My Neighbour Totoro*, Satsuki and Mei lived in the middle of a secondary forest, where they communed with the spirits of the woods. It was an obvious call for the producers of the expo to try to bring Miyazaki on board. Initially, he was resistant. 'I can't cooperate with an expo'. Eventually, though, a replica of Satsuki and Mei's house appeared in the wooded southern part of the main site, with admission restricted by lottery to fifty people at a time, sixteen times a day. Those allowed in found themselves back in the mid-1950s, 'all five senses' alert to the possibility of living so 'close to nature'. It is tempting to condemn such a combination, of entertainment capital and nativist allure, as simply more ersatz schmaltz. It is worth pausing, however. To make good on its commitment to love the earth, the expo needed sources to draw on and spaces to experience. Miyazakiworld provided both. The house remained standing, as most of the expo was dismantled, and ten years after the expo continued to attract well over 100,000 annual visitors. Come 2022, the attraction expanded. The expo memorial park was still branded as a Morikoro wonderland but was now overlaid by 'Ghibli Park'.[75]

[74] *Yomiuri Shimbun* 2006/9/15, Chūbu, 28; 2015/3/24, Chūbu, 34.

[75] *Banpaku o Tsukuru*, 164; *Yomiuri Shimbun* 2004/10/2, Chūbu, 30; 2005/9/25, am, 24; 2015/3/22, Chūbu, 29; 2017/6/2, Chūbu, 28. See Susan Napier, *Miyazakiworld: A Life in Art* (Yale University Press, 2018), for Satsuki and Mei's house (xi–xi), and the *satoyama* in which *My Neighbour Totoro* is set (101–21). For the park, Ghibli Park Co., Ltd., 'Jiburi Pāku' (2024), https://ghibli-park.jp/.

Conclusion

Back at the turn of the millennium, it seemed, expos were becoming history. The old European bastions, London and Paris, had been out of the game for at least half a century. The US had begun to drag its feet over funding for American pavilions in the early 1990s, and would withdraw from the Bureau Internationale des Exhibitions at the beginning of the following decade. In 1996, the Internet World Exposition, 'a world's fair for the information age', had tried to suggest a possible afterlife online, but in the real world, Hanover 2000 was hopeless and Aichi 2005 on the rocks. In Japan, the flood of regional expos, which had surged during the bubble, was also dwindling to a trickle, delighting the leftish *interi-tachi* (the bien-pensant) in Tokyo, who had long been convinced that expos were one of many fronts for state and market. Now, they said, the time had come to say goodbye. The sentiment was a global one. By the time Shanghai 2010 rolled around, Anglo observers from across the political spectrum were united in insisting that expos were a 'lacklustre . . . brand', wishing that they would 'lie down and die', and wondering why they still went on.[1]

[1] John Judis, 'Show and Tell', *The New Republic* 206.3 (1992/1), 15, and Colum Lynch and Robbie Gramer, 'Will Anyone Pay for Uncle Sam's World Expo Pavilion?' *Foreign Policy*, 2019/12/13, https://foreignpolicy.com/2019/12/13/pompeo-congress-dubai-world-expo-pavilion-funding-dispute-china/, for American funding issues. Steve Charnovitz, 'Why the International Exhibitions Bureau Should Choose Minneapolis for Global Expo 2023', 2017/10/15, GWU Law School Public Law Research Paper No. 2017-82, https://ssrn.com/abstract=3053623, 3, for withdrawal from the Bureau Internationale des Expositions (BIE). Carl Malamud, *A World's Fair for the Global Village* (MIT Press, 1997), for the Internet World Exposition. Mizuno, 'Sayonara', and Oda 2004 for two

The commentariat turned out to be an unreliable witness, however. Aichi turned into an unexpected success, suggesting a new direction, at least for international exhibitions. Shanghai triumphed too, albeit by turning back the clock. Reports of the death of expos were an exaggeration, based on a misreading of history and a misunderstanding of the form, as well as more than a smidgeon of first-world, metropolitan snobbery. Understanding Japan's experience of expos on its own terms allows us to rewrite the story of exhibitions in general: it can help us specify the way in which they are both symptomatic of and sometimes a requisite for development, and so grasp why they continue to this day.

Putting things on show is an ancient habit. Displaying rare things, animals, and people from far away, acquired through trade or conquest, is an old way of flaunting wealth, reach, and status. It burgeoned in new forms in early modern cities, as populations grew, networks expanded, and the possibilities for profiting from spectacle spread. Modern exhibitions, however, were initially an attempt by anxious states to impose order on the fast-changing world of the mid nineteenth century. The first industrial revolution was finally taking material shape in the form of railway networks, shipping canals, and iron warships. The second was coming into being with underwater telegraph cables and cast steel cannon. These and other innovations made it possible for the powers to accumulate wealth, to project force at a distance, not least in East Asia, and to assemble new kinds of empire, most visibly in Africa; but they also increased the instability of the present and uncertainty about the future. Other powers could use the same technologies to the same ends. Meanwhile the domestic population, victims to the disruptions of industry, were demanding less of its costs and more of its benefits, if not a wholesale transformation of the constitution of the state.

Great exhibitions were useful, therefore, for anxious hegemons (England in London in 1851 and 1862), old rivals (France in Paris in 1855, 1867, 1889, and 1900), and new pretenders (Austro-Hungary in Vienna in 1873, the US in Philadelphia in 1876, Chicago in 1893, and St Louis in 1904). They were a medium through which to take stock of recent developments, to anchor them in ancient birthright and/or manifest destiny, to convince

examples of Japanese intellectual disdain. 'Living the Dream,' *The Economist*, 2010/4/29, and Rowan Moore, 'British Pavilion; Chelsea Barracks', *The Observer*, 2010/4/25, for Anglo condescension and Lauren Hilgers, 'The World, Shanghaied', *Harper's Magazine* 321.1925 (2010), 7–9, for a longer example.

the citizenry that the present was as it should be and the future would be bright, and to parade their achievements and claims to global rivals, thereby bolstering their own self-belief, international standing, and trade. Japanese exhibits and exhibitors showed up early, with official observers returning home in the 1860s having grasped both the problem and the promise of the medium. Exhibitions overseas revealed that Japan was behind the industrial curve, which meant that it was vulnerable in a world dominated by white imperial powers; but also that it possessed resources and capabilities with which to compete in some global markets. Exhibitions at home would educate domestic producers in their comparative deficiencies, encourage them to compete and upgrade their offerings, and thereby contribute to the wealth and power of the new nation-state.

Domestic exhibitions proved effective, up to a point. Beginning in Tokyo in the 1870s, they provided space and time for organizers and exhibitors to gather and catalogue, for juries and press to inspect and evaluate, the products of agriculture and industry, alongside the technologies through which they made their way into the world, and the art and craft by which they were complemented. Exhibitions also spawned a bureaucratic infrastructure, through which the form spread to the provinces, and over time a range of offshoots, from museums through agricultural and other concours to product promotion halls. They have therefore proved convenient, fertile ground to investigate the birth of many modern things, from industrial policy and the patent system, through machinery and technology, to self-consciously Japanese art. Exhibitions never managed to discipline organizers, exhibitors, or visitors in the way that their bureaucratic promoters imagined, however. From the first, different places turned exhibitions to their own local ends; exhibitors sought to use the opportunity for self-promotion rather than civic pedagogy; and the average visitor was attracted rather by the novel, the mysterious, and the grotesque, than by didactic display. Similarly, overseas, the government never lost its belief in the utility of international exhibitions, as a way of promoting exports, providing an authorized account of the country's past and present, and thereby proving that they belonged in the ranks of the powers. But they were powerless to prevent foreign organizers and visitors consigning Japan to its customary place, as the pre-industrial fairyland, with manicured gardens, wooden architecture, handmade craft, and green tea, that they had first encountered at exhibitions; or being suspicious when it broke the mould, with its own evidence of industry and empire. Both at home and abroad, the medium of exhibition could never guarantee that the message intended by the authorities was that received by the audience.

It is tempting, however, when studying exhibitions to assume that the message is all: to see the exhibition as the simple expression of a coherent set of beliefs; to mistake the representation of these beliefs at the exhibition for the things they purport to describe; and to proceed as if they emanate wholesale from the dominant interests of the time and arrive without question in the minds of the credulous visitor. On the surface, this might explain an apparent shift of focus in international exhibitions in the West, as the preoccupations of the powerful shifted, from industry in the mid nineteenth century, to empire at the turn of the twentieth, to consumerism before and after World War II. Such an approach, however, overestimates both the coherence of the message itself and the ability of an exhibition to communicate it clearly. Lingering on any one exhibition reveals the extent to which its professed intent – to promote industry (London 1851, Tokyo 1877), to commemorate the nation (Philadelphia 1876, Tokyo 1912), to celebrate empire (Paris 1931, Tokyo 1940), or to envision the corporate world of tomorrow (New York 1939, Osaka 1970) – was usually patched together and frequently undone on site, as its simple syllogisms took material form, accommodating the various interests of sponsors, organizers, exhibitors, and visitors. Assuming that exhibitions can do efficient ideological work also fails to explain the extent to which they were the product of particular local contexts, determined as much by pragmatic considerations of geography, resource, and immediate need, as any grand vision of an abstract future. Understanding instead how exhibitions reconciled the conflicting demands of any one time and place also makes it possible to understand how they changed over time, which Japan's history of exhibitions helps to reveal.

The state remained eager to police Japan's appearance at exhibitions overseas, trying both to regulate turn-of-the-century impresarios, who were eager to make money from the white public's desire for Japanese exotica (St Louis 1904, London 1910), and to calibrate the representation of the empire, as Western suspicion mounted and the Japanese military ran amok (Chicago 1933, New York 1940). Following World War II, official exhibits at international exhibitions reverted to a familiar form: the pre-industrial trinity of manicured nature, traditionalist architecture, and meticulous craft, which had been formulated at exhibitions a century earlier and which provided an indigenous foundation for the subsequent triumph of Japanese industry, now dedicated to pacific, harmonious ends (Brussels 1958, Osaka 1970).

At home, however, already by the 1890s the state had begun to lose its interest in exhibitions as a tool of industrial policy. With the economy on a sound footing, exhibitions were taken up instead by a succession of subnational actors, who could turn them to their own particular ends: local governments used them to secure their place within the ever-shifting economic geography of the archipelago, and to justify the municipal improvements that would help them 'image up'; colonial governments to argue for their contribution to the imperial project, to advertise their products – Taiwanese sugar, Korean rice, Manchurian coal – and to solicit the immigration and investment through which they would develop; railway companies, department stores, and newspaper companies to cultivate the middle-class customers they needed to make their profits and grow their business; the army and the navy to garner admiration for their achievements, solicit support for their endeavours, and cultivate the sense of emergency that would allow them to do more; and an emerging exhibition industry to make good money by building the sites and creating the attractions to accommodate and enchant the crowds, turning on a dime from immersive recreations of battles in China in the late 1930s to celebrations of middle-class American life in the early 1950s. For over a century, from Kyoto in 1871, trying to recover from the loss of the emperor, to Yokohama, Nagoya, Hiroshima, and Fukuoka in 1989, describing their various futures in the last days of the bubble, exhibitions proved a mainstay for local authorities and private companies trying to chart their course.

It was this potential of exhibitions, not so much to prescribe meaning as to aggregate interest and so catalyse development, which finally brought the national government back to the game. Following the Tokyo Olympics of 1964, the Osaka Expo of 1970 proved a turning point, providing a pretext for investing in the infrastructure that would begin to address the unbalanced growth of the previous fifteen years, and helping to turn the economy from its focus on production, in part for exports, to an embrace of consumption, largely at home. Big international events became the preferred medium not for national pedagogy but for state planning. Sure enough, a stream of expos followed, first turning neglected peripheries into networked satellites (Okinawa 1975, Tsukuba 1985), then transforming industrial orphans into leisure centres if not ecotopias (Osaka 1990, Aichi 2005, Osaka 2025). The model was a compelling one, soon embraced by Japan's neighbours as they moved up the development ladder and the geopolitical hierarchy (Daejeon 1993, Shanghai 2010). In Japan itself, though, exhibitions were revealing a richer story than the simple mantras of state-led growth. First, again, international exhibitions were only the most visible manifestations of an enthusiasm that swept through archipelago, from Hokkaido to Kyushu, with expos and exhibits running the gamut, from corporate paeans to new technology, through municipal reinventions of urban possibility, to regional invocations of local tradition. Second, even as the economic bubble swelled, so exhibitions provided space for alternative models, of environmental mitigation and citizen participation, with which to address its excesses. Both would come to the fore in the 1990s and 2000s, once the bubble had popped.

Exhibitions as experienced by Japan can therefore also tell us something about the nature of development itself. First, the transformation of the material environment since the middle of the nineteenth century, together with the concomitant remaking of political economy, society, and culture, has prompted efforts by those responsible for these things to prescribe their direction, to reassure us that things are on track, and to distract us when they are not. The efforts have been multiple, but they might be usefully, loosely understood as various forms of spectacle: demarcating time and space, and choreographing people, places, and things, so as to provide an immersive account of a possible world. Exhibitions were the earliest of these, taking a variety of forms and spawning a range of spin-offs, many of which have forsaken the world in which we live in favour of an alternative world to which we can escape. In the form examined here, though, they have been somewhat honest, providing at least a partial take on the world as it is, while insisting that the past still counts and the future might be known. Second, by the same token, exhibitions cannot but reveal the extent to which development is always uneven and necessarily uncertain. Any exhibition has to reconcile the competing interests of those driving development forwards, and to provide a show that can assuage those exposed to its disparate effects. Exhibitions in the aggregate reveal how development takes different forms in different places and, over time, how it continues to surprise and unsettle. Finally, Japan suggests, exhibitions can be useful in mitigating those effects: justifying the capital spending with which to address the distortions of growth; even acknowledging that it is possible to leave less wreckage behind.

True, by the 2010s, there was beginning to be a drought of expos in Japan. Part of the problem was demand. After a third, more-or-less lost decade, the familiar litany of problems – demographic decline, economic stagnation, social fragmentation, political sclerosis, cultural drift – was bearing down hard on local governments. There was little municipal appetite or fiscal capacity for expos. But the geography of the drought also mapped the way in which they had left their mark. There were the vast exhibition halls in and around the big cities, which staged trade fairs and popular attractions, frequently branded as expos. There were a few long-running series, which trundled reliably on: quadrennial food expos in Osaka and confectionery expos nationwide, as well as the annual national urban greening fair.[2] Elsewhere, though, expos had morphed. Already, come the end of the previous century, they had begun to evolve, from single sites and static exhibits to a broad umbrella for events and experiences, which dispersed visitors throughout a city or a region. Now, they were subsumed within the strategic plans for regional development to which they had long been tied. One sole survivor illustrates how.

Back in the boom time, even Ehime, on the western edge of Shikoku, the smallest of the four main islands, was swept up in the enthusiasm for exhibitions, with a Technopia Expo in 1986 sketching a utopian, high-tech future. By the end of the century, though, the prefecture was beginning to confront the fact of depopulation, with Nan'yo, in its south-western corner, predicted to lose a third of its residents in the next four decades. Still, it seemed, the new, dispersed kind of expo might help. In 2004, the Ehime Machinami Expo focused on the rows of traditional housing in Nan'yo's towns and villages, trying to encourage the region's residents to turn their town-making efforts to tourist ends. Two more in the 2010s met the ongoing demand for 'mood regulation'. A first Healing Expo in 2012 highlighted Nan'yo's curative properties, in terms of its 'rich nature', outdoor recreation, and historical culture. A second in 2016 added food and cycling to the earlier menu. Three hundred events over eight months, including 'premium dining' in ten towns and an eight-day 'Tour de Nan'yo' attracted nearly 1 million additional visitors, producing 9 billion yen of additional demand, and created a 'sense of purpose' for the local community. Unprecedented rain and catastrophic flooding in 2019, however, meant that the work had to continue. A 2021 Expo of Ties (*Kizuna Haku*) was pushed back a year, due to the pandemic, but touted as a way for the region to reverse its long-term decline by targeting not only tourists but also volunteers, short-term and long-term workers, and U- and I-turn migrants, who may have been born or sought their fortune in the big city, but might now be tempted to make the countryside their home. In the promotional videos, young Tokyo creatives documented their delight in their new rural lives and promote

[2] *Yomiuri Shimbun* 2017/9/13, Chūbu, 27; 2020/11/24, Osaka, 27; 2021/1/5, am, 24.

the possibility of an 'Ehime shift'. Ehime may be the only local authority still using an 'expo' to organize and brand its outreach initiatives, but doing so points both back, to a century of regional expos, and to the continuing resonance of the term.[3]

Meanwhile, even if regional expos in Japan are thin on the ground, Aichi 2005 seems to have granted international exhibitions a new lease of life – and confirmed Japan's mastery of and commitment to the form. For old industrial heartlands, expos are a way to reinvent themselves as ecologically minded global citizens, however feeble their efforts (Hanover 2000, Milan 2015). For places on the margins, they are a way to attract attention (Zaragoza 2008, Yeosu 2012, Astana 2017). And for emerging powers and global cities, they are a way to demand that the world pay court. Shanghai 2010 and Dubai 2020 were for the most part throwbacks, testament to the path that Japan had laid. The former was an opportunity for China, which overtook Japan as the world's second-largest economy in the same year, to demonstrate its geopolitical ambitions, and for Shanghai to demonstrate its ambitions as a global city, converting premium waterfront to post-industrial use and pushing the city's marginal population aside, with the promise that a 'better city' would create a 'better life'. There were some cursory gestures to the eco-logic of Aichi 2005, but the model and the target was Osaka 1970, with a 523-hectare site (versus Osaka's 330) and 189 participating countries (versus 78). It hit its mark, too, with 73 million visitors (versus 64).[4]

[3] For 2016, Ehime Chiiki Seisaku Kenkyū Sentā, *Ehime Ibento BOX 2016* (Ehime Chiiki Seisaku Kenkyū Sentā, 2016), 70–82; Ehime-ken, '"Ehime Iyashi no Nan'yo Haku 2016" Keizai Hakyū Kōka nado ni tsuite', www.pref.ehime.jp/h14500/nanyo/documents/siryo-nanyohaku.pdf; and *Yomiuri Shimbun* 2016/12/26, Osaka, 22. For 2021, 'Ehime Nan'yo Kizuna Haku', https://kizunahaku.com; Ehime-ken, 'Ehime Nan'yo Kizuna Haku: Jisshi Keikaku', www.pref.ehime.jp/nan54180/kizunahaku/documents/jissikeikaku.pdf; and *Yomiuri Shimbun* 2020/10/21, Osaka 29. For the late 1990s healing boom, Paul Roquet, 'Ambient Literature and the Aesthetics of Calm: Mood Regulation in Contemporary Japanese Fiction', *The Journal of Japanese Studies* 35.1 (2009), 88–89. For I- and U-turn migration, Susanne Klien, *Urban Migrants in Rural Japan: Between Agency and Anomie in a Post-Growth Society* (State University of New York Press, 2020).

[4] For the expo as the culmination of a long history of Chinese exhibitionism, Sō Kennan, 'Chūgoku ni Okeru Hakurankai Būmu no Tanjō', in Sano Mayuko (ed.), *Bankoku Hakurankai to Ningen no Rekishi* (Shibunkaku Shuppan, 2015), 711–28. For an overview, Tim Winter, *Shanghai Expo* (Routledge, 2012). For the national narrative, Jennifer Hubbert, 'Back to the Future: The Politics of Culture at the Shanghai Expo', *International Journal of Cultural Studies* 20.1 (2017), 48–64. For the worldview, Astrid Nordin, 'Space for the Future: Exhibiting China in the World at the Shanghai Expo', *China Information* 26.2 (2012), 235–49. For the urban vision, Susan Fernsebner, 'Expo 2010: A Historical Perspective', *The Journal of Asian Studies* 69.3 (2010), 669–76, Sophie Houdart, 'A City

Eleven years later – the expo eventually opened from October 2021 to March 2022 – Dubai 2020 made a more concerted effort to acknowledge the looming climate crisis, with an embrace of sustainable development. For the most part, though, it too followed the older pattern established by Osaka, promising to 'connect minds' and 'create the future', laying the groundwork for a 'human-centric smart city', and enhancing the Emirati brand of 'openness' and 'tolerance', while relying on forced migrant labour to create the show.[5]

Japan, of course, keeps showing up, with new variations on the old theme of harmony, between nature and culture, tradition and technology, country and world. In Shanghai in 2010, on the far eastern edge of the Asia zone, Purple Silkworm Island was advertised as a 'living breathing organism', covered with solar panels and designed to pump out less carbon than it took in (Figure C.1). Inside, the exhibit culminated in a low-tech, twenty-minute mash-up of Chinese Kunqu opera and Japanese Noh theatre, which recounted the collaboration that had re-established the crested ibis, using Chinese stock, in Japan, but proved less media-worthy than the concurrent protests roiling Chinese cities over the Diaoyu/Senkaku islands.[6] In Milan in 2015, the exhibit riffed on the idea of *satoyama* to illustrate its commitment to 'feeding the planet', with a wooden lattice entryway and a paean to the health and sustainability of traditional food ways, which was realized by an immersive projection to conjure the harmony of rice fields, and a digital waterfall, to hymn the

without Citizens: The 2010 Shanghai World Expo as a Temporary City', *City, Culture and Society* 3.2 (2012), 127–34, and Ehara Noriyoshi, 'Toshika o Tēma toshita Shanhai Banpaku: Banpaku o Meguru Chūgoku no Kako to Mirai', in Sano Mayuko (ed.), *Bankoku Hakurankai to Ningen no Rekishi* (Shibunkaku shuppan, 2015), 681–710.

5 'Expo 2020 Dubai UAE', www.expo2020dubai.com/en, for the official story; Bashar Taha and Andrew Allan, 'Hosting a Mega Event, a Drive towards Sustainable Development: Dubai's Expo 2020', in Stan Geertman et al. (eds.), *Computational Urban Planning and Management for Smart Cities* (Springer, 2019), 223–47, for sustainable development; and *EXPOsed: Discrimination and Forced Labour Practices at Expo 2020 Dubai* (Equidem, 2022) for forced labour.

6 CPDExpo 2010, 'Japan at Expo 2010 Shanghai', www.youtube.com/watch?v=O236Z7goyvA, for the official line, and Iwata Yasushi, 'Shanhai Banpaku: Reisui Banpaku Nihonkan mara Mita Nihon no Hakurankai Gyōsei', in Sano Mayuko (ed.), *Bankoku Hakurankai to Ningen no Rekishi* (Shibunkaku Shuppan, 2015), 505–29, for admin. Andrea Chin, 'Japanese Pavilion at Shanghai Expo 2010', *Designboom* (2009), www.designboom.com/architecture/japanese-pavilion-at-shanghai-expo-2010/, for design details. Urso Chappell, 'Japan Pavilion, Expo 2010', www.youtube.com/watch?v=USinA8RjUYg, for exhibits and the performance. *Yomiuri Shimbun* 2008/2/25, Chūbu, 7, 2009/4/21, 33, 2010/4/21, 33, and 2010/6/2, Seibu, 25 for preparations and description; 2010/10/29, 6 for diplomatic kerfuffle; and 2010/11/2, 6 for a summing up.

FIGURE C.1 Ecological aspiration and diplomatic marginalization. The Japanese Pavilion at Shanghai 2010.
Source: Qilai Shen/Bloomberg via Getty Images.

diversity of Japanese cuisine.[7] Finally, in Dubai in 2021, an origami lattice cooled the waiting visitor, while inside miniature dioramas drew on the old practice of allusion (*mitate*) to rehearse Japan's environmental commitments and export prowess. Tiny camels dotted the dunes of a desert made out of an Emirati *kandura*, the ankle-length, long-sleeved robe worn throughout the Arab world, which in turn was made out of fabric manufactured in Japan. The exhibit ended by inviting visitors to share 'your ideas and messages' for the next world expo, in 2025 in 'the exciting Kansai' (sic). Neither the city of Osaka nor the Japanese state has given up on the utility of international exhibitions as a catalyst for development.[8]

It has not quite happened yet, but Expo 2025 already echoes Aichi 2005, with its origins in the overconfidence of the bubble and the anxiety of its aftermath. The ground was initially laid back in the 1950s, with the

[7] teamLab, 'Expo Milano 2015', www.teamlab.art/e/expomilano2015/, for the exhibits. Danna, 'Expo 2015 in Milan', for one take on the whole.

[8] Public Relations Office, Government of Japan, 'Where Ideas Meet: Japan Pavilion at Expo 2020 Dubai' (2022/2), www.gov-online.go.jp/eng/publicity/book/hlj/html/202202/202202_09_en.html, and Expo 2020 Dubai, 'Japan Pavilion' (2022), www.expo2020dubai.com/en/understanding-expo/participants/country-pavilions/japan, for design and exhibits.

reclamation of the first of what would eventually become three offshore islands just south of the mouth of the Yodo River in Osaka Bay. The third, Yumeshima (Dream Island) began to appear in 1977, initially as a place to dump construction waste. By 1988, it was being reimagined, following Tokyo, as a teleport, but the collapse of the bubble meant the authorities had to scramble to fill the hole. The 2008 Olympics was their first choice, but Osaka's bid lost in the first round of voting in 2001. Osaka's share of the domestic economy was also sliding. Companies were leaving for Tokyo at a rate of knots.[9] Enter the populist. Hashimoto Tōru, a Tokyo-born, Osaka-raised, right-wing lawyer and media personality, was elected as Governor in 2008 and Mayor in 2011, with neoliberal designs to 'restore' the city and region, through a mix of privatization, outsourcing, and annexation of the surrounding municipalities.[10]

A couple of years later, the shine was coming off, but the way forward was becoming clear. An 'integrated resort' would be built on Yumeshima, aiming to win one of the casino licences made possible by a new law. A world expo in 2025 would announce Osaka's restoration, as advised by expo enthusiast Sakaiya Taichi, and make its case for the extension of the promised maglev line from Nagoya. By 2016, the two initiatives had come together, with Yumeshima as the preferred site for both resort and expo, even though only a third of the island had been reclaimed. METI stepped in to coordinate the expo bid, broadening the theme, expanding the site, and dividing it into three zones: 'pavilion world', with the familiar crowd of national, corporate, and NGO pavilions; 'water world', incorporating a hotel and guest facilities; and 'green world', devoted mainly to transport and parking, wrapping around and including the existing solar plant. When the final BIE vote was held in 2018, Osaka came out comfortably ahead.[11] Two years later, the masterplan made the

[9] *Yomiuri Shimbun* 2017/4/12, am, 13; 2017/5/14, Osaka, 25; and 2018/11/28, Osaka, 38. For context, Gavan McCormack, 'Paving over the Kansai', JPRI Occasional Paper (1994), https://jpri.org/wp-content/uploads/2021/11/Occasional-Paper-01.pdf, and Hiroshi Watanabe, 'Osaka: The Altered Landscape', *Japan Quarterly* 42.3 (1995), 317–31.

[10] Charles Weathers, 'Reformer or Destroyer? Hashimoto Tōru and Populist Neoliberal Politics in Japan', *Social Science Japan Journal* 17.1 (2014), 77–96.

[11] For early reporting, *Yomiuri Shimbun* 2014/8/7, Osaka, 38; 2015/1/20, Osaka, 38; 2015/4/29, Osaka, 27; 2015/7/8, Osaka, 31; 2016/5/24, Osaka, 29; 2016/10/29, Osaka, 35. For METI, ibid., 2017/2/14, Osaka pm, 1. For the BIE application and winning the vote: 2017/9/22, Osaka pm, 1; 2018/6/8, Osaka, 17; and 2018/11/24, 1.

vision clear. The 28 million expected visitors will enter the expo from a new subway station, surmounted by a skyscraper, funnel between Japanese government and corporate pavilions, then pass up onto a huge circular roof, half a kilometre in diameter, or under it, into the international and theme zone. The expo will focus on 'Designing Future Society for Our Lives', in line with the United Nations (UN)'s seventeen sustainable development goals and Japan's own plans for Society 5.0, both of which hold out the possibility of living within the planet's means. Eight theme pavilions will explore what this might mean, from amplifying life 'by integrating it with technology' to 'better co-being', courtesy of a 'forest of tranquility'. A 'people's living lab' will produce solutions from the bottom-up, including citizens not just as observers but as participants and producers too.[12]

Of course, not everyone is happy. The connection to the casino was a question mark in the early going, and its ethical propriety, social effects, and financial viability remain usefully indeterminate targets for critics. More material concerns remain too, about the stability of the ground and problems of access, given what an earthquake might to do to a site built in part out of construction waste, rather than close-packed composite, and to the single bridge and subway line (Figure C.2). Most immediate, a year ahead of the opening ceremonies, is the eternal expo equation of time, budget, and popular interest. Local companies were never enthusiastic about bearing their chunk of the predicted costs, and criticism of the near doubling of the construction budget, from 125 to 235 billion, has been exacerbated by the delayed construction of the foreign pavilions. The authorities protest that all this is par for the course and point to the 161 countries and regions that will participate. More worrying is the increasing public indifference, attributed in part to the overly intellectual theme. Nearly nine out of ten Japanese know that it is happening, but the number interested in going has slid, from over 50 per cent at the end of 2021, to just over 40 in 2022, to the low 30s in December 2023. Advance tickets went on sale with 500 days to go, but early interest was desultory.

12 *2025-nen Nihon Kokusai Hakurankai Kihon Keikaku* (2025-nen Nihon Kokusai Hakurankai Kyōkai, 2020), 9 for SDGs and Society 5.0, 28–38 for the elaboration of the theme, and 51–70 for the site plan. *Yomiuri Shimbun* 2019/3/28, Osaka, 32, for the skyscraper. See also 'Expo 2025: Osaka, Kansai, Japan', www.expo2025.or.jp/. For Society 5.0, 'Society 5.0', www8.cao.go.jp/cstp/english/society5_0/index.html, and Hitachi-UTokyo Laboratory, *Society 5.0: A People-Centric Super-Smart Society* (Springer Singapore, 2020).

FIGURE C.2 The container berths and the solar plant were in place, but eighteen months before the expo, construction was just beginning.
Source: Richard A. Brooks/AFP via Getty Images.

All of this is irrefutable evidence for some commentators, who insist that, if it is not cancelled, the expo will destroy Osaka.[13]

Time will tell. On past performance, it is hard to believe that either the promises of its promoters or the cautions of its critics will entirely come to pass. Expos have always been spectacles of development. Osaka 2025, too, will tell a story where we have been, where we are now, and where we might go next. But it will also make it possible to see things otherwise. It will do something for the economy. It will leave a space behind, in which to gamble away one's cares, or to mourn a future betrayed. And when the circus leaves town, there will be other expos in other cities, following in the footsteps of Osaka, Okinawa, Tsukuba, Aichi, and the rest. There will be a horticultural expo in Yokohama and a specialized

[13] For early corporate indifference, *Yomiuri Shimbun* 2015/7/29, Osaka, 23, for the casino, 2016/9/24, Osaka, 33, and for costs, 2016/9/30, Osaka, 1. For a compendium, 2017/4/25, Osaka, 1; 2017/4/26, Osaka, 3; and 2017/4/27, Osaka, 31. For increasing costs and delayed foreigners, 2023/12/3, 31. For disinterested public and slow ticket sales, 2023/12/15, 25, and 2024/2/1. For one recent take-down, Saitō Minako, 'Ōsaka Banpaku o Chūshi ni Subeki, Kore Dake no Riyū' (2024/2), www.webchikuma.jp/articles/-/3415, who draws on Sakurada Teruo, Takayama Shin, and Yamada Akira, *Kajino, Banpaku de Ōsaka ga Kowaweru: Ishin ni yoru Keizai, Seikatsu Daihakai* (Akebi Shobō, 2022).

expo in Belgrade, both in 2027. There will be a world expo in Riyadh in 2030. Busan lost out but is eyeing a bid for 2035. Even the US has been trying to get back in the game.[14] And wherever there is an expo, Japan will most likely be there.

[14] For Busan, *Korea Times*, 2023/11/29, www.koreatimes.co.kr/www/nation/2024/03/113_364104.html. For the US, US Department of State, 'Hosting the World's Fair', www.state.gov/hosting-the-worlds-fair/.

Select Bibliography

This bibliography includes only the major expo-related primary and secondary materials on which the book draws. All other references can be found in the footnotes.

DATABASES, NEWSPAPERS, AND PERIODICALS

Akahata
Asahi Jānaru
Asahi Shimbun. Accessed via https://database.asahi.com/
Chōsen oyobi Manshū
Chūō Kōron
Ekonomisuto
Fūzoku Gahō
Gazette des Beaux Arts
Gendai no Me
Gijutsu to Ningen
Illustrated London News
InterCommunication
Japan Quarterly
The Japan Times
Kenchiku to Shakai
Kikan Geijutsu
New York Times
Nikkei Asia
Nomura, *Hakurankai Shiryō Collection*. Accessed via www.nomurakougei.co.jp/expo/
Rejā Sangyō Shiryō
Sankei Nyūsu
Shin Kenchiku
Shisō

Tsūsan Jānaru
Yomiuri Shimbun. Accessed via https://database.yomiuri.co.jp/

CATALOGUES, GUIDES, AND REPORTS

London 1851

Official Catalogue of the Great Exhibition of the Works of Industry of All Nations, 1851 (Spicer Brothers, 1851)

London 1862

Alcock, Rutherford, *Catalogue of Works of Industry and Art, Sent from Japan* (Clowes & Sons, 1862)
International Exhibition 1862: Official Catalogue of the Industrial Department (Truscott & Simmons, 1862)
International Exhibition, 1862: Reports by the Juries on the Subjects in the Thirty-Six Classes into Which the Exhibition Was Divided (Clowes & Sons, 1863)
Shaffner, Taliaferro P., and William Owen, *The Illustrated Record of the International Exhibition of All Nations, in 1862* (Ward and Lock, 1862)

Paris 1867

Chevalier, Michel, *Rapports du Jury International* (Paul Dupont, 1868)
Commission Impériale, *Rapport sur l'Exposition Universelle de 1867 à Paris* (Imprimerie Impériale, 1869)
Exposition universelle de 1867 à Paris, *Catalogue Officiel des Exposants Recompensés par le Jury International* (Dentu, 1867)
Exposition universelle de 1867 à Paris, *Guide Officiel à l'Exposition Universelle de 1867* (Dentu, 1867)
Reports on the Paris Universal Exhibition, 1867, Volume 1, Containing the Report by the Executive Commissioner, and Appendices (1868)

Vienna 1873

Commission Impériale Japonaise, *Notice sur l'Empire de Japon et sur Sa Participation à l'Exposition Universelle de Vienne, 1873* (Lévy, 1873)
Tanaka, Yoshio, and Hirayama Narinobu, *Ōkoku Hakurankai Sandō Kiyō* (Meiji Bunken Shiryō Kankōkai, 1964)

Philadelphia 1876

Imperial Japanese Commission to the International Exhibition at Philadelphia, *The Empire of Japan: Brief Sketch of the Geography, History and Constitution* (1876)

Imperial Japanese Commission to the International Exhibition at Philadelphia, *Official Catalogue of the Japanese Section, and Descriptive Notes on the Industry and Agriculture of Japan* (1876)
Japanese Department of Education, *An Outline History of Japanese Education* (Appleton, 1876)

Tokyo 1877

Naikoku Kangyō Hakurankai Jimukyoku, *Meiji Jūnen Naikoku Kangyō Hakurankaijō Annai* (1877)
Meiji Jūnen Naikoku Kangyō Hakurankai Shuppin Kaisetsu (Meiji Bunken Shiryō Kankōkai, 1962)
Meiji Jūnen Naikoku Kangyo Hakurankai Shuppin Hōkokusho (Meiji Bunken Shiryō Kankōkai, 1964)
Nakamura, Juntarō, *Naikoku Kangyō Hakurankai Michishirube* (Itō Shōnosuke, 1877)
Ōkubo, Toshimichi, *Naikoku Kangyō Hakurankai Shokisoku* (Tsuruda Shin'ya, 1876)

Paris 1889

Huard, Lucien, *Livre d'Or de l'Exposition* (Boulanger, 1889)
L'Exposition de Paris (1889) (La Librairie Illustrée, 1889)
Nōshōmushō, *Fukoku Pari Bankoku Dai Hakurankai Hōkokusho* (Fujimi Shobō, 1890)

Tokyo 1890

Itō, Ken'ichi, *Dai Sankai Naikoku Hakurankai Kaijō Annai* (Meikōsha, 1890)
Ōkubo, Tsunekichi, *Dai Sankai Naikoku Kangyō Hakurankai Kaijō Michishirube* (Kōgyokudō, 1890)

Chicago 1893

Holley, Marietta, *Samantha at the World's Fair* (Funk & Wagnalls, 1893)
Imperial Japanese Commission of the World's Columbian Exposition, *History of the Empire of Japan* (Dai Nippon Tosho, 1893)

Paris 1900

Le Livre d'Or de l'Exposition de 1900 (Édouard Cornély, 1900)
Le Panorama: Exposition Universelle 1900 (L. Baschet, 1900)

Osaka 1903

Daigokai Naikoku Kangyō Hakurankai Kyōsankai, *Ōsaka to Hakurankai* (Daigokai Naikoku Kangyō Hakurankai Kyōsankai, 1902)
Daigokai Naikoku Kangyō Hakurankai Kinen Shashinchō (Gyokumeikan, 1903)

Hasegawa, Nyogen, *Daigokai Naikoku Kangyō Hakurankai Shokisoku* (Nagoya Tsūshinsha, 1902)
Kadota, Masatsune, *Daigokai Naikoku Kangyō Saikai Benran* (Taiwan Kyōkai, 1903)
Miura, Ippei, *Daigokai Naikoku Kangyo Hakurankai Yōran* (Meiji Bunken Shiryo Kankokai, 1973)

St Louis 1904

Hanson, J. W., *The Official History of the Fair, St. Louis, 1904* (Monarch Book Co., 1904)
Holley, Marietta, *Samantha at the St. Louis Exposition* (GW Dillingham Co., 1904)
Hoshi, Hajime, *Handbook of Japan and Japanese Exhibits at World's Fair, St. Louis, 1904* (1904)
Imperial Japanese Commission to the Louisiana Purchase Exposition, *The Exhibition of the Empire of Japan, Official Catalogue* (International Exposition, St. Louis, MO, 1904)

London 1910

Japan-British Exhibition, 1910, Shepherd's Bush, London: Official Guide (Bemrose and Sons, 1910)
Mutsu Hirokichi, *The British Press and the Japan-British Exhibition of 1910* (Routledge, 2001)
Nichiei Hakurankai Jimukyoku Jimu Hōkoku (Nōshōmushō, 1912)
Official Report of the Japan British Exhibition 1910 at the Great White City, Shepherd's Bush, London (Unwin Brothers, 1911)

Tokyo 1912

Takushoku Hakurankai Jimu Hōkoku (Takushoku Hakurankai Zanmu Toriatsukaijo, 1913)

Osaka 1913

Meiji Kinen Takushoku Hakurankai Annaiki (Yūshinsha, 1913)
Meiji Kinen Takushoku Hakurankai Hōkoku (Meiji kinen takushoku hakurankai, 1913)

Seoul 1915

Chōsen Sōtokufu, *Chosen Sōtokufu Shisei Nenpō, 1915* (Kuresu Shuppan, 1991)

Seoul 1929

Chōsen Hakurankai Kinen Shashinchō (Chōsen Sōtokufu, 1930)

Taipei 1935

Taiwan o Daihyōsuru Mono (Taiwan Shinbunsha, 1935)

Nagoya 1937

Nagoya Han Taiheiyō Heiwa Hakurankai kaishi (Han Taiheiyo Heiwa Hakurankai, 1938)

Paris 1937

Livre d'Or Officiel de l'Exposition Internationale des Arts et des Techniques Dans la vie Moderne, Paris, 1937 (Éditions Spec, 1937)

New York 1939

The Japanese Pavilion, New York World's Fair, 1939 (Japanese Commission to the New York World's Fair, 1939)

San Francisco 1939–1940

Golden Gate International Exposition (Japanese Committee for the Golden Gate International Exposition, 1939)

Official Guide Book: Golden Gate International Exposition on San Francisco Bay (Crocker, 1939)

Tokyo 1940

Kigen 2600-nen Kinen Nihon Bankoku Hakurankai Gaiyō (Kigen 2600-nen Kinen Nihon Bankoku Hakurankai Jimukyoku, 1938)

Brussels 1958

Exposition Universelle et Internationale de Bruxelles, 1958, 9 vols. (Établissements Généraux d'Imprimerie, 1958)

Osaka 1970

Expo '70: Nihon Bankokuhaku Shidō no tame ni (Monbushō, 1970)

Japan World Exposition, Osaka, 1970: Official Report, 3 vols. (Commemorative Association for the Japan World Exposition, 1972)

Nihon Bankoku Hakurankai no Keisatsu Kiroku (Ōsakafu Keisatsu Honbu, 1971)

Nihon Bankoku Hakurankai Kōshiki Gaido (Nihon Bankoku Hakurankai Kyōkai, 1970)

Nihon Bankoku Hakurankai Kōshiki Kiroku, 3 vols. (Nihon Bankoku Hakurankai Kinen Kyōkai, 1972)

Nihon Bankoku Hakurankai Kōshiki Kiroku Shiryōshū (Nihon Bankoku Hakurankai Kinen Kyōkai, 1971)

Nihon Bankoku Hakurankai Kōshiki Kiroku Shiryōshū Bessatsu, 46 vols. (Nihon Bankoku Hakurankai Kinen Kyōkai, 1971)
Nihon Bankokuhaku Jiten (Marunouchi Risāchi Sentā, 1968)
Nihonkan (Tsushō Sangyōshō Kigyōkoku Bankokuhaku Junbishitsu, 1970)

Okinawa 1975

Expo '75: Official Report (Japan Association for the International Ocean Exposition, 1976)
Kaiyōhaku Handobukku (Nihon Kōgyō Shinbunsha, 1973)
The Sea We Would ike to See: The International Ocean Exposition, Okinawa, 1975 (Japan Association for International Ocean Exposition, 1975)

Kobe 1981

Kobe Port Island Exposition: Official Guide (Kobe Port Island Exposition Association, 1981)
New Cultural City on the Sea: Port Island, Kobe City (Kōbe-shi Kaihatsu-kyoku, 1979)

Tsukuba 1985

Expo '85: Dwellings and Surrounding – Science and Technology for Man at Home (Japan Association for the International Exposition, Tsukuba, 1985, 1981)
Tsukuba Expo '85: Official Guide Book (Japan Association for the International Exposition, 1985)

Fukuoka 1989

Asian-Pacific Exposition Fukuoka '89 (Association for the Asian-Pacific Exposition, 1987)

Hiroshima 1989

Sea and Islands Expo Hiroshima 1989 (Sea and Islands Expo Association, 1989)

Yokohama 1989

Uchū to Kodomotachi: Yokohama Hakurankai (Yokohama Hakurankai Kyōkai, 1989)

Osaka 1990

Expo '90, Osaka, Japan: Official Guidebook (Japan Association for the International Garden and Greenery Exposition, 1990)

Aichi 2005

Banpaku o Tsukuru (2005-nen Nihon Kokusai Hakurankai Kyōkai, 2005)
Maeda, Eisaku, *Kyoshoku no Aichi Banpaku* (Kōbunsha, 2005)
Zaidan Hōjin 2005-nen Nihon Kokousai Hakurankai Kyōkai, *Expo Japan 2005* (Zaidan Hōjin 2005-nen Nihon Kokousai Hakurankai Kyōkai, 2000)

Osaka 2025

2025-nen Nihon Kokusai Hakurankai Kihon Keikaku (2025-nen Nihon Kokusai Hakurankai Kyōkai, 2020)

BOOKS, ARTICLES, AND DISSERTATIONS

Altick, Richard D., *The Shows of London* (Belknap Press, 1978)
Asakura, Kamezō, *Misemono Kenkyū* (Heibonsha, 1992)
Aso, Noriko, *Public Properties: Museums in Imperial Japan* (Duke University Press, 2014)
Auerbach, Jeffrey, *The Great Exhibition of 1851: A Nation on Display* (Yale University Press, 1999)
Awazu, Kiyoshi, and Hariu Ichirō, 'Genzō no Naka de', *Kikan Geijutsu* 4.2 (1970), 26–39
Bennett, Tony, *The Birth of the Museum: History, Theory, Politics* (Routledge, 1995)
Blanchard, Pascal, Nicolas Bancel, Gilles Boetsch, Eric Deroo, and Sandrine Lemaire, *Human Zoos: Science and Spectacle in the Age of Colonial Empires* (University of Chicago Press, 2009)
Blaxell, Vivian, 'Preparing Okinawa for Reversion to Japan: The Okinawa International Ocean Exposition of 1975, the US Military and the Construction State', *The Asia-Pacific Journal: Japan Focus* 8.29.2 (2010)
Bokovoy, Matthew, *The San Diego World's Fairs and Southwestern Memory, 1880–1940* (University of New Mexico Press, 2005)
Borggreen, Gunhild, 'Ruins of the Future: Yanobe Kenji Revisits Expo 70', *Performance Paradigm* 2 (2006), 123–36
Bromfield, David, 'Japanese Representation at the 1867 Paris International Exhibition and the European Response to It', in Alan Rix and Ross E. Mouer (eds.), *Japan's Impact on the World* (Japanese Studies Association of Australia, 1984), 133–50
Bureau International des Expositions, *Commémoration Du 75ème Anniversaire de La Convention de Paris* (BIE, 2004)
Christ, Carol Ann, '"The Sole Guardians of the Art Inheritance of Asia": Japan at the 1904 St. Louis World's Fair', *Positions: East Asia Cultures Critique* 8.3 (2000), 675–709

Coaldrake, Kimi, 'Fine Arts versus Decorative Arts: The Categorization of Japanese Arts at the International Expositions in Vienna (1873), Paris (1878) and Chicago (1893)', *Japan Forum* 25 (2013), 174–90

Conant, Ellen, 'Refractions of the Rising Sun: Japan's Participation in International Exhibitions, 1862–1910', in Toshio Watanabe and Tomoko Sato (eds.), *Japan and Britain: An Aesthetic Dialogue, 1850–1930* (Lund Humphries, 1991), 79–92

Danna, Daniela, 'Expo 2015 in Milan: The Power of the Machine', *Journal of Political Ecology* 24.1 (2017), 901–20

Debord, Guy, *Comments on the Society of the Spectacle* (Verso Books, 1988)

The Society of the Spectacle (Zone Books, 1994)

Ehara, Noriyoshi, 'Toshika o Tēma toshita Shanhai Banpaku: Banpaku o meguru Chūgoku no Kako to Mirai', in Sano Mayuko (ed.), *Bankoku Hakurankai to Ningen no Rekishi* (Shibunkaku Shuppan, 2015), 681–710

Ericson, Mark, 'Sovereignty on Display: The Tokugawa Bakufu and the Paris Universal Exposition of 1867', *The Journal of Japanese Studies* 47.2 (2021), 277–311

Fernsebner, Susan, 'Expo 2010: A Historical Perspective', *The Journal of Asian Studies* 69.3 (2010), 669–76

'When the Local Is the Global: Case Studies in Early Twentieth-Century Chinese Exposition Projects', in David Raizman and Ethan Robey (eds.), *Expanding Nationalisms at World's Fairs: Identity, Diversity, and Exchange, 1851–1915* (Routledge, 2017), 173–94

Figal, Gerald, *Beachheads: War, Peace, and Tourism in Postwar Okinawa* (Rowman & Littlefield, 2012)

Freestone, Robert, and Marco Amati, *Exhibitions and the Development of Modern Planning Culture* (Routledge, 2014)

Frost, Warwick, and Jennifer Laing, *Exhibitions, Trade Fairs and Industrial Events* (Routledge, 2017)

Fujiki, Takao, and Toyoda Kentarō, '1937-nen Pari Bankoku Hakurankai Nihonkan no Sekkei Ikisatsu ni tsuite', *Nihon Kenchiku Gakkai Keikakukei Ronbunshū* 514 (1998), 217–23

Fujioka, Rika, 'Takashimaya Sōsōki ni okeru Hakurankai no Yakuwari', *Keiei Kenkyū* 50.1–2 (1999), 249–66

Fujita, Yoshihisa, '"Japan Ekisupo" ni yoru Chihō Hakurankai o megutte', *Aidai Shigaku: Nihonshi, Ajiashi, Chirigaku* 11 (2002), 61–82

Fujitani, Takashi, *Splendid Monarchy: Power and Pageantry in Modern Japan* (University of California Press, 1996)

Fukuma, Yoshiaki, 'Senji Hakurankai to "Seisen" no Hokorobi', in Fukuma Yoshiaki, Nanba Kōji, and Tanimoto Naho (eds.), *Hakuran no Seiki: Shōhi Nashonariti Mejia* (Azusa Shuppansha, 2009), 277–306

Fukuoka, Maki, *The Premise of Fidelity: Science, Visuality, and Representing the Real in Nineteenth-Century Japan* (Stanford University Press, 2012)

Fukuzawa, Yukichi, *Seiyō Jijō* (Keio Gijuku Daigaku Shuppankai, 2002)

Furukawa, Takahisa, *Kōki, Banpaku, Orinpikku: Kōshitsu Burando to Keizai Hatten* (Yoshikawa Kōbunkan, 1998)

Furuta, Ryō, 'Nihon no Bijutsu Tenrankai: Sono Kigen to Hattatsu', *Museum* 545 (1996), 29–56

Gardner, William O., 'The 1970 Osaka Expo and/as Science Fiction', *Review of Japanese Culture and Society* 23 (2011), 26–43

Gilbert, James, *Whose Fair?: Experience, Memory, and the History of the Great St. Louis Exposition* (University of Chicago Press, 2009)

Greenhalgh, Paul, *Ephemeral Vistas: The Expositions Universelles, Great Exhibitions, and World's Fairs, 1851–1939* (Manchester University Press, 1988)

Gregory, Breandan, 'The Spectacle Plays and Exhibitions of Imre Kiralfy, 1887–1914' (PhD thesis, University of Manchester, 1988)

Ha, Sae-bong, 'Mohyeongui Jeguk', *Tongyang Sahak Yŏn'gu* 78 (2002), 165–99

'Taiwan and Its Self-Images: The Case of the Osaka Exhibition in 1903', *Taiwan Shi Yanjiu* 14.7 (2007), 1–39

Hamada, Yō, 'Shūkyō to Hakuran no Kindaishi', in Fukuma Yoshiaki, Nanba Kōji, and Tanimoto Naho (eds.), *Hakuran no Seiki: Shōhi Nashonariti Mejia* (Azusa Shuppansha, 2009), 131–60

Hariu, Ichirō, 'Hanpaku: Hansen Undō no Shikō Sakugo', *Gendai no Me* 10.10 (1970), 126–35

Harris, Neil, 'All the World a Melting Pot? Japan at American Fairs, 1876–1904', in Akira Iriye (ed.), *Mutual Images: Essays in American-Japanese Relations* (Harvard University Press, 1975), 24–54

Haryū, Ichirō, 'Expo '70 as the Ruins of Culture (1970)', *Review of Japanese Culture and Society*, 23 (2011) 44–56

Hashizume, Setsuya, *Dai-Ōsaka Imēji: Zōshokusuru Manmmosu, Modan Toshi no Genzō* (Sōgensha, 2007)

Hashizume, Shinya, 'Nenchū Gyōji to Shinbun Kigyō: Ryōgoku Kokugikan no Nōryō Ibento o megutte', in Toshihiro Tsuganesawa (ed.), *Kindai Nihon no Media Ibento* (Dōbunkan, 1996), 169--91

Jinsei wa Hakurankai: Nihon Rankaiya Retsuden (Shōbunsha, 2001)

Nihon no Hakurankai: Terashita Tsuyoshi Korekushon (Heibonsha, 2005)

Hatanaka, Akihiro, *Gorin to Banpaku: Kaihatsu no Yume, Honrō no Rekishi* (Shunjūsha, 2020)

Hayashi, Yoko, 'Les Enfants de l'Exposition Universelle d'Osaka: La Génération-Symbole de la Scène Artistique Japonaise des Années 90', in Éric Mézil (ed.), *Donai Yanen! Et maintenant!* (École nationale supérieure des beaux-arts, 1998), 39–59

Hedinger, Daniel, 'Fighting a Peaceful War: Japan at World Exhibitions in the 1860s and 1870s', in Volker Barth (ed.), *Innovation and Education at International Exhibitions* (Bureau international des expositions, 2007), 71–94

Henry, Todd, 'Celebrating Empire, Fighting War: The 1940 Exposition in Late Colonial Korea', *Asea Yŏn'gu* 51.4 (2008), 72–112

Assimilating Seoul: Japanese Rule and the Politics of Public Space in Colonial Korea, 1910–1945 (University of California Press, 2014)

Herbert, James D., *Paris 1937: Worlds on Exhibition* (Cornell University Press, 1998)

Hirai, Seiji, 'Hakurankai kara mita Kamagasaki no Rekishi: Kokusaku no Kage ni tsukai Sterareru Hiyatoi Rōdōsha', *Shin Nihon Bungaku* 43.1 (1988), 110–15

Holt, Elizabeth, *The Art of All Nations, 1850–73: The Emerging Role of Exhibitions and Critics* (Princeton University Press, 1982)
Hopson, Nathan, '"A Bad Peace?": The 1937 Nagoya Pan-Pacific Peace Exhibition', *Japanese Studies* 38.2 (2018), 137–51
Hotta, Yoshihiro, *Yoshida Hatsusaburō no Chōkanzu o Yomu: Egakareta Kindai Nihon no Fūkei* (Kawade Shobō Shinsha, 2009)
Hotta-Lister, Ayako, *The Japan-British Exhibition of 1910: Gateway to the Island Empire of the East* (Japan Library, 1999)
Hotta-Lister, Ayako, and Ian Nish (eds.), *Commerce and Culture at the 1910 Japan-British Exhibition: Centenary Perspectives* (Global Oriental, 2013)
Houdart, Sophie, 'L'Image ou sa Dissolution au Moment de la Préparation de l'Exposition Internationale Japonaise de 2005', *Autrepart*, 4 (2002), 141–66
'"Un Chaos Savamment Ordonné...": L'Élaboration Conceptuelle de l'Exposition Internationale Japonaise en 2005', *Ateliers d'anthropologie* 30 (2006), 35–62
'A City without Citizens: The 2010 Shanghai World Expo as a Temporary City', *City, Culture and Society* 3.2 (2012), 127–34
Hubbert, Jennifer, 'Back to the Future: The Politics of Culture at the Shanghai Expo', *International Journal of Cultural Studies* 20.1 (2017), 48–64
Hur, Nam-lin, 'Invitation to the Secret Buddha of Zenkōji: Kaichō and Religious Culture in Early Modern Japan', *Japanese Journal of Religious Studies* 36.1 (2009), 45–63
Ikawa, Mitsuo, 'Manshū Jihen Zengo no Nagoya Shinbun no Ibento', in Tsuganesawa Toshihiro and Ariyama Teruo (eds.), *Senjiki Nihon no Media ibento* (Sekai Shisōsha, 1998), 113–27
'Genshiryoku Heiwa Riyō Hakurankai to Shinbunsha', in Tsuganesawa Toshihiro (ed.), *Sengo Nihon no Media Ibento, 1945–1960-nen* (Sekai Shisōsha, 2002), 247–65
Ikeguchi, Kotarō, *Nihon no Bankoku Hakurankai: Sono Igi, Keikaku, Kōka* (Tōyō Keizai Shinpōsha, 1968)
Bankokuhaku to Mirai Senryaku (Daiyamondosha, 1970)
Inoue, Shōichi, *Senjika Nihon no Kenchikuka: Āto, Kitchu, Japanesku* (Asahi Shinbunsha, 1995)
'Konpanion ga Jokanshu to Yobareta Koro: Hakurankaijō ni okeru Josei Setsugūin no Seiritsu to Tenkai', in Sano Mayuko (ed.), *Bankoku Hakurankai to Ningen no Rekishi* (Shibunkaku Shuppan, 2015), 385–405
Inuzuka, Takaaki, 'Pari Bankoku Hakurankai to Satsuma Gaikō: "Puropaganda" no Gaikō Senryaku', in Kagoshima Junshin Joshi Daigaku Kokusai Bunka Kenkyū Senta (ed.), *Shin Satsuma Gaku, Vol. 1, Sekai no Naka no 'Satsuma'* (Nanpō Shinsha, 2002), 13–52
Ishida, Ayū, 'Shōwa Zenki to Kokusan Yōfuku Hakurankai', in Fukuma Yoshiaki, Nanba Kōji, and Tanimoto Naho (eds.), *Hakuran no Seiki: Shōhi Nashonariti Mejia* (Azusa Shuppansha, 2009), 72–85
Ishikawa, Atsuko, 'Shiryō Kara Miru Rankaiya to Sōshokugyō no Rekishi', in Sano Mayuko (ed.), *Bankoku Hakurankai to Ningen no Rekishi* (Shibunkaku Shuppan, 2015), 351–83

Isono, Naohide, 'Yakuhinkai, Bussankai Nenpyō (Zōteiban)', *Keiō Gijuku Daigaku Hiyoshi Kiyō, Shizen Kagaku* 29 (2001), 55–65

Isozaki, Arata, 'Recalling the Days of Expo Art (?) (2001)', *Review of Japanese Culture and Society* 23 (2011), 72–80

Itō, Mamiko, '1915-nen Panama Taiheiyō Banpaku to San Furanshisuko Hainichi Undō', *Mediashi Kenkyū* 23 (2007), 25–41

Meiji Nihon to Bankoku Hakurankai (Yoshikawa Kōbunkan, 2008)

'Hakurankai to Shokuminchi: Shokuminchi Teikoku Nihon no Kenjiyoku', in Ugai Masashi and Kawaguchi Akihiro (eds.), *Kinō no Nihon: Kindai Shakai to Bōkyaku Sareta Mirai* (Yūshisha, 2012), 64–74

Kal, Hong, *Aesthetic Constructions of Korean Nationalism: Spectacle, Politics, and History* (Routledge, 2011)

Kanda, Kōji, 'Nanki Kumano Taikenhaku to Kumano no hyōshō', in Sano Mayuko (ed.), *Bankoku Hakurankai to Ningen no Rekishi* (Shibunkaku Shuppan, 2015), 321–47

Kargon, Robert H., Karen Fiss, Morris Low, and Arthur P. Molella (eds.), *World's Fairs on the Eve of War: Science, Technology, & Modernity, 1937–1942* (University of Pittsburgh Press, 2015)

Kasahara, Kazuto, 'Rekishi, Kankō, Hakurankai: Dai Yonkai Naikoku Kangyō Hakurankai to Heian Sento Senhyakunen Kinensai no Toshi Kūkan', 10+1 36 (2004), 164–77

Kawaguchi, Hitoshi, 'Meiji Sue no Chihō ni okeru Kodomo Hakurankai ni tsuite', *Kyūshū Zōkei Tanki Daigaku Kiyō* 22 (2000), 33–46

Kawasaki-shi Okamoto Tarō Bijutsukan (ed.), *Okamoto Taro, EXPO '70: Taiyō no Tō kara no Messeji* (2000)

Kawashima, Yūsuke, 'Hitsuzen Demo Gūzen Demo Naku: 1995-nen "Sekai Toshi Hakurankai", Chūshi no Seijigakuteki Bunseki', *Nagoya Daigaku Hōsei Ronshū* 269 (2017), 309–28

Kim, Tae-woong, 'Industrial Exhibitions ("Gongjinhoe") and the Political Propaganda of Japanese Imperialism in the 1910s', *International Journal of Korean History* 3 (2002), 179–223

Kiyokawa, Yukihiko, 'Shokusan Kōgyō toshite no Hakurankai, Kyōshinkai no Igi', *Keizai Kenkyū* 39.4 (1988), 340–59

Komatsu, Sakyō, 'Nippon 70 Nendai Zengo: Dokyumento – Orinpikku Kara Bankokuhaku e', *Bungei Shunjū* 49.2 (1971), 220–70

Kyodai Purojekuto Ugoku (Kōsaidō, 1994)

Koresawa, Yūko, 'Meijiki ni okeru Jidō Hakurankai ni tsuite', *Tōkyō Kasei Daigaku Kenkyū Kiyō* 35.1 (1995), 159–65, and 37.1 (1997), 127–37

Kornicki, Peter, 'Public Display and Changing Values: Early Meiji Exhibitions and Their Precursors', *Monumenta Nipponica* 49.2 (1994), 167–96

Kuni, Takeyuki, *Hakurankai no Jidai: Meiji Seifu no Hakurankai Seisaku* (Iwata Shoin, 2005)

Hakurankai to Meiji no Nihon (Yoshikawa Kōbunkan, 2010)

KuroDalaiJee, 'Performance Art and/as Activism: Expo '70 Destruction Joint-Struggle Group', *Review of Japanese Culture and Society* 23 (2011), 154–73

Kushima, Tsutomu, *Maboroshi Bankoku Hakurankai* (Shōgakkan, 1998)

Kwon, Hyeokhui, 'An Analysis of Korean Intellectual Responses to the Exhibition of Koreans at Japanese Expositions: Nationalism and the Discourse on Northeast Asian Solidarity at the Turn of the Century', *Sungkyun Journal of East Asian Studies* 17.1 (2017), 19–40

Kyōto Hakuran Kyōkai, *Kyōto Hakurankai Enkakushi* (Fujimi Shobo, 1997)

Lacambre, Geneviève, 'La Découverte de l'Art Japonais à Travers les Expositions Universelles de Paris, 1867, 1878, 1889, 1900', in Terukazu Akiyama and Tōru Haga (eds.), *L'Age du Japonisme. Le IIème Colloque Franco-japonais* (Société Franco-Japonaise d'Art et d'Archéologie, 1983), 33–46

Langlois, Lisa Kaye, 'Exhibiting Japan: Gender and National Identity at the World's Columbian Exposition of 1893' (PhD thesis, University of Michigan, 2004)

Levitt-Pasturel, Deborah, 'Critical Response to Japan at the Paris 1878 Exposition Universelle', *Gazette Des Beaux-Arts* 119 (1992), 68–80

Li, Tae-Moon, '1915-nen "Chōsen Bussan Kyōshinkai" no Kōsei to Naiyō', *Keiō Gijuku Daigaku Hiyoshi Kiyō, Gengo, Bunka, Komyunikēshon* 30 (2003), 25–61

Lockyer, Angus, 'The Logic of Spectacle c. 1970', *Art History* 30.4 (2007), 571–89

'Expo Fascism? Ideology, Representation, Economy', in Alan Tansman (ed.), *The Culture of Japanese Fascism* (Duke University Press, 2009), 276–95

'The Problem of Sovereignty in an Age of Empire: Representing Japan in Paris in 1867', *Critical Asian Studies* 45.4 (2013), 615–42

Low, Morris, *Visualizing Nuclear Power in Japan: A Trip to the Reactor* (Palgrave Macmillan, 2020)

Malamud, Carl, *A World's Fair for the Global Village* (MIT Press, 1997)

Manita, Yukio, *Chiiki o Tsukuru Yume Sōchi: Hakurankai Kara Chiiki o Miru* (Seibundō Shinkōsha, 1991)

Marcon, Federico, *The Knowledge of Nature and the Nature of Knowledge in Early Modern Japan* (University of Chicago Press, 2015)

Markus, Andrew, 'Shogakai: Celebrity Banquets of the Late Edo Period', *Harvjasiastud Harvard Journal of Asiatic Studies* 53.1 (1993), 135–67

'The Carnival of Edo: Misemono Spectacles from Contemporary Accounts', *Harvard Journal of Asiatic Studies* 45.2 (1995), 499–541

Markwyn, Abigail, 'Economic Partner and Exotic Other: China and Japan at San Francisco's Panama-Pacific International Exposition', *Western Historical Quarterly* 39.4 (2008), 439–65

Maruyama, Hiroshi, 'Meiji Shoki no Kyōto Hakurankai', in Yoshida Mitsukuni (ed.), *Bankoku Hakurankai no Kenkyū* (Shibunkaku Shuppan, 1986), 221–48

Mashiyama, Kazushige, 'Maboroshi no Hakuran Toshi Keikaku: Tōkyō Tsukishima, Nihon Bankoku Hakurankai', in Sano Mayuko (ed.), *Bankoku Hakurankai to Ningen no Rekishi* (Shibunkaku Shuppan, 2015), 267–95

'Kigen 2600-nen Kinen Nihon Banpaku no Keikaku to Sono Shūhen: 1930 Nendai no Kokusai Hakurankai Nihon Tenji o Meguru Renzokusei', in Sano Mayuko (ed.), *Banpakugaku: Bankoku Hakurankai toiu, Sekai o Haakusuru Hōhō* (Shibunkaku Shuppan, 2020), 160–90

Matsuda, Kyōko, *Teikoku no Shisen: Hakurankai to Ibunka Hyōshō* (Yoshikawa Kobunkan, 2003)
Matsumiya, Hideharu, 'Bankoku Hakurankai to Myūjiamu', in Nishikawa Nagao and Matsumiya Hideharu (eds.), *'Beiō Kairan Jikki' o Yomu: 1870-nendai no Sekai to Nihon* (Hōritsu Bunkasha, 1995), 229–78
Mattie, Erik, *World's Fairs* (Princeton Architectural Press, 1998)
Mitchell, Timothy, 'The World as Exhibition', *Comparative Studies in Society and History* 31.2 (1989), 217–36
Mizuta, Miya Elise, '"Fair Japan": On Art and War at The Saint Louis World's Fair, 1904', *Discourse* 28.1 (2006), 28–52
Nadis, Fred, 'Nature at Aichi World's Expo 2005', *Technology and Culture* 48.3 (2007), 575–81
Nagoya-shi Hakubutsukan, *Nagoya no Hakurankai* (Nagoya-shi Hakubutsukan, 1982)
Nakagawa, Dōji, *Rankaiya Ichidai: Waga Hakurankai 100 Nenshi* (Kōdansha, 1969)
Nakamaki, Hirochika, 'Chūkū Kōzō de Toku Senri Nyū Taun to Ōsaka Banpaku', in Sano Mayuko (ed.), *Bankoku Hakurankai to Ningen no Rekishi* (Shibunkaku Shuppan, 2015), 301–8
Nakamori, Yasufumi, 'Criticism of Expo '70 in Print: Journals Ken, Bijutsu Techō, and Dezain Hihyō', *Review of Japanese Culture and Society* 23 (2011), 132–44
Nakamura, Kiyoko, 'Kindai Nihon ni okeru "Kodomo" Imēji to Kodomo Hakurankai', *Bijutsu Kyōiku Gaku* 18 (1997), 215–24, and 19 (1998), 223–34
Nakase, Toshikazu, *Bankokuhaku to Jōhō Fashizumu* (Azekura Shobō, 1970)
Nanba, Kōji, 'Hachijū Nendai no Chihōhaku "Ransai" to Sono Ato', in Fukuma Yoshiaki, Nanba Kōji, and Tanimoto Naho (eds.), *Hakuran no Seiki: Shōhi Nashonariti Mejia* (Azusa Shuppansha, 2009), 5–36
Nishio, Rintarō, 'Kokusai Hakurankai toshite no Nagoya Han-Taiheiyō Heiwa Hakurankai: Sono Hikari to Kage', *Hibarino* 23 (2001), 49–60
Nonaka, Katsutoshi, 'Meiji Shoki ni Shiroato de Kaisai sareta Hakurankai ni kansuru Kenkyū', *Toshi Keikaku Ronbunshū* 41 (2006), 911–16
Nordin, Astrid, 'Space for the Future: Exhibiting China in the World at the Shanghai Expo', *China Information* 26.2 (2012), 235–49
Oda, Masanori, 'Sayonara Banpaku, San Tabi / Nippon Banpaku 70 Nyūsu', 10 +1 36 (2004), 112–15
Okamoto, Tarō, 'Bankokuhaku ni Kaketa Mono', in Tange Kenzō and Okamoto Tarō (eds.), *Nihon Bankokuhaku: Kenchiku, Zōkei* (Kōbunsha, 1971)
Ōkubo, Toshimichi, *Ōkubo Toshimichi Monjo* (Tokyo Daigaku Shuppankai, 2014)
Pluvinage, Gonzague, *Expo 58: Between Utopia and Reality* (Editions Racine, 2008)
Raizman, David, and Ethan Robey (eds.), *Expanding Nationalisms at World's Fairs: Identity, Diversity, and Exchange, 1851–1915* (Routledge, 2017)
Roche, Maurice, *Mega-Events and Modernity: Olympics and Expos in the Growth of Global Culture* (Routledge, 2020)

Rydell, Robert W., *All the World's a Fair: Visions of Empire at American International Expositions, 1876–1916* (University of Chicago Press, 1985)
World of Fairs: The Century-of-Progress Expositions (University of Chicago Press, 1993)
Saito, Asato, 'Global City Formation in a Capitalist Developmental State: Tokyo and the Waterfront Sub-Centre Project', *Urban Studies* 40.2 (2003), 283–308.
Saitō, Gesshin, *Zōtei Bukō Nenpyō* (Heibonsha, 1968)
Sakaiya, Taichi, *Chijō Saidai no Gyōji: Bankoku Hakurankai* (Kōbunsha, 2018)
'Banpaku ni Sengoshi o Yomu: Okinawa Kaiyōhaku (1975 nen) o Chūshin ni', in Sano Mayuko (ed.), *Banpakugaku: Bankoku Hakurankai toiu, Sekai o Haakusuru Hōhō* (Shibunkaku Shuppan, 2020), 369–86
Sakazaki, Otsurō, 'Hanpaku no Shisō', *Kikan Geijutsu* 4.2 (1970), 58–61
Sand, Jordan, *House and Home in Modern Japan: Architecture, Domestic Space, and Bourgeois Culture, 1880–1930* (Harvard University Asia Center, 2003)
Tokyo Vernacular: Common Spaces, Local Histories, Found Objects (University of California Press, 2013)
Sanderson, Rachel, 'Queensland Shows the World: Regionalism and Modernity at Brisbane's World Expo 88', *Journal of Australian Studies* 27.79 (2003), 65–75
Sano, Mayuko, *Bankoku Hakurankai to Ningen no Rekishi* (Shibunkaku Shuppan, 2015)
Banpakugaku: Bankoku Hakurankai toiu, Sekai o Haakusuru Hōhō (Shibunkaku Shuppan, 2020)
Saotome, Kenji, '1970 nen Ōsaka Banpaku no Kihon Rinen: "Bankokuhaku o Kangaeru Kai" ni yoru Sōan Sakusei no Haikei to Keii', in Sano Mayuko (ed.), *Banpakugaku: Bankoku Hakurankai toiu, Sekai o Haakusuru Hōhō* (Shibunkaku Shuppan, 2020), 246–65
Satō, Doshin, *'Nihon Bijutsu' Tanjō: Kindai Nihon no 'Kotoba' to Senryaku* (Kōdansha, 1996)
Sawada, Yūji, 'Aichi Banpaku Zen'ya: Hakurankai no Kikaku Seisaku Genba kara', in Sano Mayuko (ed.), *Bankoku Hakurankai to Ningen no Rekishi* (Shibunkaku Shuppan, 2015), 475–503
Sawaragi, Noi, *Sensō to Banpaku* (Bijutsu Shuppansha, 2005)
Servant, Claude, and Ippei Takeda, *Study on the Impact of International Expositions* (Bureau International des Expositions, 1996)
Shepherdson-Scott, Kari, 'Conflicting Politics and Contesting Borders: Exhibiting (Japanese) Manchuria at the Chicago World's Fair, 1933–34', *The Journal of Asian Studies* 74.3 (2015), 539–64
Shibusawa, Eiichi, *Shibusawa Eiichi Taifutsu Nikki* (Nihon Shiseki Kyōkai, 1928)
Shigematsu, Kiyoshi, *Towairaito* (Bungei Shunjū, 2002)
Shirayama, Mari, 'Senji Senden to Shashin Hekiga: Yamahata Shōgyoku to 1939-nen Nyū Yōku Bankoku Hakurankai no "Yakushin Nihon" o Chūshin ni', in Sano Mayuko (ed.), *Banpakugaku: Bankoku Hakurankai toiu, Sekai o Haakusuru Hōhō* (Shibunkaku Shuppan, 2020), 191–207
Sigur, Hannah L., 'A Neoclassical Translation: The Hōōden at the 1893 World's Columbian Exposition', in David Raizman and Ethan Robey (eds.),

Expanding Nationalisms at World's Fairs: Identity, Diversity, and Exchange, 1851–1915 (Routledge, 2017), 92–108

Sumi, Yutaka, *Tokugawa Akitake: Banpaku Tonosama Ichidaiki* (Chūō Kōronsha, 1984)

Suzuki, Hiromune, 'Dai Sankai Naikoku Kangyō Hakurankai to Ajia Dai Hakurankai Keikaku', *Hōsei Shiron* 20 (1993), 44–52

Tada, Osamu, *Okinawa Imēji no Tanjō: Aoi Umi no Karuchuraru Sutadīzū* (Tōyō Keizai Shinbunsha, 2004)

Takai, Masashi, 'Hakurankai to Jendā, Kaisō: "Konpanion" o Jirei toshite', in Fukuma Yoshiaki, Nanba Kōji, and Tanimoto Naho (eds.), *Hakuran no Seiki: Shōhi Nashonariti Mejia* (Azusa Shuppansha, 2009), 161–81

Takara, Yūsei, *Jiritsu no Keizaigaku: Fukki to Kaihatsu no Seisaku Dokuhon* (Roman Shobō, 1988)

Taki, Kōji, 'Banpaku Hantairon: Dezain ni okeru Ningen Kaifuku o Mezasu', *Tenbō* 121 (1969), 172–77

'Kindai no Shūmatsu', *Ken* 1 (1969), 64–69

Tange, Kenzō, 'Ikutsuka no Keiken', *SD* 184 (1980/1), 176–90

Teasley, Sarah, *Designing Modern Japan* (Reaktion Books, 2022)

Teramoto, Noriko, *Pari Bankoku Hakurankai to Japonisumu no Tanjō* (Shibunkaku Shuppan, 2017)

Tōkyō Kokuritsu Hakubutsukan, *Tōkyō Kokuritsu Hakubutsukan Hyakunenshi* (Tokyo Kokuritsu Hakubutsukan, 1973)

Tomio, Kentaro, 'Visions of Modern Space: Expositions and Museums in Meiji Japan', in Helen Hardacre and Adam L. Kern (eds.), *New Directions in the Study of Meiji Japan* (Brill, 1997), 719–33

Tseng, Alice, *The Imperial Museums of Meiji Japan: Architecture and the Art of the Nation* (University of Washington Press, 2008)

Modern Kyoto: Building for Ceremony and Commemoration, 1868–1940 (University of Hawai'i Press, 2018)

Tsuganesawa, Toshihiro (ed.), *Kindai Nihon no Media Ibento* (Dōbunkan, 1996)

Sengo Nihon no Media Ibento, 1945–1960-nen (Sekai Shisōsha, 2002)

Tsunoyama, Yukihiro, *Uīn Bankokuhaku no Kenkyū* (Kansai Daigaku Keizai Seiji Kenkyūjo, 2000)

Umesao, Tadao, *Umesao Tadao Chosakushū*, 23 vols. (Chūō Kōronsha, 1989–94)

Urushima, Andrea, 'Genesis and Culmination of Uzô Nishiyama's Proposal of a "Model Core of a Future City" for the Expo 70 Site (1960–73)', *Planning Perspectives* 22.4 (2007), 391–416

'The 1970 Osaka Expo: Local Planners, National Planning Processes and Mega Events', *Planning Perspectives* 26.4 (2011), 635–47

Ward, Max, 'Displaying the Worldview of Japanese Fascism: The Tokyo Thought War Exhibition of 1938', *Critical Asian Studies* 47.3 (2015), 414–39

Wilson, Sandra, 'Exhibiting a New Japan: The Tokyo Olympics of 1964 and Expo '70 in Osaka', *Historical Research* 85.227 (2012), 159–78

Winter, Tim, *Shanghai Expo* (Routledge, 2012)

'Auto-Exoticism: Cultural Display at the Shanghai Expo', *Journal of Material Culture* 18.1 (2013), 69–90

Winther-Tamaki, Bert, 'To Put on a Big Face: The Globalist Stance of Okamoto Tarō's Tower of the Sun for the Japan World Exposition', *Review of Japanese Culture and Society* 23 (2011), 81–101

Yamada, Atsushi, 'Ou Seiei Nanjing Seifu Shita no Dai Tōa Sensō Hakurankai', in Shibata Tetsuo and Yamada Atsushi (eds.), *Chūgoku to Hakurankai: Chūgoku Nisenjūnen Shanhai Bankoku Hakurankai ni itaru Michi* (Seibundo, 2014), 95–112

Yamaji, Katsuhiko, *Kindai Nihon no Shokuminchi Hakurankai* (Fūkyōsha, 2008)

Chihō Toshi no Kakusei: Taishō Shōwa Senzenshi, Hakurankai hen (Kansei Gakuin Daigaku Shuppankai, 2017)

Yoshida, Mitsukuni, *Bankoku Hakurankai: Gijutsu Bunmeishi-teki ni* (NHK Books, 1985)

Bankoku Hakurankai no Kenkyū (Shibunkaku Shuppan, 1986)

Yoshii, Takao, 'Daigokai Naikoku Kangyōhaku no Disupurē', in Yoshida Mitsukuni (ed.), *Bankoku Hakurankai no Kenkyū* (Shibunkaku Shuppan, 1986), 221–48

Yoshimi, Shun'ya, *Hakurankai no Seijigaku: Manazashi no Kindai* (Chūō kōronsha, 1992)

Banpaku Gensō: Sengo Seiji no jubaku (Chikuma Shobō, 2005)

'A Drifting World Fair: Cultural Politics of Environment in the Local/Global Context of Contemporary Japan', in Tomiko Yoda and Harry Harootunian (eds.), *Japan after Japan: Social and Cultural Life from the Recessionary 1990s to the Present* (Duke University Press, 2006), 295–414

'Mō Hitotsu no Media toshite no Hakurankai: Genshiryoku Heiwa Riyō Haku no Juyō', in Tsuchiya Yuka and Yoshimi Shunya (eds.), *Senryō Suru Me, Senryō Suru Koe: CIE/USIS Eiga to VOA Rajio* (Tokyo Daigaku Shuppankai, 2007), 291–316

Yoshimoto, Midori, 'Textiles Pavilion: An Anomaly and Critique of Expo '70', *Review of Japanese Culture and Society* 23 (2011), 113–31

Yoshimura, Tomohiro, '1900 Nendai Shotō no Nishihama Buraku to Hikakugyō: Daigokai Naikoku Kangyō Hakurankai ni miru Seihin o Chūshin ni', *Osaka Jinken Hakubutsukan Kiyō* 4 (2000), 61–73

Ziomek, Kirsten, *Lost Histories: Recovering the Lives of Japan's Colonial Peoples* (Harvard University Asia Center, 2019)

Zwigenberg, Ran, '"The Coming of a Second Sun": The 1956 Atoms for Peace Exhibit in Hiroshima and Japan's Embrace of Nuclear Power', *The Asia-Pacific Journal: Japan Focus* 10.6.1 (2012)

Index

Individual exhibitions that are discussed in detail in the main text are referenced here by city and year, with their most commonly used title in parentheses: e.g. Osaka 1970 (Japan World Exposition). Exhibitions that are mentioned in passing are referenced by city only.

Printed by Integrated Books International,
United States of America